Basin Futures

Water reform in the Murray-Darling Basin

Basin Futures

Water reform in the Murray-Darling Basin

Edited by Daniel Connell and R. Quentin Grafton

THE AUSTRALIAN NATIONAL UNIVERSITY

E PRESS

E PRESS

Published by ANU E Press
The Australian National University
Canberra ACT 0200, Australia
Email: anuepress@anu.edu.au
This title is also available online at: http://epress.anu.edu.au

National Library of Australia Cataloguing-in-Publication entry

Title: Basin futures : water reform in the Murray-Darling basin / edited by Daniel Connell and R. Quentin Grafton.

ISBN: 9781921862243 (pbk.) 9781921862250 (ebook)

Subjects: Darling River Watershed (Qld. and N.S.W.)--Management.
Murray River Watershed (N.S.W.-S. Aust.)--Management.

Other Authors/Contributors:
Connell, Daniel, 1947-
Grafton, R. Quentin, 1962-

Dewey Number: 333.730994

Cover design and layout by ANU E Press

Cover image: © MDBA; Mouth of the Murray River, by Michael Bell

Printed by Griffin Press

Contents

Part III. Legal issues

Part IV. Economics

Part V. Governance

To Donna Brennan

a significant contributor to public policy in rural Australia

List of figures

List of tables and boxes

Boxes

Basin Futures: An introduction

This book represents a remarkable collection of evidence, opinions, proposals and remedies for the troubles that ail the Murray–Darling Basin. It grew out of a workshop and conference held on 14 and 15 October at The Australian National University shortly after the release of the *Guide to the proposed Basin Plan* by the Murray–Darling Basin Authority (MDBA) on 8 October 2010. At the time there was a wave of protest about what was proposed, and what might happen should water reform continue its course. This protest may have lost its radiance somewhat in 2011 but there is still no shortage of heat. This book sheds some much-needed light over these controversial issues.

The workshop participants and others who have been invited to contribute to this book were asked to respond in their own ways, and with their own expertise, to the questions: what led to where we are today? And what can we do about it? We owe a great debt to our fellow contributors. They, like us, strongly believe a better future is possible, although they might not agree about the path(s) we should take.

The Basin's troubles have been a long time in the making. They arise from the inherent trade-offs that result from multiple and rival demands for a scarce resource. The challenges that have resulted have shaped our nation, our constitution and who we are as a people. What we decide to do in the Basin—for good or ill—in the coming years will be watched by generations to come. It will help create the sense of not only what Australia is, but how Australia is perceived by onlookers overseas who see our part of the world as the proverbial 'canary in the coalmine' in terms of future water scarcity.

All told, there are 27 chapters in this book that cover the issues of landscape; communities and stakeholders; Indigenous issues; law; economics; and governance—written by some of Australia's leading thinkers on water issues. The common purpose of all the chapters is to 'shine a light' on water reform in the Basin. We stress emphatically that we do not agree with everything that is written in every chapter. Neither is there a consensus across all the chapters. But we would not expect this from a book the two editors of which have tried to be as pluralistic as they can be when they requested diverse thoughts and ideas from many different disciplines and experiences. Nevertheless, between these pages you can find a robust debate, whether it be about the amount of water that should be reallocated from consumptive to non-consumptive purposes or the process of reform and what might be missing, and what should be included in the proposed Basin Plan.

The chapters discussing community issues emphasise the importance of process and good communications to achieve positive change in the way natural resources are managed in the Murray–Darling Basin (MDB). Leith Boully and Karlene Maywald draw on their experience working in the MDB to comment that although over-allocation is widely acknowledged as a problem by irrigators, the fact that many people subsequently made major investment decisions based on those over-allocations creates a serious dilemma, which will need careful negotiation to resolve. They also stress the community dimension of the challenge. This means that programs to promote acceptance of reform need to go much wider than merely compensating individual irrigators who are willing to sell their entitlements. These points are supported by Catherine Gross, who draws on recent episodes of conflict in south-western New South Wales and central Victoria to stress that whether or not people perceive the processes used to make decisions as fair and reasonable will influence their thinking about the decisions that result. The need to connect and consult better was a theme or sub-theme of many chapters. Part of the problem was the way in which the release of the draft Plan was treated by the media—a result that was at least partly due to the approach of governments and the MDBA, according to Åsa Wahlquist, who adds her valuable thoughts about how it should have been done.

Chris Miller's chapter supports these recommendations by drawing on the large body of available national and international literature to outline what would be involved in developing a strong community-engagement strategy. Key elements include retention of local capital, attraction of new capital investment, fostering of the development of new products and services for a diversified economy, extension of community ownership of local services and enterprises, encouragement of new migration, retention of young people, building of human, social and cultural capital, collaboration with other basin communities, partnerships with local, State and Federal governments and integration of community development with local natural-resource management. This discussion is complemented by Martin Mulligan's examination of the concept of 'community' and of a number of Australian examples of where community development has been successful.

Most recent public discussion about the constitutional foundations of the *Water Act 2007* has not gone beyond rather simplistic discussion of whether social and economic considerations should be seen as equal to the imperative to achieve environmental sustainability or whether they are secondary. The chapter by Douglas Fisher pushes beyond that public debate to examine the complex mesh of 'protectable rights and enforceable duties' and the concept of 'ecologically sustainable development' that are central to the way in which the courts will interpret the *Water Act*. In addition, he expands on the application of the rules contained within the Act and discusses how its compliance provisions will work. Central to what happens in the courts will be the way in which scientific evidence is used and weighed; that subject is the theme of the chapter by Jennifer McKay.

In 2004 the National Water Initiative (NWI) placed a high priority on restoring Indigenous rights to water in the MDB, but since then there has been considerable confusion about what this could mean in practice. The two chapters on Indigenous issues complement each other neatly. One discusses what the waterways of the MDB mean to Indigenous people and the other considers what can be done to take account of their interests in the water-reform process. Jessica Weir's chapter explains that water management in the past—with its acceptance of increasing allocations for production and the resulting ongoing environmental decline—was in effect an act of dispossession similar to that experienced by Indigenous people when they lost access to their land in the nineteenth century. What can be done about incorporating Indigenous issues into water decisions is the theme of Sue Jackson's chapter. It outlines work done by the CSIRO and others at the behest of the MDBA as part of the preparation of the draft guidelines to the Basin Plan, and links the commitments of the NWI and the requirements of the *Water Act* to a number of international agreements to which Australia is bound. It also explores the concept of a 'cultural flow', which involves using water to improve the social, cultural and economic conditions of Indigenous people in addition to the riverine environment.

The potential economic impacts of the Basin Plan are discussed in a number of chapters. John Quiggin approaches these issues from the perspective of risk, examining in particular the interaction between uncertainty and property rights—a relationship that is particularly fraught in the MDB because, arguably, property rights have been allocated to more water than is sustainably available. He also addresses the issues of risk assignment. Quentin Grafton argues that social, economic and environmental outcomes in the MDB could all be improved if the $5.8 billion now allocated to infrastructure investment was divided between the buybacks fund and community-orientated investments in those regions that will be affected by the reduction in water for irrigation. (Under the current Plan, the $5.8 billion for irrigation-infrastructure investment will supplement the $3.1 billion in payments for water purchases, thus, focusing the public investment to promote change in the MDB solely on irrigators rather than the communities of which they are part.) John Quiggin provides similar comments in terms of the cost effectiveness and the ability of the current spending priorities and current funding for water reform to sufficiently achieve a 3000–4000 gigalitre per year reallocation. Grafton also makes the point that there is no economic evidence provided in the Guide or supporting technical documentation to justify the assertion that an increase in environmental flows of more than 4000 GL/yr would generate socioeconomic costs that are too high.

Central to the debates about the potential impacts of the Basin Plan are the results of modelling exercises used to predict the outcomes of different buyback and investment scenarios. Qiang Jiang provides a chapter that examines the ways the various models are being used. The impacts on irrigation are central to this debate. The attitudes of irrigators to what is taking place are examined by Henning Bjornlund, Sarah Wheeler and Jeremy Cheesman, who reveal that opposition to a significant redirection of water to the environment has become stronger in recent years. Their research also indicates that much of the money gained from the sale of water entitlements could be used to retire debt or will be invested away from the MDB, thus highlighting the potential negative impact of current policies on irrigation-based communities. There is also a chapter from the late and distinguished agricultural economist Donna Brennan. She assesses the proposed new regime from the perspective of its impact on water markets and deduces that the focus on environmental outcomes will cause significant anomalies in water-trading transactions in regions such as the upper Goulburn River. Taking a different perspective, Darla Hatton Macdonald, Rosalind Bark, Dustin Garrick, Onil Banerjee, Jeff Connor and Mark Morrison focus on the many costs and benefits that are left out of standard assessments, which, if taken into account, would radically transform public understanding of the benefits of water reform in the MDB.

Much of the public debate so far has focused on the perceived lack of consideration of social and economic issues in developing the Basin Plan. Jamie Pittock and Max Finlayson redress this imbalance somewhat by arguing that the Basin Plan falls seriously short of meeting the requirements of the international treaties dealing with the environment that provide much of the constitutional underpinning for the *Water Act*. One of the major sources of pressure making delivery of those commitments increasingly difficult is climate change. As part of the preparation of the draft Basin Plan, the MDBA commissioned a substantial suite of research projects to investigate its potential impact. These are summarised in the chapter written by Nick Schofield. He explains that while the planetary trajectory of global warming is now well defined, the ability of science to provide reliable regional forecasts for catchments such as the MDB will need considerable further development before it can reliably underpin policy.

Also in this section discussing how science should inform policy is the chapter by Richard Norris, who argues that there has been a growing tendency in recent years to treat environmental benefits as an optional extra to be provided only to the degree that major interest groups find convenient and not too disruptive to their established activities. In opposition to that view, Norris restates the case that the environment is the source of a wide range of ecological services of high value, both monetary and non-monetary. He explains that continued access to such services depends on the maintenance of key ecological functions in the riverine system through the provision of adequate water and the maintenance of essential patterns of flow, and that this view accords with the stated aims of the *Water Act*.

As John Williams explains in his chapter, water policy is about managing the interaction between human aspirations, demands and impacts and the wider environment. Williams summarises the main elements at play in the context of the MDB and explains why management success will require more than just balancing off the competing claims of human stakeholders. The compromises that are negotiated also have to be within the long-term limits of what the environment can provide. This is a key issue to be considered when discussing how great should be the shift of water away from production back to the environment. At the very least it should be sufficient to halt the ongoing environmental decline that characterised the past century. In discussions about water policy in the MDB, there is a tendency to concentrate on surface water, but the NWI requires that it be managed in coordination with groundwater where systems are linked. How best to do this is discussed in the chapter by Stuart Richardson, Ray Evans and Glenn Harrington, who draw on a number of recent examples to show how both science and stakeholders can be effectively involved.

The book contains a wide range of chapters discussing water governance in the MDB. Neil Byron is highly critical of the underlying design of the *Water Act* and calls for a fundamental reassessment of the sort of institutional arrangements that are needed to best manage the MDB. The call to go back to basics is also supported by Ray Ison and Philip Wallis, who discuss approaches that could be used to re-conceptualise the task and encourage institutional innovation. A reminder that water reform in Australia is part of an international movement is provided Dustin Garrick and Rosalind Bark, who argue that the growing occurrence in the western United States of negotiation between stakeholders to avoid the courts is a development worthy of more attention in Australia.

Other chapters accept the *Water Act* as a given and discuss how it could be made to work to produce the best possible outcomes under current circumstances. Daniel Connell writes about the Commonwealth Environmental Water Holder—an organisation that could be used to implement large-scale rehabilitation even if the Basin Plan is indefinitely delayed by political conflict and litigation. Also working within the *Water Act* is Mike Young, who proposes that the process for producing the Basin Plan should be redesigned to allow for more effective engagement with affected communities. He recommends that regions should play a greater role to allow for increased innovation and more flexibility to adapt to local circumstances. Drawing on her research into water reform in Europe, Katherine Daniell also discusses this theme. She argues that 'the centralised technocratic approach' used to develop the Basin Plan was always likely to alienate people in the MDB. Instead, governments should draw on the large body of experience now available for the promotion of community participation. Communities and the regional sub-catchment level are also the focus for Mark Hamstead, who stresses that there will need to be much more effective water planning if the goals of the *Water Act* are to be realised.

We trust that you will value the insights in these pages as our nation struggles with a 'wicked water problem'. If you have the time to read this book cover to cover or even if you 'cherry pick', you will soon realise the real issues are about not water *per se*, but water and its interactions with people. Key questions we and our fellow contributors have struggled with include: what are the values people hold about agriculture and about the environment? How can we 'balance' competing demands? Who should bear the costs of transition to a sustainable future? Who should pay?

All these questions can, and should, be debated, and this is what this book does. Whether you are a farm worker living in Griffith worried about your job or a retiree in Sydney who cares passionately about the Coorong, the chapters in this book have much to offer. The book tells a clear story that business as usual is not working in the Basin. Contrary to what some people think, these problems have most certainly not been resolved by the floods of 2010–11, just as they were not created simply by the Millennium Drought. Decisions must be made, and made soon, and these cannot be half-measures or we risk irreparably damaging the future of all those who live, work and care about our basin, our home.

Daniel Connell and R. Quentin Grafton
Centre for Water Economics, Environment and Policy
Crawford School of Economics and Government
The Australian National University

Part I. Water and the Basin

1. Understanding the Basin and its Dynamics

John Williams

Introduction

The competing tensions between water extraction for immediate human use and water essential to the long-term ecological function and sustainability of the rivers and groundwater systems in the Murray–Darling Basin (MDB) sit at the centre of public policy debate on water reform. Yet it is much more than this. The people of the Basin are faced with the enormous challenge of transforming themselves into more resilient communities. This requires managing and reconstructing the conflict between the climatic and biophysical realities of the Basin and the earlier private and public policy aspirations of the European settlers that have dominated for the past 150 years. Water reform is, then, a social process by which communities work to align land use and economic industries so that they work more in harmony and within the capacity of the hydrological and ecological processes operating in the landscape and thereby are able to harvest a wider range of ecosystem services than they currently do. For water reform to be embedded in such a process, it is critical, however, that the dynamics of the biophysical processes operating within the geological and geomorphic form of the ancient Basin be fully appreciated and understood. Without an understanding of the Basin's form and functionality, water-reform implementation will probably solve one problem while creating several others. This must be done against a backdrop of climate change which is impacting on the very high climatic variability that over the past decade has seen a severe, nine-year drought and a year of large floods. These extreme events — both very wet and very dry — are what characterise the Basin of the past and they can be expected to be an increasing part of a climate-change future (Francis and Hengeveld 1998; Min et al. 2011; Pall et al. 2011).

The biophysical nature of the Basin's rivers and groundwater systems, coupled with this climatic variability and change, calls for water reform to be implemented in innovative ways that will test the fabric of our society and stretch our scientific knowledge to the limit. For it is management of greater extremes — the floods and the droughts — in accord with the ecological functional requirement of these rivers and the increasing demand for water extraction to satisfy human need that call attention to the need for radical water reform in the light of failure of current policy and practice. Present-day problems that confront the Murray–Darling Basin (MDB) can be related to the way the societal narrative, cultural values and knowledge have clashed with the climatic and geological history of the ancient Australian continent.

The MDB heritage we have today is the result of an unfortunate coincidence between human action and the set of geological and climatic forces that formed the Basin. Human activity over the past 150 years has exacerbated geological, hydrological and ecological processes driven by a history of changing and highly variable climate through time and across the Basin (Williams and Goss 2002). The Basin is ancient. What we see today bears residual features and the overprinting of a long history of climate changes, involving many sequences and oscillations between very humid and extremely arid environments. Clearing of forest and woodland vegetation, in conjunction with the application of irrigation water, has produced in less than 200 years a change in groundwater equilibrium and river flow regimes that in many ways mimic the changes that have resulted in the past from climatic oscillations (Bowler 1990). These changes brought by Europeans through grazing, clearing of forests and woodlands and the development of irrigation have resulted in the return of conditions that existed about 18 000 years ago — once again we have high saline water tables discharging to rivers. The flow regimes of the rivers have been drastically changed so that the floodplain ecosystems that drive much of their ecological function are disconnected, and the flows to flush out salt and refresh the system are far too infrequent. Innovation, problem solving, and the managerial capacity of farmers have sustained an impressive productivity growth through the twentieth century, particularly in cereal production. Great wealth from the production of food and fibre has been fundamental to the wealth and wellbeing all Australians now share. The Basin has yielded much and has a heritage of place and natural history that is very important to Australians everywhere. The Basin is our heartland and holds symbols of our rural heritage, upon which Australian identity has been built. But now we see much of what has been built under threat from economic, social and environmental change and decline.

Water reform in the Basin is cast against this background of the ancient bio-geophysical processes that must be understood and managed while finding new expressions and narratives within which the Basin's communities recast and rebuild more resilient futures. This chapter seeks to examine the nature of the Basin's geology, hydrology and ecology, and to weave into this the interaction of the human aspirations, values and visions that have shaped our communities and that generate the human and physical heritage within which water reform must take place.

Biophysical Background

Geological History and Basin Structure

The Murray–Darling Basin's streams and rivers sit in a shallow basin, which is very old, very flat, contains large stores of salt, and with respect to groundwater is very nearly blind in that it has no outlet to the sea.

In geological terms, the Basin has a very ancient foundation. The oldest rocks (pre-Cambrian), which outcrop in the western margins, date back at least some 600 million years. Most of the Basin has a basement of ancient (Palaeozoic age of 230–540 million years) rocks that were eroded to a palaeo-plain. Over this ancient platform, sedimentary rocks formed basins during the Mesozoic age (some 60–250 million years ago) in the case of the Great Artesian Basin (GAB), and, later, during the Caenozoic age (less than 60 million years), the Murray Basin was laid down (Ollier 1995). These two basins are separated by a system of tectonic warp axes that corresponds to the drainage divides. These are the two major sedimentary groundwater basins over which the Murray–Darling Basin catchment is located (Ollier 1995). Within both basins there has been broad down-warping, subsidence or sinking of these sedimentary rocks. This has resulted in sedimentary rocks infilling a low-lying, saucer-shaped depression (Evans et al. 1990), rimmed and underlaid by folding and partly metamorphosed ancient basement rocks. These ancient (Palaeozoic) rocks now form the subdued mountain ranges around most of the Murray–Darling Basin — apart from the south-western rim, where concealed basement rocks just beneath the surface form the Padthaway Ridge. This separates the MDB from the Southern Ocean.

Whilst the Murray Basin is very large, the sedimentary rocks are relatively thin. The maximum thickness is 600 m, found over the region of most subsidence, while at the margins of the Basin the thickness of the sediments is less than 200 m. They form a pancake-like veneer over the older basement rocks (Evans et al. 1990). Because the sedimentary rocks are quite thin, the Basin has a relatively small capacity for groundwater storage. This saucer-shaped structural configuration with subsidence just south of the centre, covered by a thin layer of sediments, has important implications for the nature of the Basin and the way surface and groundwater must be managed. The MDB is essentially a closed groundwater basin within which groundwater drainage is directed internally towards the central subsidence and thicker sediments, rather than towards the side where the Murray connects to the sea.

Because the Basin is blind and because the sediments in which groundwater can be stored are relatively thin they offer a relatively small storage capacity as the sedimentary rocks are largely water saturated. Thus, there is little capacity in the groundwater system for the storage of additional recharge. Thus, if the groundwater systems receive increased recharge as they have since European settlement, the additional recharge cannot escape. The water tables must rise.

These features all point to a most important conclusion. Minimal groundwater recharge will drive a rapid water table rise, and because the Basin is essentially blind and therefore has a small discharge capacity, the response of groundwater levels to reductions in recharge rate will be very slow. When the additional recharge is reduced, the water table fall will be very slow largely determined by the small discharge capacity via the Murray or by evaporation from land-surface discharge regions in the depressions and lakes of the landscape. Thus, groundwater systems can be filled easily, but must empty and discharge very slowly. This is a most unfortunate fundamental fact about the MDB and it is essential to understand that the Murray River needs to have large flushing flows to carry salt to the oceans where it came from (Evans et al. 1990).

About 40–60 million years ago, the central area of the MDB subsided as the eastern highlands were uplifted. This subsidence formed two distinct regions of sedimentation and later groundwater accumulation. The southern area is known as the Murray Groundwater Basin, which is not fully synonymous with the catchment but it does underlie a great deal of it. The northern area, over which the Darling River and its tributaries now flow, is the southern part of the Great Artesian Basin. The climate of the early Tertiary (40–60 million years ago) was very much wetter than at present and the Murray Basin then contained large swamps and bogs, and thick sediments that were laid down in broad valleys. With increasing subsidence and eastern highlands uplift, stream dissection and incision in the highlands resulted in sand and gravel deposition in fans as the rivers entered the plains.

During the Miocene (26–7 million years ago) the sea level rose relative to the land, and the inland sea covered the south-western corner of the Murray Groundwater Basin. Marine materials were deposited in sand sheets. In the past two million years, the sea retreated, leaving a succession of stranded beach ridges and relic coastlines. Following the sea's retreat, a huge freshwater lake developed, as there was a blockage across the Murray. During the Quaternary glacial period about two million years ago, the climate became very arid, with dry and windy conditions prevailing. Another set of dunes — this time Aeolian — was built by the wind action. In the past 30 000 years, a thick blanket of fine alluvium has been laid down over coarse sediments in the old bedrock of the central area of the Murray Basin. A similar process took place in the Darling.

The sea once occupied the Mallee and most of the Murray Basin, extending to Balranald in New South Wales, with thin reaches stretching to Kerang, Victoria, at its peak, before retreating from about three to four million years ago. Whilst the salt associated with this intrusion has long since left the Basin through leaching, the retreat of the sea established the ultimate gradient and outlet for the Basin and the modern (past 500 000 years) landscape development of the Basin. This retreat of the sea had a number of other important consequences.

Not only did the climate dry from the extensive wet rainforest period (12–30 million years ago), but earth movements dammed the Murray outlet to result in the huge Lake Bungunnia. The lake formed about 2.5 million years ago and continued to exist for about two million years, until about 500 000–700 000 years ago when the outflow point was deepened sufficiently to drain the lake and permit the Murray River to cut a deep gorge through earlier sediments to provide an outlet to the sea.

Modern Features of the Landscape, Waterways and Vegetation

Within the time frame of the Murray–Darling's origins, there are four factors (Evans et al. 1990) that control the modern landscape features

1. the low level of tectonic activity over long periods
2. a strong east–west gradient of increasing aridity
3. the marine influence on the south-western corner of the Basin
4. the prevailing south-westerly winds.

Compared with other continents, Australia has been remarkably free of volcanic or mountain-building activity in recent time. While the Australian continent has drifted north from Antarctica over the past 60–80 million years, very minor changes in topography have occurred (Bowler 1990; Ollier 1995). The Great Dividing and Flinders ranges and the extensive plains between were already present from at least 20 million years ago. These ranges are very subdued features compared with the mountains of other continents. The late-Quaternary (past million years) history of the Murray–Darling Basin has been of minor tectonic movement and the evolution of landforms under increasingly arid conditions (Wasson 1987). The major subdivisions such as the Eastern Upland, Cobar Plains, Murray and Upper Darling basins have largely remained as they are, but within these, landform changes have occurred to produce the rivers, dunes, alluvial plains and slope colluvium as they are today (Wasson 1987).

The closed nature of the Murray Basin results in a strong interaction between groundwater and surface water. In the west, the River Murray is an efficient drain providing the natural pathway for removing groundwater and its dissolved salts. In fact, the lower sections of the river have always been a salt drain. The changes that have occurred in the groundwater system of the southern Murray Basin over the past 150 years of European settlement appear to mirror in magnitude the changes that would have occurred over thousands of years as a result of climate oscillations that characterised periods over the past 500 000 years.

The result is that old groundwater recharge structures and mountains of salt sitting dormant for maybe 18 000 years are being reactivated. This implies that an increase in groundwater flow to streams and rivers as has happened in the geological past will again occur — with an accompanying increase in movement of salt to land and rivers. Large quantities of saline groundwater enter the Murray as it moves through the channel that is deeply incised into the sediments underlying the Mallee. Reversal of the process that caused the rising water tables will not halt the discharge of salt to land and river until the high water tables are able slowly to discharge. This is a critical matter to take note of when determining the sustainable-flow regimes of these river systems under any water-reform agenda. Failure to give attention to these issues will further damage the river function and further reduce the options for those communities dependent on healthy river and floodplain function.

The cycles of saline/non-saline associated with rising/falling water tables were driven by oscillations in climate, and current 50-year cycles can be seen in shallow groundwater today (Rancic et al. 2009). Within an otherwise geologically stable basin, a central feature of the landscape was the erosion/deposition/transportation of salt and sediments within the Basin. Very little material left the Basin. The sediments and salts were recycled, reworked and accumulated. These oscillations in climate to both wetter and drier than at present were sufficient to move sediments from shallow groundwater and saline environments to non-saline environments. This is a salutary characterisation of the Basin to be better managed under water reform.

While the onset of regional salinity was a relatively rapid response to changing climatic conditions, the recovery from the saline lakes and rivers was a much slower process. About 13 000 years ago, trees and shrubs returned to the landscape, and shallow groundwater levels fell, enabling other, more bio-diverse vegetation to establish on the once salinised land. Although the evidence of the recovery is clear, the mechanism that brought the recovery remains unclear (Bowler 1990).

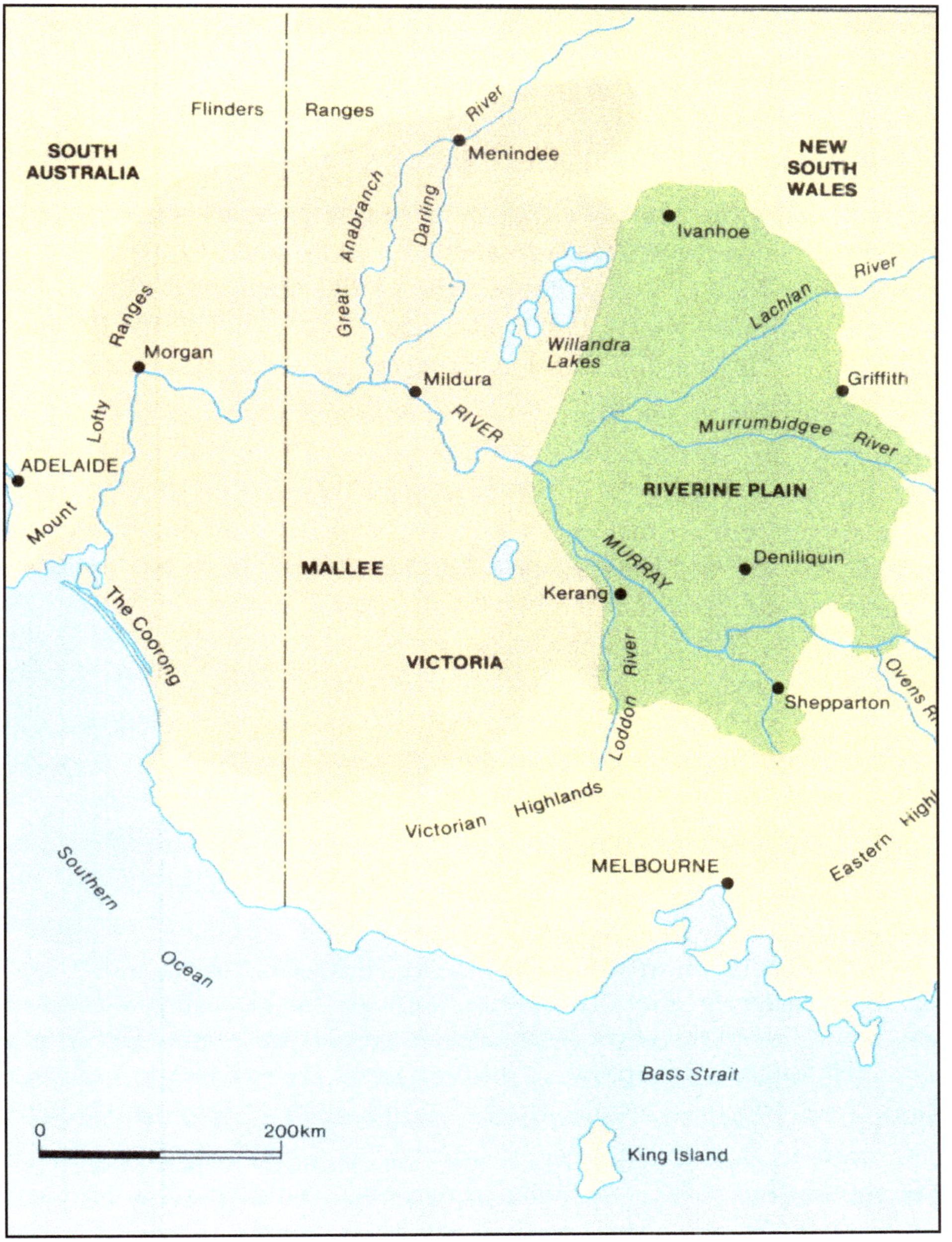

Figure 1.1 The large geomorphic features of the Murray–Darling Basin

Source: Rutherford (1990:18).

River Form, Hydrology and Ecological Functionality

The rivers in the Murray–Darling Basin as we know it today occupy — and are substantially controlled by — a catchment formed by the interactions of geology, erosion and depositional processes driven by oscillating humid and arid climates over many millions of years. An appreciation of these processes and the history is critical to understanding the context in which water reform must be cast. If water reform is not cast to address how the Basin works, it will fail and will need to be redone in years to come.

The MDB encompasses some 14 per cent of the Australian continent, stretching from the subtropics of central Queensland to the southern alps of Victoria, across the extensive floodplains of the Murray River to the Lower Lakes at its mouth and, ultimately, to the Great Southern Ocean (see Figure 1.1).

The Basin's boundaries to the east and south are provided by the Great Dividing Range including the Australian Alps while in the north, west and south-west, the boundaries are much less distinct. For instance, in the Wimmera in the south-east and the Bulloo Basin in the north-west, where the Darling rises, the boundaries are subdued watersheds with catchments dominated by internal drainage, which thus contribute very little to surface-water flow. Elsewhere low to medium-altitude ranges mark the Basin's limits: the Mount Lofty Ranges are in the south-west, the Grey and Barrier ranges to the west and the Chesterton and Warrego ranges in the north. Extensive plains and low uplands that are less than 200 m above sea level dominate most of the Basin. Thus, low relief and very low gradients dominate the flow regimes and the movement of floods and pulses of flow in the rivers for the greater part of their length. The three largest rivers of the Basin — the Murray, the Murrumbidgee and the Darling — are not only the three longest rivers on the Australian continent, but more importantly they are central components of our folk law and our history for both Aboriginal and European settlers. These rivers — their red gums, the wetlands and the arid lands of the west — are welded into the array of Australian icons and culture.

The MDB is essentially a shallow bowl almost full of deposited material (Rutherford 1990). Two large depositional geomorphic forms dominate the Basin. In the west, the rivers have carved channels and anabranches over ancestral channels and troughs of the ancient Mallee plains (Figure 1.1). This Mallee plain in the west is characterised by extensive sequences of sand ridges of both marine and Aeolian origin with no tributary stream junctions except those of the Murray and the Darling. To the east, a large proportion of the Basin consists of very large riverine floodplains of 200 m or less in elevation, characterised by meandering river systems of multiple channels, anabranches, billabongs and wetlands (Rutherford 1990). The riverine plain is essentially a complex network of river channels and floodplains overprinting on ancient river channels and floodplains, with alluvial deposits interspersed with widespread and complex patterns of Aeolian depositions. Present-day drainage often breaks up into complex distributor systems extending across the plain to form a mosaic of ancient and modern channels, which generally rejoin forming the Murray trunk stream near the junction between the riverine plain and the Mallee sands. All the streams of the Riverina are characterised therefore by very low gradients with enormous year-to-year and decade-to-decade variability in flow, and the movement of water, nutrients and sediments between the floodplain and the network of ancient and modern channels.

The oldest of the ancient river systems (MDBMC 1987:10) are the *'deep leads'* — deeply buried channels now filled with sands and gravels. The next are the 15–30 000-year-old structures known as *'prior streams'*, and their sediment-filled channels are slightly elevated above the surface of the current floodplain. The most recent are the *'ancestral rivers'* because of their close relationship with the present drainage networks — showing up as winding depressions across the landscape. The interlinking of these systems of sands and gravels, which now contain groundwater, is critical to the present hydrology of the Basin.

The riverine floodplain of the Murray and the Murrumbidgee has a remarkable network of drainage lines, channels and wetlands. Many other, much older systems have evidence that these streams carried much more water and sediment than they do today. The earlier streams were responsible for the depositions of the alluvial plains. In the post-glacial period (the past 15 000 years), stream discharge has moderated and sediment loads have become finer, muddier and with slower stream velocities. Stream banks, slopes and dunes have stabilised, groundwater has fallen and lakes have dried up. The windblown *parna* and riverine alluvium have been deposited during the Quaternary across the riverine plain and along the eastern slopes (Butler 1958) — a consequence of a wet, stable period of soil formation — followed by wind erosion and deposition of the previously mentioned material to be reworked once again in the soil-forming process that characterises the Riverina floodplains.

Middle Murray Rivers carried sandy bed loads and built larger meadow scrollbars, a process which is no longer occurring. Flow from the rivers was very much greater than at present, probably fed by snow-melt from glaciers and large areas of permanent snow in the eastern uplands. There were periods of much greater flow than this prior to extraction for irrigation.

The river channels have evolved to be connected regularly to these over-bank ecosystems via a range of flood events interspersed with long periods of droughts under historical and recent highly variable rainfall sequences, as set out in Figure 1.2.

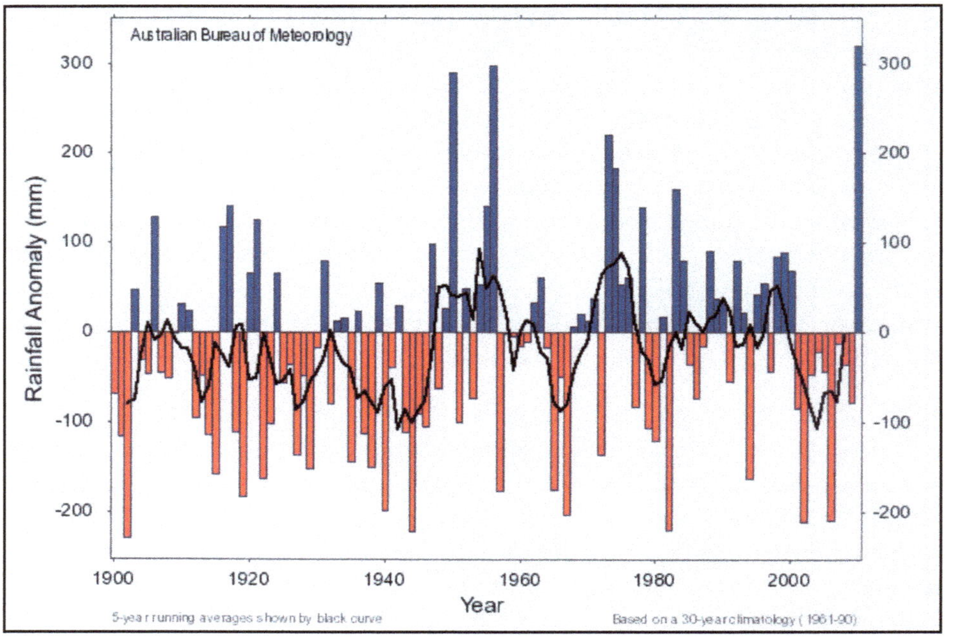

Figure 1.2 Annual rainfall anomaly for the Murray–Darling Basin, 1899–2010

Source: <http://www.bom.gov.au/cgi-bin/climate/change/timeseries.cgi?graph=rranom&area=mdb&season=0112&ave_yr=5>

The highly variable climate means that water availability varies greatly from season-to-season as does its quality during the seasons and between seasons. The highly variable flow of the rivers has led to large storages being built to reduce the variability of supply. The impact of such large storages on the river-flow regime, water quality and ecological functioning of the rivers have been very great indeed (NRC 2009b).

For the rivers, groundwater, wetlands, floodplains, lakes and estuaries to regain their ecological function and become healthy, they must regularly have over-bank flows that connect the channel to the floodplains, billabongs and wetlands. Plants and algae in these places transfer and enrich the river water with energy, carbon, nutrients and food-web elements, which then move back over time to the channel and drive the ecological activity along the length of the river (Overton and Saintilan 2010). These critical exchanges are represented for red-gum communities in Figure 1.3.

This process might be repeated many times along the path of these floodplain rivers. The flow of water through all these components of the river system is fundamental to the ecological health and function of the Basin. River regulation and over-allocation have severed these vital connections. It is in the interests of all water users that these functions be restored (NRC 2009b).

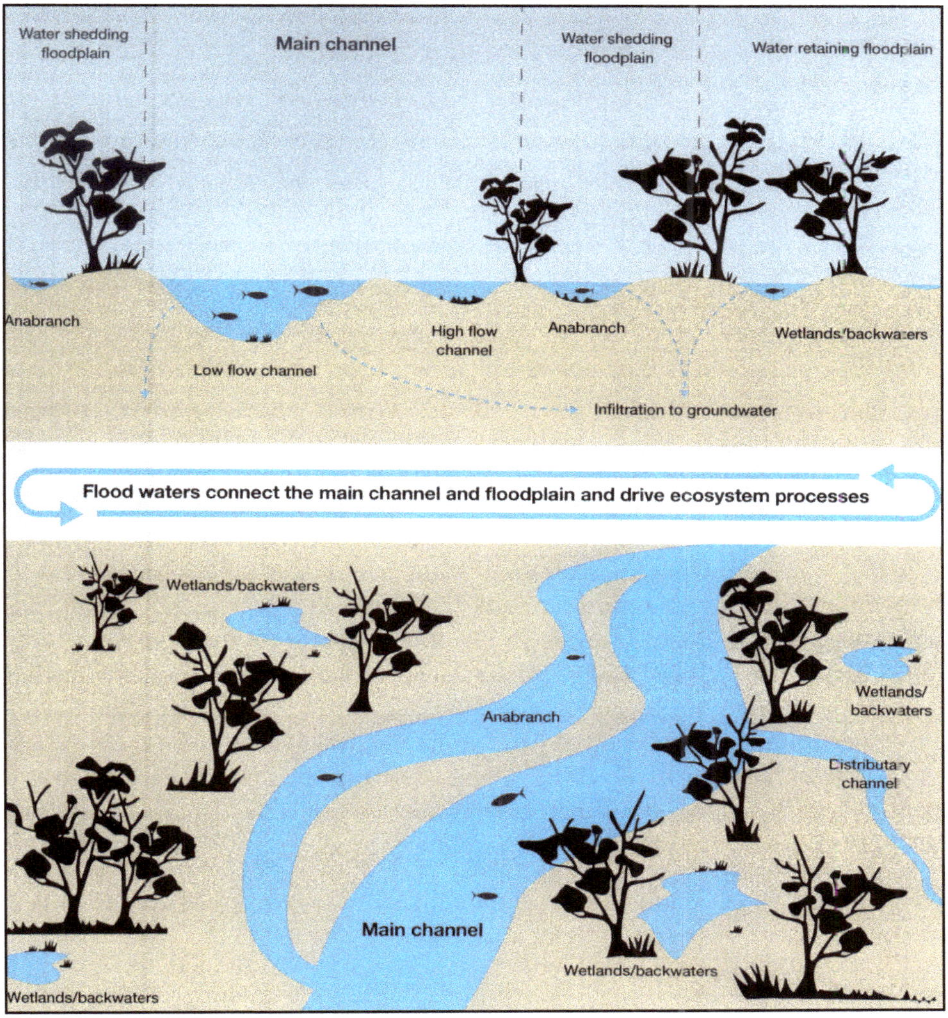

Figure 1.3 Cross-section and oblique views, respectively, of ecological function and the floodplain hydrology of river red-gum forests

Source: NRC (2009b:165).

This is how our Murray–Darling floodplain rivers work. They must be connected regularly to the over-bank ecosystems (Doody et al. 2009). Not only do these surface-dependent ecosystems in billabongs, floodplains and wetlands supply to the river much of its photosynthetic-driven ecosystem function, these places are also where many of the groundwater aquifers are recharged. Floods over floodplains are important to the recharge of groundwater. Our river channels, floodplains and groundwater systems are nearly always interconnected. Water flows from flooding rivers into the underground water and quite often flows through these groundwater aquifers back to the river to provide the base flow of our rivers in dry times. These connections and flow to and from the river channel are critical to the health of the river system (NRC 2009b). They need to be regularly fed and connected.

This floodplain geomorphology has shaped the ecosystem function of the river, maintained key habitats and ensured bird and fish-breeding events. The sequences of flooding events across these floodplains created over 30 000 wetlands in the MDB, 11 of which are listed under the Ramsar Convention on Wetlands of International Importance.

The environmental challenge is that current extractions are such that the frequency of low-flow years associated with closure of the Murray mouth has increased from about 5 per cent prior to European settlement to more than 60 per cent. The biggest impact is in the Lower Lakes, the Coorong and Murray mouth at the end of the system, but assets throughout the Basin are in a major state of decline (Senate Standing Committee on Rural and Regional Affairs and Transport 2008:39). In addition to reductions in flows due to high rates of extraction and drought, river regulation has reduced the variability and frequency of low to medium over-bank or flood events so that floodplains are now much drier and more saline as a result of evaporation from off-river pools (NRC 2009b:102–99). Far fewer regular flood events and minimal flows during dry periods along with stream salinisation mechanisms interact to produce acid-sulphate sediments in substantial parts of the Basin (Hall et al. 2006; Lamontange et al. 2006; McCarthy et al. 2006). These can cause permanent damage to the benthic habitat, create acid waters and release heavy metals, which can result in total collapse of ecological functions in these ecosystems.

Overall, the ecological health of 20 of the 23 river valleys in the Basin is classified as either poor or very poor (Davies et al. 2008).

Managing erosion and deposition processes to maintain healthy refuges within the channels and the floodplain through floods and droughts is a critical issue which must be addressed in successful water reform.

The Murray–Darling rivers need both floods and drying regimes that restore healthy ecological function across the whole Basin. Peter Cullen wrote:

> This is a challenging area for science, but current thinking indicates that the goal of ecological management is to restore or maintain resilience so the systems can cope with the shocks of climate or other factors they experience. It takes extreme events like droughts and floods to let us see whether we have kept resilience in our systems. We are not managing these systems for some benign 'average' condition, but so they can cope with the extremes that characterize the Australian climate and our agricultural markets. (Cullen, forthcoming).

Water reform will need to pioneer ways to go forward with management that can make river systems resilient to the shocks of the droughts and massive floods that are often amplified by our engineering interventions. To do that with current climate variability will challenge our science and our society, but to add the impacts of climate change on variability and changed probability distributions for our rainfall will stretch both science and society to their limits.

Climate History and Future Change

Climate is Highly Variable in Space and Over Time

Inflows to the Darling River in the north of the Basin are derived from highly variable episodic summer rainfall, often driven by monsoon depressions. The Murray River and its tributaries have their source in the Australian Alps and receive most of their inflows from rain and snow in winter and spring (Figure 1 4).

This steep decline in rainfall features a complementary increase in evaporation. Most of the Basin west of the Wagga Wagga–Dubbo axis has an annual water deficit; it is only in the eastern margins — maybe 15 per cent of the Basin (Crabb 1997:6) — that there is a water surplus that can drive the river flows in the Basin. In fact, today some 37 per cent of the flow is driven from about 3 per cent of the Basin.

This is illustrated in Figure 1.5 (Donohue et al. forthcoming), which provides the average Budyko-modelled annual run-off across the Murray–Darling Basin for the 1981–2006 period, the distribution of basin run-off expressed as a percentage of total run-off and the percentage of basin run-off for a given percentage of basin area. Clearly, the hydrology of the Basin is characterised by eastern upland headwaters contributing most water for very long, low-gradient rivers meandering through semi-arid and arid plains. There are, however, summer flows from monsoon influences and depressions that attach to these, which can give high flows in the Darling River. The consequence is a river system with very high spatial and temporal variability in flow regimes where floods and droughts are important to the ecological systems that have evolved over long periods of geomorphic stability.

It is in this landscape of extraordinary tectonic stability that the modern climate (past 500 000 years) operated to drive the erosion, depositional and fluvial features of the Basin. Whilst the geological foundation is old and stable, the climate, in contrast, has been highly variable and has oscillated between very humid and extremely arid periods, with strong gradients across the Basin — perhaps much greater than we see today (Butler 1958).

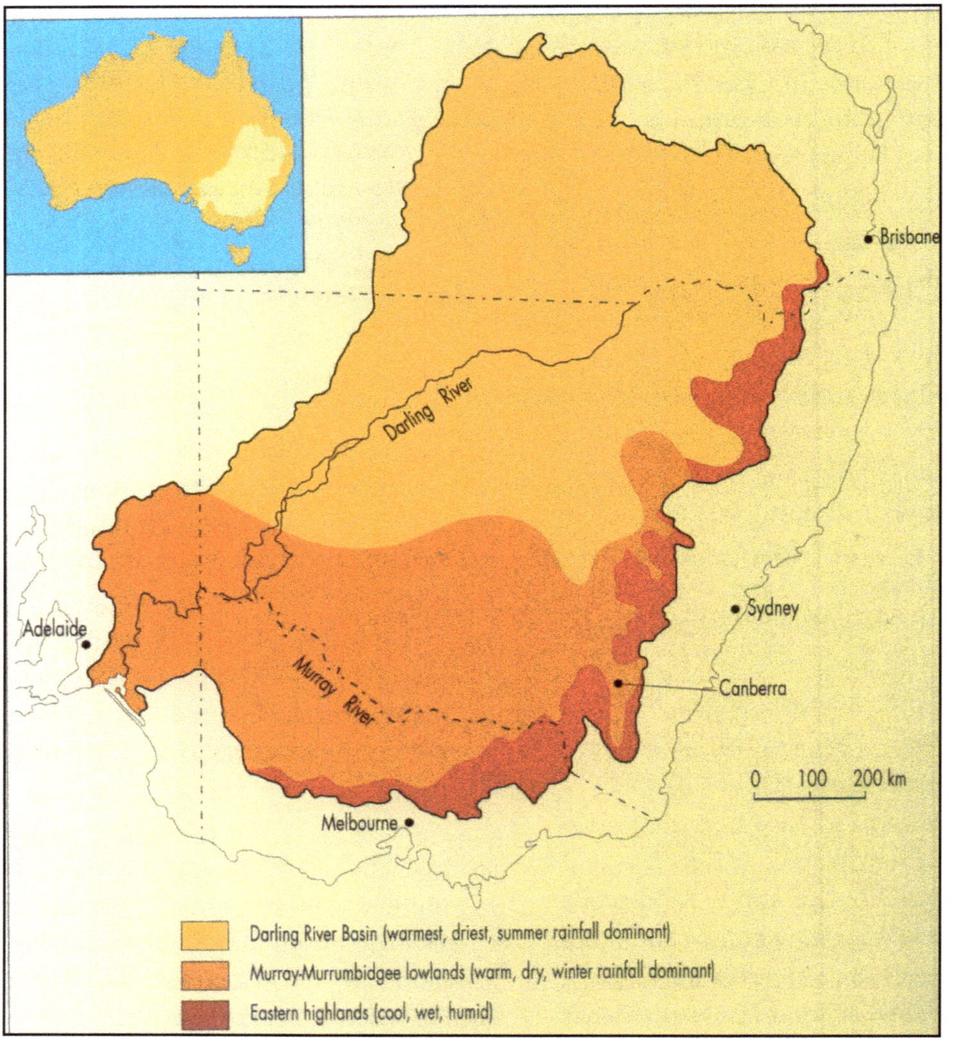

Figure 1.4 The three major climatic regimes of the Murray–Darling Basin

Source: Nix and Kalma (1982), reproduced with permission © CSIRO

Winds blowing across the continent from the south-west have been a major influence on landscape formation. Lakes have been formed by wind erosion of dry clays from valley floors. Along with the formation of lunettes, great clouds of dust laden with salt were lifted aloft by strong westerly winds to be transported and redeposited downwind to the east. Thus, the dust and salts are recycled up the valleys by the wind and the salt in clay aggregates, blown from the lake floors. Westerly winds have formed the west–east dune patterns that constitute the most distinctive feature of the Mallee region: long, low-gradient rivers with highly variable flow regimes are sourced from small upland regions.

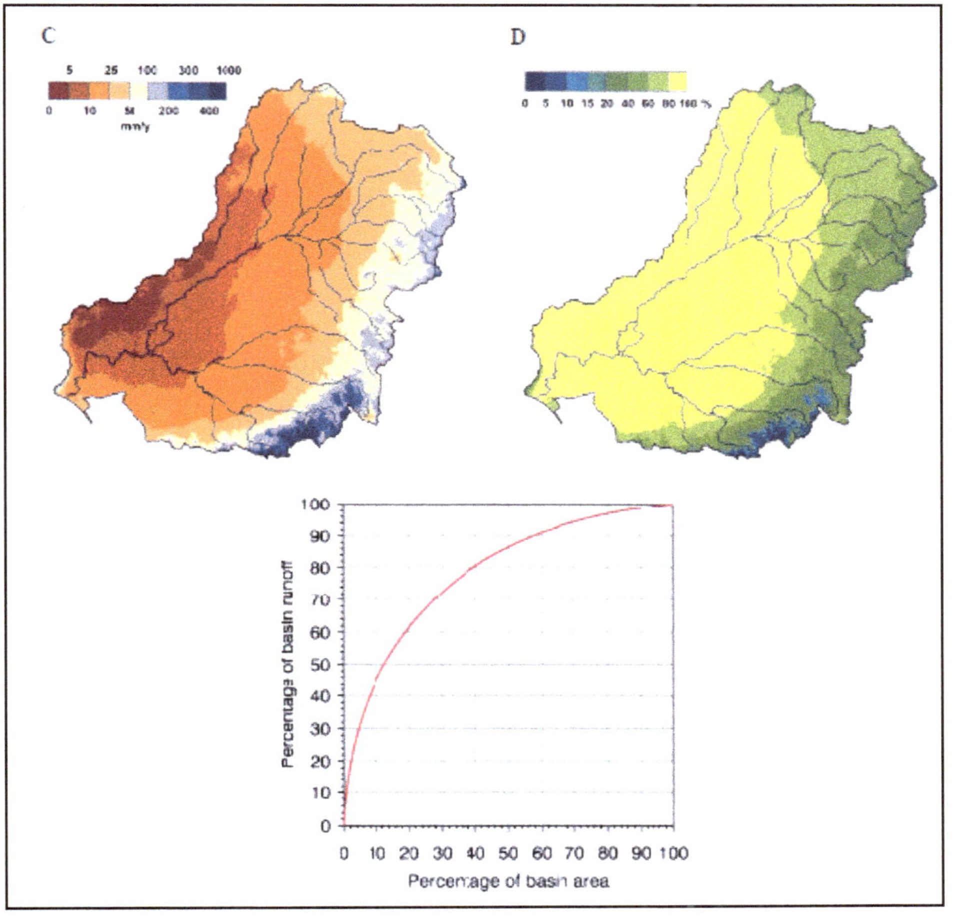

Figure 1.5 Average hydrological fluxes across the Murray–Darling Basin, 1981–2006

Notes: Plot C is Budyko-modelled annual run-off, and plot D is the distribution of basin run-off expressed as a percentage of total run-off. The lower plot shows the percentage of basin run-off for a given percentage of basin area.

Source: Donohue et al. (forthcoming).

The Murray River and its tributaries rise in the well-watered areas of the south-eastern highlands and flow westward through the dry interior. It is similar for the Darling, where the headwaters are in relatively well-watered, summer-dominant rainfall regions, in contrast with the winter-dominant snow-fed flow of the Murray Basin's rivers. In the Darling, as rivers pass from east to west and move into more arid regions, residual salts are concentrated in these arid landscapes where little surface run-off contributes to flow. The entry of groundwater — much of it very saline — to the Murray system as it moves through these arid lands is, however, significant. As mentioned earlier, this groundwater discharge to the Murray is the only mechanism by which groundwater can exit the Basin.

The MDB is large in area, yet very small in discharge, characterised by extreme events and thus very high variability. Over the period 1894–1993 the annual discharge from the Murray and the Darling ranged from a low of 1626 to 54 168 GL (Maheshwari et al. 1995).

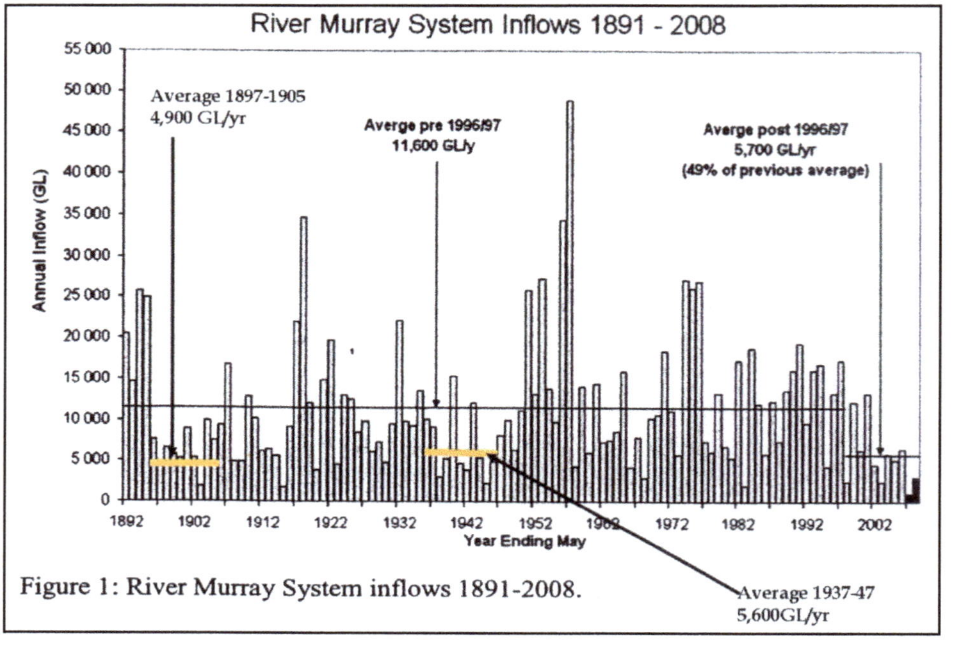

Figure 1.6 Murray River inflow, 1891–2008

Source: MDBC (2008:78).

Climate-Change Impacts on Drought, Floods and Water Use

The latest modelling indicates that Australian average temperatures are projected to rise between 0.6 and 1.5°C by 2030, and, with business as usual in terms of global greenhouse gas emissions, the warming is projected to be between 2.2 and 5°C by 2070. In the southern MDB, it is projected that there will be decreases in rainfall, especially during winter and spring, which are traditionally the times of the highest precipitation (Bureau of Meteorology 2010).

This large reduction from 10 to 20 per cent by 2030 under median climate change for the southern catchments is a trend consistent with a southward shift in the Southern Annular Mode (Meneghini et al. 2007) and is what would be expected with global warming. The central and northern river systems are expected to have smaller reductions, particularly where summer rainfall dominates the generation of water yield. It is also expected that ocean-warming patterns and the expression of La Niña influences on the northern Basin might increase summer rainfall and contribute to large flooding events across the Basin. Even in the south, where the mean is expected to decrease, the distribution of rainfall around that mean is less certain and could indicate an increase in extremes of both severe droughts and large floods (Min et al. 2011; Pall et al. 2011). In line with the work of Khan (2008), we can expect that climate-change impacts will change the statistical distribution of rainfall. While means are helpful, the task for best-practice river management will be to manage the extremes of droughts and floods and work with statistical probability distributions of both historical and predicted rainfall using models that incorporate the climate-change drivers attributed to anthropogenic greenhouse gas (Min et al. 2011; Pall et al. 2011).

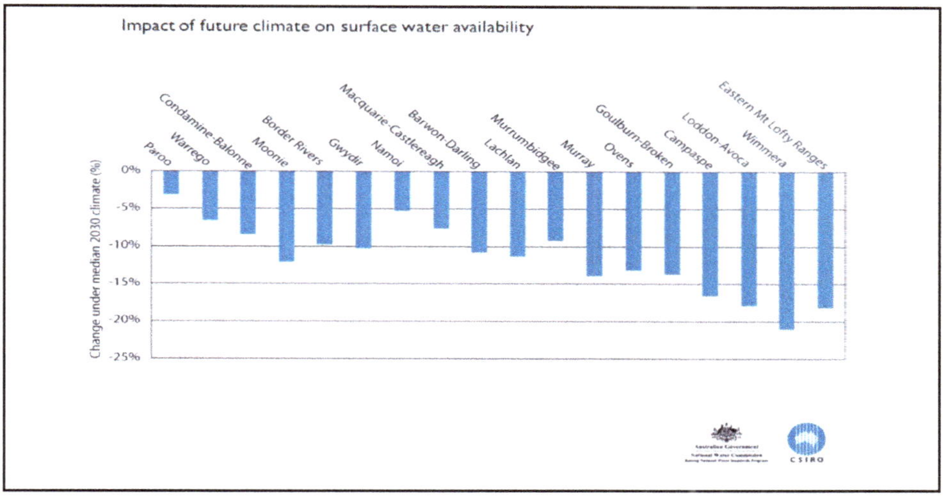

Figure 1.7 Predicted impacts of future climate on surface-water availability in the river systems of the Murray–Darling Basin

Source: CSIRO (2008:34).

Climate-change projections (CSIRO 2008:26) on run-off for the MDB are set down in Figure 1.8 and highlight the large uncertainty in future run-off projections. Averaged over the whole Basin, the median estimate under a mid-range climate-change scenario is a reduction in average annual run-off of 9 per cent by 2020, 15 per cent by 2050 and 23 per cent by 2070 (CSIRO 2008:26). Critically, the fall in surface-water availability over the past decade of the millennium drought in the southern part of the Basin has been much greater than the worst-case climate-change scenarios for 2030, as autumn rainfall has fallen by as much as 30 per cent (Proctor et al. 2009:9) and annual run-off relative to 1990 by nearly 40 per cent (CSIRO 2008:26). Overall in the Basin, we can expect a general drying, but also associated with a declining mean we can expect increases in variability and thus an increase in extreme dry periods and extreme wet periods.

The historical high variability and the likelihood that this might be further enhanced in ways unknown set a very exacting context for implementing water reform. The reform must be able to address this variability in that droughts and large floods are part of the Basin's history. The ecological functionality of the Basin has evolved to require this variation in flooding and drying. Such a pattern of flow regimes does not readily accommodate water-resource development and irrigated agriculture. This is the nub of the difficulties that will need to be addressed by water-reform policy and its implementation catchment by catchment, community by community.

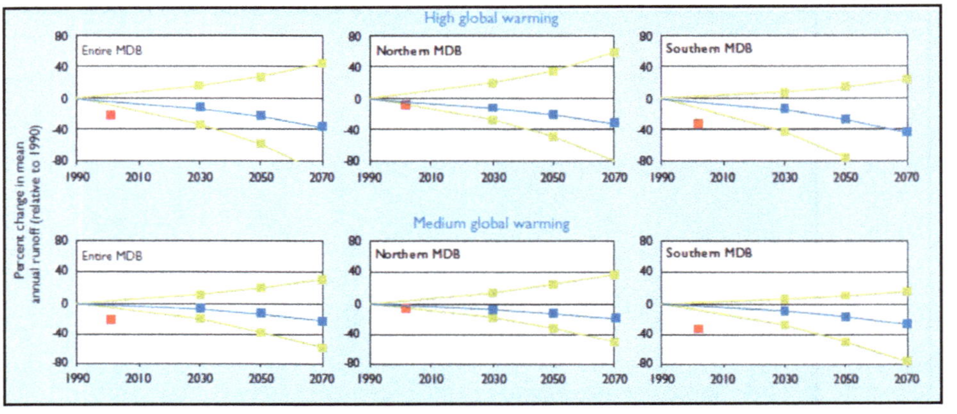

Figure 1.8 Run-off projections for 2030, 2050 and 2070 relative to 1990 for the entire Murray–Darling Basin, the northern Basin and the southern Basin under high and medium global-warming scenarios

Notes: The lines represent the upper and lower ranges and the median predictions of change to run-off. The squares locate the percentage change in run-off associated with the recent, 1997–2006 climate.

Source: CSIRO (2008:26).

In light of these circumstances, river-basin management is faced with greatly over-allocated rivers with water extraction that has not taken account of the huge climatic variability (see Figure 1.2), where a 40 per cent reduction in run-off is part of history. In addition, the climate-change projections for temperature, rainfall and run-off are alarming given the current dire straits of many environmental assets and ecosystems in the MDB. Fortunately, the millennium drought was broken by a major flood-producing rainfall event in 2010 (see Figure 1.2) — largely as the consequence of the El Niño event of 2009 transitioning into a very significant La Niña event in 2010 and 2011. If the nine-year dry had continued in the southern MDB then the sustainable diversion limits (SDLs) proposed under the Basin Plan could have been 'too little, too late'. The challenge is threefold: 1) immediately to reduce extractions to levels that prevent key ecosystems crossing critical thresholds; 2) ensuring environmental flows adjust to reduced inflows in ways that do not risk the long-term sustainability of key environmental assets; and 3) developing environmental watering that mimics pre-development frequency of flows and ensures episodic flooding events.

The European Aspirations

For the first European settlers, Australia was, indeed, a lucky country. The natural conditions were very conducive to pastoralism: extensive grasslands with minimal clearing required, mild temperatures and large areas. And the economic conditions were favourable, too: capital and convict and other cheap labour flowing from Great Britain and a rich market for the dominant product — wool. The early European settlers in Australia had no choice but immediately to adapt to the unique environment of high climate variability, low soil fertility and scarce water resources. In fact, the early settlers were arguably better adapted to Australian conditions than the assisted settlers of the twentieth century. They had large farming leases and relied heavily on what nature provided in natural-resource inputs, and it would seem they had a respect and even admiration for the natural environment (Cathcart 2009; Idriess 1993:7–238).

The frontier was born with the westward push over the Blue Mountains — opportunity mixed with risk to personal safety. By 1860, most of the economically useful land in the Murray–Darling Basin had been taken up, but, significantly, it was in the hands of a few settlers on extensive leases. The path to the natural riches of the Basin was led not only by explorers but also by squatters seeking new grasslands. Again, it was the promise of material gain that was a dominant factor, and the productivity and resource security of the Basin entered the national psyche (Powell 1993). The squatters' values and beliefs are still remnant in rural Australia today and can be traced to these times: man against the odds, heroism and scorn of authority (Connell 2007:7–47; Gray and Lawrence 2001).

In time, land administration caught up with expansion, and, under the policies of the day, agriculture was transformed to a system based on small proprietors drawing financial and labour resources from largely within the family (Gray and Lawrence 2001). Pastoral land use gave way, in part, to cropping. As in the United States, in Australia, the closer-settlement policy was built on a 'yeoman ethic'. In Powell's words:

> [T]he reassuring picture [was] of small freehold properties owned and controlled by industrious families for their own immediate and essentially non-commercial benefit. The implication of self-sufficiency entailed the requirement that each freehold be carefully cultivated to some significant extent, and the undisguised moral injunction suggested that the land be 'passed down' by successive generations, like heirlooms. This imagery has been a primary influence in [the] economic and political life of Australia since the mid-nineteenth century. (Powell 1993:24)

This imagery remains strong today.

There have been waves of closer-settlement schemes since the end of World Wars I and II. The policies of the day placed great store on the family farm as the instrument of community development and assumed viable farm sizes in times of agricultural prosperity.

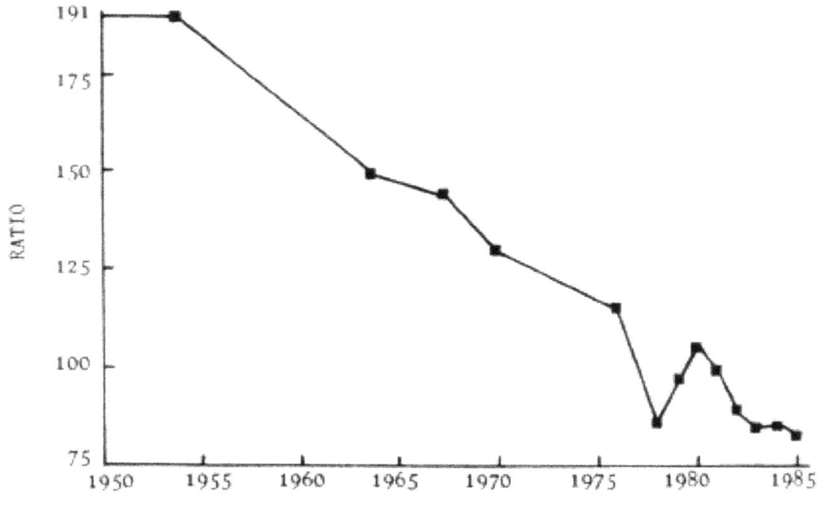

Figure 1.9 Farmers' terms of trade, 1950–85 ratio of prices received to prices paid

Note: B.A.E. data with 1980–81 = 100.

Source: Crofts (1985).

The early years of rapid expansion of both dryland and irrigated agriculture in the MDB were characterised by very favourable terms of trade, as in Figure 1.9 (Crofts 1985) — reflected in relatively cheap energy, fertilisers and agricultural chemicals coupled with rail transport systems tailored to the pattern of expansion. Most importantly, this was characterised by sequences of high rainfall that filled newly constructed dams and gave favourable seasons that favoured the establishment and utilisation of exotic pasture systems based on the greatly improved grain genetics arising from the 'green revolution'. Unfortunately, the very favorable terms of trade were not long lasting (see Figure 1.7). Wool, grain and meat prices declined rapidly relative to rising costs of production. Steeply declining terms of trade continue to plague most agricultural industries in the Basin. Only those industries with the ability to lift efficiency gains at rates that exceed the declining terms of trade have been able to build improving enterprise equity and profit. The trends in commodity prices and generally rising costs of production generate long-term financial pressure upon many farm families in broadacre dryland industries (Barr 2002, 2009).

From 1970, when commodity prices and market growth plummeted, the legacy of this policy was apparent: farms becoming marginal in profitability in the face of declining terms of trade, and, within a generation, rural reconstruction then rural adjustment schemes were introduced. This concentration in settlement and decline in farm profitability had many implications besides economic 'belt tightening' by the occupying families. Specifically, there has been a great environmental cost and a legacy where farm business units do not have the resources to address it — hence the expression 'you can't be green while you're in the red'.

But in contradiction with this policy position, Australian agriculture — from its European beginnings and quite unlike that of the United States — was largely export oriented. From first settlement, it was subjected to commercial forces both in an expectation of returns on the capital brought from abroad and in the relative lack of dependency on domestic consumption. It was not long before the natural resource and environmental constraints had to be tackled. They were successful, through raw innovation by farmers and sustained assistance by public investment in research and development, extension and education. This is where Australia did follow the US model, as early as the 1890s. Under pro-development policies of the day, science and technology were applied to 'problem solving': fallowing to conserve moisture, the application of phosphate fertilisers, or the exploitation of groundwater, which clearly was in response to recognised limits in dryland production at the time. The greatest legacy today is from the clearing of native vegetation and its replacement with introduced crops and pastures, aided by all manner of innovative engineering devices to prepare the ground for farming. The European 'land-use model' changed from extensive grazing where the adapted plant species remained largely intact, although often over-grazed, to an economic incentive to cultivate and crop all the arable area.

Innovation, problem solving, and the managerial capacity of farmers have sustained an impressive productivity growth through the twentieth century, particularly in cereal production. At the same time, evidence of the land-degradation impacts was mounting (Williams 2001). Retention of significant portions of land under native vegetation was advocated as early as the 1890s (Powell 1993). Concern within the scientific community about soil erosion grew, until the twin events of the Depression and the 'dust bowl' in the United States gained widespread attention. There were policy and institutional responses — typically, soil-conservation legislation and government research and advisory programs. But the form of land degradation that remained elusive was dryland salinity. Surprisingly, the association between vegetation clearing and the onset of salinity — in Western Australia — was first published in a scientific journal in 1924. In the Murray–Darling Basin, dryland salinity remains an ongoing threat, not only to farmland but to river water quality and water supplies, thus placing dryland and irrigation farming in an awkward relationship.

Irrigation development in the southern Basin dates back to the 1880s and was driven early by the need to overcome the variability of the Australian climate. Damming and regulating rivers are synonymous with water conservation and a new development frontier — *'making the deserts bloom'* (Cathcart 2009; Powell 1993; Williams 2003). While the earliest farmers used their own resources to exploit groundwater, this was soon followed in Victoria by State-funded infrastructure and closer-settlement schemes. Again, there was a policy objective on inland settlement, community development and, during the war years, population dispersal for national security reasons. Subsequently, the level of exploitation of River Murray water became a source of interstate conflict (Cullen 2002) and its value reached a point where a River Murray Commission's Report could state that *'water is of such inestimable value in an arid country like Australia that the State of people allowing it to go to waste or not effectively using it forfeits the moral although not of course the legal right to the enjoyment of the prospective privileges derived from these rights'* (Powell 1993:62).

The continued, then rapid expansion of irrigation development from the 1950s has resulted in a dilemma even more stark than dryland farming: enormous economic development but at major environmental cost to rivers. The Murray–Darling Basin Ministerial Council responded with some of the most interventionist natural-resource policies in Australian history: the Cap on diversions in 1995, and the Salinity and Drainage Strategy of 1989 (Connell 2007). What was overlooked post-1950 for 30 years was that all this expansion took place in the same high-rainfall climate sequence that drove the expansion of dryland agriculture (Khan 2008).

Figure 1.10 depicts this coincidence between irrigation expansion, as reflected in storage capacity and diversions in the Basin, with a large number of years with large positive rainfall anomalies over the 1950s to 1980s, with the exception of four drier years in the 1960s. This period is perhaps the wettest period in our recent history (Khan 2008).

Unfortunately, it came to a shuddering halt with the millennium drought from 2000 to 2009. Irrigated agriculture suffered a very large reduction in water availability over these nine years. But a reduction of up to 40 per cent in water use across the irrigation industries resulted in the gross value of irrigated agricultural production in the Basin falling only from $5.5 billion to $5.1 billion. This, however, masked large impacts on profit margins and enterprise equity despite the benefits of water trading, which has cushioned substantially the economic impact. These industries now have been weakened such that both the economic and the social wellbeing of the irrigation communities have deteriorated (Chapter 11).

The declining financial strength of dryland agriculture has been coupled with the impaired economic and social conditions of irrigated agriculture resulting in many communities in the MDB being under significant economic and social stress (Grafton and Jiang, 2010; Chapter 14).

Prior to the millennium drought, it was assumed that, unlike dryland farming — which we have seen is increasingly under price–cost pressures and has limited capital accumulation through profits — irrigation industries have a greater capacity to address their ecological footprint. This must now be seriously questioned.

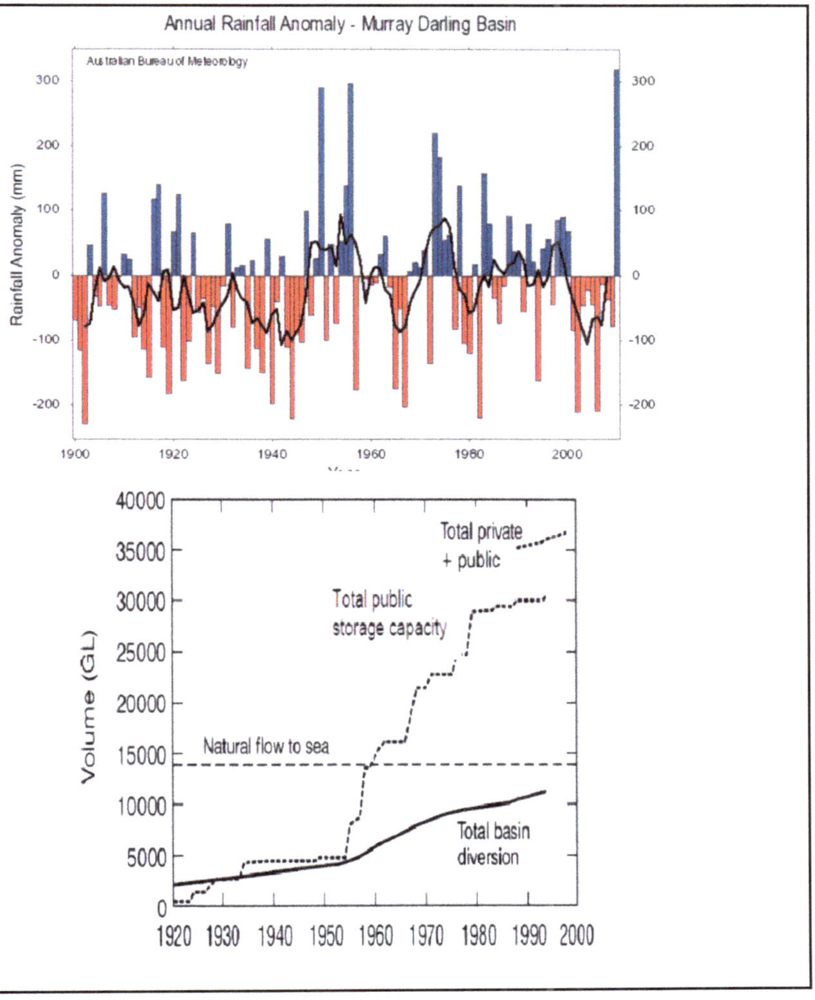

Figure 1.10 The annual rainfall anomaly for the Murray–Darling Basin, 1891–2010, and the storage capacity and diversions in the Basin, 1920–94

Source: Williams (2004).

The water-reform agenda that is urgently needed to restore some fundamental ecological functions to the river system and the groundwater now must be cast in a context of communities of the MDB suffering declining economic and social capacity. In such context, is water reform achievable? If it is to be achieved, how can this take place within the limited social and economic capacities of the MDB communities?

Human Interaction with the Bio-Geophysical Realities must be Managed in Successful Water Reform

Water Reform is About People, Policy and Practice

Irrigation industries by their nature are more intensive than others in their land use, and one of their productive resources — water — is amenable to tight management. In fact, in recent years, even with the bold national government policy (the water-reform agenda) — including the Cap on diversions in the Basin and water trading — there is still considerable scope to maintain or improve economic output from a scarce and increasingly scarcer natural-resource base. In fact, the Government can boast a pretty good track record—for instance, bringing off-farm impacts of pesticide use under control, and limiting or even reversing groundwater rise on farms. The serious impact on river health of the over-extraction of water from rivers to support these industries was, however, not addressed. This fundamental resource was taken for granted. Initially, it was believed that irrigation industries do not suffer the extremes of capital risk and income decline that the dryland industries experience. In general, they were thought to have the capacity to reinvest in their future and respond positively to the pressure to reduce their environmental footprint. The millennium drought has now thrown those assumptions into serious question. Failure to use historical data from equally severe droughts in the first 50 years of Federation (see Figure 1.10) blinded policy makers, water managers and communities alike to the fact that climate variability was very important to the water supply and how it was allocated, and that irrigation industries were just as vulnerable as the dryland industries.

Long-term average surface-water extractions in the Basin are about 12 000 GL/year, while the nominal total surface-water entitlements are approximately 16 000 GL. Long-term average surface-water availability is about 14 500 GL/yr.

The Basin is home to two million people, including 18 000 irrigators and more than 40 000 primarily dryland farmers (ABS et al. 2009), who collectively farm more than 80 per cent of land in the Basin. As we have seen, irrigated agriculture was actively encouraged through most of the twentieth century by State governments in the form of land grants and subsidised off-farm infrastructure and water storage (Connell 2007). To provide reliable flows for irrigation — especially during dry years — large public dams with total capacity in excess of two years of average flows were built and now account for about 80 per cent of water-storage capacity in the Basin (see Figure 1.10). The largest were constructed in the second half of the twentieth century — a period (as shown in Figure 1.10) that coincided with relatively wet conditions and large flooding sequences (Khan 2008; Vives and Jones 2005) — and water extractions more than doubled from the 1950s to the 1990s (Productivity Commission 2010).

Consequently, by the 1990s, the nominal volumes of surface water allocated to water entitlements that varied in terms of their reliability of supply by type, catchment and State, were about one-third greater than what was usually available. More than 90 per cent of the volume is diverted to water annual crops such as pasture, cotton and rice, and also perennials such as grapes and fruit trees (Productivity Commission 2010). By the 1980s, there was increasing concern over the effects of extractions on the health of the rivers and their ecosystems. A spectacular algal bloom in the summer of 1991–92 that extended along more than 1000 km of the Darling River gave the issue international prominence and led, ultimately, to a freeze on further growth in surface-water extractions at 11 600 GL against a long-term natural flow to the sea of approximately 14 000 GL (Chartres and Williams 2006:Figure 4). This became known as 'the Cap'. An unfortunate failure of public policy was that the Cap did not apply to groundwater extractions and groundwater use rapidly accelerated after its imposition on surface water (SKM 2003).

While the Cap limits surface-water extractions to what irrigators would have received given the infrastructure and management rules in place in 1993–94, assuming the same hydrological and climactic conditions, it does not address the over-allocation of the Basin's rivers. Simultaneous with the establishment of the Cap, the States in the Basin agreed to allow water trading by separating rights to land and water so that while total diversions were capped, individual irrigators could increase their extractions if they purchased water entitlements from others. Water trading in terms of permanent water entitlements came into being as part of the National Water Initiative in 2003 and was part of an accord to return water to over-allocated rivers. When water licences of various durations and forms were converted to indefinite water entitlements which became a tradeable water right this represented a large transfer of public assets to the private sector. The accord and social contract was in light of this transfer to be a return of water to the public to provide the water for over-allocated rivers systems of the Basin. The nature of this social contract is the foundation for the water reform process in which we are currently involved. This should not be forgotten.

Trade in these entitlements — and to a greater extent trade in the annual allocations of water assigned to each entitlement every irrigation season — has grown rapidly, with the trade in entitlements increasing more than twenty-fold since implementation of the Cap. Nevertheless, rules still remain that effectively prevent the trade of water entitlements across State boundaries (ACCC 2009).

The Cap was implemented with the water-management rules that operate separately in each State and that are counter-cyclical. Namely, they provide for a greater proportion of inflows to irrigators in dry years than in wet periods. These rules were justified, in the belief that the environment would get its 'fair share' of the water during flood events, and this would be consistent with the natural flows to which the Basin's ecosystems had evolved. A drying trend in the southern part of the Basin that began shortly after the implementation of the Cap (Vernon-Kidd and Kiem 2009), coupled with an increase in decadal mean temperature of between 0.2 and 0.3°C since 1960 in many parts of the Basin (Bureau of Meteorology 2010), has reduced water availability, and made the Cap a non-binding constraint in many parts of the Basin, as water-sharing agreements were suspended in most States. In contrast, the counter-cyclical water rules have become much more important, as they have reduced environmental flows in the MDB by much more than the actual declines in inflows during the recent millennium drought.

Agriculture and associated development in the past 213 years have contributed to economic growth and population wellbeing which is as good as you will find anywhere in the modern world; however, the exploitation of natural resources beyond their rates of replenishment and out of line with ecological functioning of the river systems has been a cost associated with this economic growth. Costs include declining river health, increased surface and groundwater use, rising salinity and acidity, loss of soil structure and condition; and environmental impacts that are quite stark — measured in the invasion of environmental weeds and feral animals, in flora and fauna species loss, and in ecosystem breakdown.

For industries that are largely dependent on natural resources, the over-allocation or overuse of resources gives very masked signals because of the long-run nature of the response to development and the year-to-year variability obscuring the trends. On the other hand, there are much sharper signals of environmental impact and these are readily observed and communicated widely and forcefully by conservation advocates on the fringe of these mainstream industries.

These environmental impacts are an early indicator of a loss of resilience in ecosystem function in the natural resources base that underpins our land-based industries — that is, the biophysical processes and systems are moving from the complex to the simple. These changes are foretelling a decline in the natural-resource base and ultimately in economic productivity.

It is clear that this negative interaction between people and their land, water and biodiversity is fundamental to the task that must be resolved for water reform to take place. Water reform is only one component of a whole complex of issues confronting communities in the MDB. From the brief and inadequate analysis here, it is clear that the social, economic and political fabric of the MDB communities is at a critical stage of change. The economic foundations and social fabric for both dryland and irrigated agriculture are no longer as they were when settlement was established. There is evidence that many communities are suffering serious financial and social difficulties. Many sense a serious threat to their existence. Water reform, then, is cast at a time in the history of the Basin when it is extremely difficult to understand and analyse the tensions and contradictions between the cultural values and social and economic aspirations of communities both in the Basin and outside with the necessary future actions, dictated by the bio-geophysical realities, necessary to restore resilience in the function of ecosystems upon which all ultimately depend. The narrative — the mix of values, aspirations and visions — for the MDB can no longer be cast as the one that served the past, but the new narrative for the Basin and its communities has yet to emerge.

This clash between bio-geophysical reality, economic reality, and social and cultural values demands a policy response. It is difficult and progress is slow. Australians have been at it in the Murray–Darling Basin for nearly 150 years. Why has it come to this? Why are we incapable of turning around the long-run environmental impacts of food and fibre production and associated development, measured symptomatically as declining river health and ecological function of rivers, the loss of wetlands, dryland salinity, decline in the quality of our soils and increasing loss of native species?

The challenge to us all is quite confronting now.

Our Future

For more than 100 years, Australians have fought over the waters of the Murray–Darling Basin. Water is a scarce resource, and, as we have developed the extraction industries of the Basin, we have not left enough water in the rivers to sustain a healthy river system. This will challenge our science and our society to find management solutions that can yield river systems resilient to the shocks of drought and massive floods which are often intensified by our engineering interventions. This is daunting enough with current climate variability, but then to add to this mix the impacts of climate change on climate variability and changed probability distributions for our rainfall will stretch us to our limits.

A new Basin Plan is being developed to address this blight on our history. Under the *Water Act 2007*, the job of the Plan is to provide for limits on the quantity of water that may be taken from the Basin water resources as a whole. It is part of a broader set of water reforms designed to bring balance back to the Basin (Wentworth Group 2010).

As we have seen, irrigated agriculture is the biggest consumer of the water we take out of the Basin. It uses the water to produce high-quality food and fibre. It produces valuable exports. It also provides jobs and is a foundation industry in the economies of communities along the rivers of the Basin. For the river to work, however, there must be enough water in the system to connect the wetlands and floodplains, and flush the salts, nutrients and sediments through the lakes, estuaries and the Murray mouth.

The changes the new Basin Plan brings can be delivered in different ways. They can be done badly, as we have seen in the past. Or they can be achieved in ways that provide the water to meet the environmental needs of the river system and at the same time help businesses and communities to optimise opportunities and adapt to a future with less water.

Perhaps the narrative we need is that the current crisis in the Murray–Darling Basin provides the best opportunity since Federation for Australians to work together to rebuild our Murray–Darling heartland, resulting in more resilient communities and healthier rivers. We must accept that we have a future with less water and a system that is currently over-allocated and is also confronted by climate change.

In the past 50 years, the majority of Australians — whether we live in the city or the country — have benefited greatly from the development of irrigated agriculture in the Murray–Darling Basin. In a relatively short time, we have developed an industry that produces much of the top-quality food and fibre we all enjoy at a cost the majority of us can easily afford. This growth in irrigation has been achieved largely as a result of families and individuals investing their time, money and aspirations into their farms, infrastructure and businesses. Industries that support irrigation have developed and employed people. In turn, service industries — ranging from the pub to the newsagent — have been set up, or expanded, to service the needs of the growing population. People have built their livelihoods; they have fed and clothed their families, paid school fees and mortgages — all off the back of irrigation and the industries it supports. Regional towns have grown as a result of the people irrigation brought. More people has meant social groups and sporting teams have been set up or expanded. Trophies have been won and lost and community spirit and identity have evolved.

The gold that fed this rush was water.

But we took too much and we did not take note of our history of droughts that could bring these industries to a halt. We now must do that: look to our climate history and to our climate future and rebuild our communities so that they have resilient futures in light of our climate and bio-geological heritage.

The current approach to the adjustment — although beginning to return some water to the system — has a fundamental flaw. Irrigation communities already suffering see it as another attack on their livelihoods. People cannot see a positive vision for themselves and their community in a future with less water.

In some areas, those who can see a vision for themselves and are willing to sell must struggle with the negative judgements of the rest of the community many of whose members are conscious of the importance of the water and the income it potentially but indirectly generates for them. A large proportion of the community that relies on irrigation-related activities for its income has no water to sell and does not benefit from its sale out of their region.

The social implications of the adjustment are massive, and if we do not address them we will not make the adjustment. We will instead fight amongst ourselves to protect our livelihoods. Rules limiting trading volumes, embargoes on trade and bickering between States highlight this reality.

In this challenge, however, lies opportunity.

With the right social processes in place, the irrigation communities of the Murray–Darling Basin could develop a new vision for their future. For some communities, the vision might be a future without irrigation. These local visions could integrate into a broader vision for a sustainable and profitable Murray–Darling Basin with healthy rivers, wetlands and floodplains.

Implementation of part of this process of water reform, and the $12.9 billion the Government has allocated, could be the catalyst to deliver a new future rather than the threat to communities it is currently seen as.

Other, existing nation-building and rural-development programs could be integrated with the water reforms to deliver services and infrastructure to help communities develop new opportunities. To deal with this, we will need a well-balanced, three-legged-stool approach to water reform. Currently, we have only two legs: buyback and infrastructure improvement to lift efficiency. Without the third leg of support to help regional communities plan for a future with less water and structurally adjust, the stool will fall over. This third leg is missing, and our communities are being expected to make these huge adjustments with little support from government.

Australian society as a whole has played a role in the development of this catastrophe through our government's over-allocation of water extraction from our rivers and groundwater. It seems only fair that all Australians take responsible action to assist our communities to make the required adjustment so that water extraction is in line with the capacity of the rivers and groundwater.

Ultimately, this will give us all an assurance of a more sustainable future. To support communities in the Basin and build on this legacy it will be increasingly important that Australians build a regulatory framework around food and fibre production in the Basin that enshrines its sustainability credentials as the water reform is implemented. Williams and McKenzie (2008) argue for a regulatory framework in Australia that ensures that all food reaching the consumer is produced in ways that minimise the damage to natural resources and the environment. Environmental management systems and proper labelling of food and its footprint are first steps and are currently maturing. But this alone is not sufficient. A regulatory framework is required that establishes that, for food and fibre to be marketed, it must have been produced by means which meet an Australian standard for sustainable food or fibre products. Such a standard must apply to both Australian grown and imported products. The water reforms when implemented have the potential to make the MDB one of the most sustainable food and fibre production basins on the planet. Australians will need to then maximise the benefits of such an achievement by ensuring we have a regulatory framework which enables Basin communities to capture of this competitive advantage on local and overseas markets.

For communities to begin to shape futures, it is so important that there be honesty and transparency in the magnitude of the reduction in water extraction that is compatible with a healthy Murray–Darling. Most regional cities, towns and communities within the Murray–Darling Basin face massive social and economic impacts of a water-reform agenda designed to improve the health of over-allocated rivers and groundwater. This upheaval comes at a time of severe drought and against a backdrop of climate change. Communities are faced with making tough and painful decisions.

There is ample evidence that regional communities and industry are actively taking responsibility for planning to live with less water and accept the need to return water to the river.

Certainly, the government buyback of water allocations and entitlements is a critical part of the solution, as is the government investment in water and irrigation infrastructure. But there is an urgent need to bring together these two elements in the water-reform agenda with a third element involving a strong focus on and commitment to community and industry planning as part of a package for regional development.

Governments have put some $12 billion on the table to address water reform in the Murray–Darling Basin. This investment should be a key plank in the regional development, rebuilding and revitalisation of the communities of the Murray–Darling. It is a magnificent opportunity to support, facilitate and resource our communities to find their own solutions for a more resilient future. Elsewhere in this book (Chapter 11; Hoggett et al. 2008; Miller 2010; Wentworth Group 2010) a number of writers explain how to empower individuals and groups of people by providing the skills they need to effect change in their own communities. These skills are often concentrated around building social cohesion through the formation of large social groups working for a common agenda.

We must support regional communities in a number of different ways to help them plan for a future with less water and provide the structural-adjustment support that will be required. History suggests that many attempts to assist autonomous adjustment backfire. Structural adjustment can be done very well or very badly. Funds skilfully applied to target areas can greatly speed up adjustment processes, especially if there are substantial public benefits at stake.

The whole water-reform package could be seen as an opportunity for major regional development based on community assistance for planning, building new futures and making the necessary structural adjustment. With this focus, the most effective use can then be made of water buybacks coupled with investment in infrastructure and on-farm innovation to drive water-use efficiency. Putting the focus on community development and the assistance required by communities who are faced with major change and adjustment could turn the current crisis into an opportunity for Australians to work together to rebuild our Murray–Darling heartland, resulting in more resilient communities and healthier rivers.

Acknowledgments

This chapter draws heavily on concepts developed in earlier work by myself, Kevin Goss and Daniel Connell. I am most grateful to Quentin Grafton and Chris Miller for access to unpublished work as well as their time to discuss issues raised in this chapter. I am also indebted to Tim Stubbs and Peter Cosier along with other colleagues in the Wentworth Group of Concerned Scientists for use of concepts and material from various drafts of our collective work and thinking. Tim McVicar and CSIRO colleagues made available data and figures from unpublished work, and I am most grateful. Belinda Crozier gave generously of her time against very short time frames to get the manuscript into a form suitable to hand to our editors.

Bibliography

Australian Bureau of Statistics (ABS), Australian Bureau of Agricultural and Resource Economics (ABARE) and Bureau of Rural Sciences (BRS) 2009, *Socio-economic context for the Murray–Darling Basin*, Descriptive report, MDBA Technical Report Series: Basin Plan, BP02, Murray–Darling Basin Authority, Canberra.

Australian Competition and Consumer Commission (ACCC) 2009, *Water Trading Rules: Draft advice*, Australian Competition and Consumer Commission, Canberra.

Barr, N. 2002, 'Structural change in Australian agriculture: implications for natural resource management', in *People Strategies and Outcomes 1986–96*, Australian Natural Resources Atlas, Department of Sustainabililty, Environment, Water, Population and Communities, Canberra, <http://www.anra.gov.au/topics/people/strategies-outcomes/index.html>

Barr, N. 2009, *The House on the Hill: The transformation of Australia's farming communities*, Land & Water Australia, Canberra, <http://lwa.gov.au/products/pn20461>

Bowler, J. 1990, 'The last 500,000 years', in N. Mackay and D. Eastburn (eds), *The Murray*, Murray–Darling Basin Commission, Canberra, pp. 96–110.

Bureau of Meteorology 2010, *State of the Climate*, Bureau of Meteorology, Canberra.

Butler, B. E. 1958, *Depositional systems of the riverine plains of southeastern Australia in relation to soils*, Soils Publication No. 10, CSIRO Division of Soils, Canberra.

Cathcart, M. 2009, *The Water Dreamers—The remarkable history of our dry continent*, Text Publishing, Melbourne.

Chartres, C. J. and Williams, J. 2006, 'Can Australia overcome its water scarcity problems?', *Journal of Developments in Sustainable Agriculture*, vol. 1, pp. 17–24.

Commonwealth of Australia 2007, *Water Act 2007*, No. 137 (2007), Department of the Attorney-General, Canberra.

Commonwealth of Australia 2004, *Intergovernmental Agreement on a National Water Initiative*, Commonwealth of Australia, Canberra.

Commonwealth Scientific and Industrial Research Organisation (CSIRO) 2008, *Water availability in the Murray–Darling Basin*, A report from CSIRO to the Australian Government, Murray–Darling Basin Sustainable Yields Project, CSIRO, Canberra.

Connell, D. 2007, *Water Politics in the Murray–Darling Basin*, The Federation Press, Sydney.

Crofts, F. C. 1985, 'Thirty years of pasture research which changed Australia', *Proceedings of 3rd Agronomy Conference 1985 Hobart, Tasmania, Australia*, <http://www.regional.org.au/au/asa/1985/invited/p-01.htm#TopOfPage>

Crabb, P. 1997, *Murray–Darling Basin Resource*, Murray–Darling Basin Commission, Canberra.

Cullen, P. W. 2002, 'The common good', in D. Connell (ed.), *Unchartered Waters*, Murray–Darling Basin Commission, Canberra, p. 138.

Cullen, P. (forthcoming), 'Facing up to the water crisis in the Murray–Darling Basin', in V. Cullen (ed.), *This Land Our Water*, ATF Press, Adelaide.

Davies, P., Harris, J., Hillman, T. and Walker, K. 2008, *Sustainable rivers audit: a report on the ecological health of rivers in the Murray–Darling Basin*, Prepared by the Independent Sustainable Rivers Audit Group for the Murray–Darling Basin Ministerial Council, Canberra.

Donohue, R. J., Roderick, M. L. and McVicar, T. R. (Forthcoming), 'Assessing the differences in sensitivities of runoff to changes in climatic conditions across a large basin', *Journal of Hydrology*.

Doody, T., Overton, I. and Pollock, D. 2009, 'Floodplain inundation mapping', in I. C. Overton, M. J. Colloff, T. M. Doody, B. Henderson and S. M. Cuddy (eds), *Ecological outcomes of flow regimes in the Murray–Darling Basin*, Report prepared for the National Water Commission by CSIRO Water for a Healthy Country Flagship, Canberra.

Evans, R., Brown, C. and Kellett, J. 1990, 'Geology and groundwater', in N. Mackay and D. Eastburn (eds), *The Murray*, Murray–Darling Basin Commission, Canberra, pp. 75–95.

Francis, D. and Hengeveld, H. 1998, *Extreme Weather and Climate Change*, Climate and Water Products Division, Atmospheric Environment Service, Canadian Ministry of Supply and Services, Ontario.

Grafton, R. Q. and Jiang, Q. 2010, *Economics of drought, water diversions, water recovery and climate change in the Murray–Darling Basin*, CWEEP Research

Paper 10-01, Centre for Water Economics, Environment and Policy, The Australian National University, Canberra, <http://cweep.anu.edu.au/pdf/publications/research_papers/10-01_WaterEconomics.pdf>

Gray, I. and Lawrence, G. 2001, *A Future for Regional Australia: Escaping global misfortune*, Cambridge University Press, UK.

Hall, D., Baldwin, S., Rees, G. and Richardson, A. 2006, 'Distribution of inland wetlands with sulfidic sediments in the Murray–Darling Basin, Australia', *The Science of the Total Environment*, vol. 370, pp. 235–44.

Hoggett P., Mayo M. and Miller C. 2008, *The dilemmas of development work: Ethical challenges in regeneration*, Policy Press, Bristol.

Idriess, I. 1993, 'The cattle king', in *Ion Idriess's Greatest Stories*, HarperCollins, Sydney, pp. 1–238.

Khan, S. 2008, 'Managing climate risks in Australia: options for water policy and irrigation management', *Australian Journal of Experimental Agriculture*, vol. 48, pp. 265–73.

Lamontange, S., Hicks, W. S., Fitzpatrick, R. W. and Rogers, S. 2006, 'Sulfidic materials in dryland river wetlands', *Marine and Freshwater Research*, vol. 57, pp. 775–88.

McCarthy, B., Conalin, A., D'Santos, P. and Baldwin, D. 2006, 'Acidification, salinisation and fish kills at an inland wetland in south-eastern Australia following partial drying', *Ecological Management and Restoration*, vol. 7, pp. 218–23.

Maheshwari, B. L., Walker, K. F. and McMahon, T. A. 1995, 'Effects of regulation on the flow regime of the River Murray, Australia', *Regulated Rivers: Research and Management*, vol. 10, no. 1, pp. 15–38.

Meneghini, B., Simmonds, I. and Smith, I. N. 2007, 'Association between Australian rainfall and the Southern Annular Mode', *International Journal of Climatology*, vol. 27, pp. 109–21.

Miller, C. 2010, 'Developing capacities and agency in complex times', in *Challenging Capacity Building*, eds. M. Clarke and S. Kenny, Palgrave Macmillan, Melbourne.

Min, S.-K., Zhang, X., Zwiers, F. W. and Hegerl, G. C. 2011, 'Human contribution to more-intense precipitation extremes', *Nature*, vol. 470 (17 February), pp. 378–81.

Murray–Darling Basin Authority (MDBA) 2010, *Guide to the proposed Basin Plan*, Murray–Darling Basin Authority, Canberra, <http://thebasinplan.mdba.gov.au/guide/guide.php?document=the-murray-darling-basin>

Murray–Darling Basin Commission 2008, Annual Report 2007-2008, Murray-Darling Basin Commission, Canberra, p.78. <http://publications.mdbc.gov.au/download/MDBC-AR-0708.pdf>

Murray–Darling Basin Ministerial Council (MDBMC) 1987, *Murray–Darling Basin Environmental Resources Study*, NSW State Pollution Control Commission, Sydney.

Natural Resources Commission (NRC) 2009a, *Riverina Bioregion Regional Forest Assessment: River red gum and other woodland forests*, NSW Natural Resources Commission, Sydney, <http://www.nrc.nsw.gov.au/content/documents/Red%20gum%20-%20FAR%20-%20Complete.pdf>

Natural Resources Commission (NRC) 2009b, 'Implications of changes in climate for water availability and flooding regimes', in *The Final Assessment Report on Riverina Bioregion Regional Forest Assessment: River red gums and other woodland forests*, NSW Natural Resources Commission, Sydney, <http://www.nrc.nsw.gov.au/content/documents/Red%20gum%20-%20FAR%20-%20Ch%208.pdf>

National Water Commission (NWC) 2009a, *National Water Commission Australian Markets Report 2008–2009*, National Water Commission, Canberra.

National Water Commission (NWC) 2009b, *Australian Water Reform 2009: Second biennial assessment of progress in implementation of the National Water Initiative*, National Water Commission, Canberra.

Nix, H. A. and Kalma, J. D. 1982, 'The climate of the Murray-Darling Basin.' in *Murray–Darling Basin Project Development Study*, Stage 1 Working Papers, CSIRO Division of Water and Land Resources, Canberra.

Ollier, C. D. 1995, 'Tectonic and landscape evolution in Southeast Australia', *Geomorphology*, vol. 12, no. 1 (April), pp. 37–44.

Overton, I. and Saintilan, N. (eds) 2010, *Ecosystem Response Modelling in the Murray–Darling Basin*, CSIRO Publishing, Melbourne.

Pall, P., Aina, T., Stone, D. A., Stott, P. A., Nozawa, T., Hilberts, A. G. J., Lohmann, D. and Allen, M. R. 2011, 'Anthropogenic greenhouse gas contribution to flood risk in England and Wales in autumn 2000', *Nature*, vol. 470 (17 February), pp. 382–6.

Powell, J. M. 1993, *The Emergence of Bioregionalism in the Murray–Darling Basin*, Murray–Darling Basin Commission, Canberra.

Proctor, W., Hosking, K., Carpenter, T., Howden, M., Stafford Smith, M. and Booth, T. 2009, *Future research needs for climate change adaptation in the Murray–Darling Basin*, Report to the Murray–Darling Basin Authority, Canberra.

Productivity Commission 2010, *Market Mechanisms for Recovering Water in the Murray–Darling Basin*, Productivity Commission, Melbourne.

Rancic, A., Salas, G., Kathuria, A., Acworth, I., Johnston, W., Smithson, A. and Beale, G. 2009, *Climatic Influences on Shallow Fractured-Rock Groundwater Systems in the Murray–Darling Basin, NSW*, NSW Department of Environment and Climate Change, Sydney South, <http://www.environment.nsw.gov.au/resources/salinity/09108GroundwaterMDB.pdf>

Rutherford, I. 1990, 'Ancient river, young nation', in N. Mackay and D. Eastburn (eds), *The Murray*, Murray–Darling Basin Commission, Canberra, pp. 17–36.

Senate Standing Committee on Rural and Regional Affairs and Transport 2008, *Water Management in the Coorong and the Lower Lakes*, Parliament of Australia, Canberra.

Sinclair Knight Merz (SKM) 2003, *Projections of groundwater extraction rates and implications for future demand and competition for surface water*, Murray–Darling Basin Commission Publication 04/03, Sinclair Knight Merz, Melbourne.

Vernon-Kidd, D. C. and Kiem, A. S. 2009, 'Nature and causes of protracted droughts in southeast Australia: comparison between the federation, WWII, and big dry droughts', *Geophysical Research Letters*, vol. 36, no. L22707, <DOI:10.1029/2009GL041067>

Vives, B. and Jones, R. 2005, *Detection of abrupt changes in Australian decadal rainfall (1890–1989)*, Technical Paper No. 73, CSIRO Department of Atmospheric Research, Aspendale, Vic.

Wasson, R. J. 1987, 'Geology, geomorphology and mineral resources of the Murray–Darling Basin', in *Murray–Darling Basin Project Development Study*, Stage 1 Working Papers, CSIRO Division of Water and Land Resources, Canberra, pp. 1–9.

Wentworth Group 2003, *Blueprint for a national water plan*, A report from the Wentworth Group of Concerned Scientists, Sydney, <http://www.clw.csiro.au/new/WWF0703-02%20A4%20Horiz.pdf>

Wentworth Group 2010, *Sustainable Diversions in the Murray–Darling Basin—An analysis of the options for achieving a sustainable diversion limit in the Murray–Darling Basin*, Wentworth Group of Concerned Scientists, Sydney, <http://www.wentworthgroup.org/blueprints/sustainable-diversions-in-the-murray-darling-basin>

Williams, J. 2001, 'Farming without harming—can we do it?', *Agricultural Science*, vol. 14, no. 1, pp. 20–4; vol. 14, no. 2, pp. 34–7.

Williams, J. 2003, 'Can we myth-proof Australia?' *Australasian Science*, vol. 24, no. 1 (January/February).

Williams, J. 2004, 'Wealth from water: a national perspective', in G. Batten and J. Kent (eds), *Proceedings of the Wealth from Water Conference*. Wagga Wagga, NSW, 14 March 2003, Charles Sturt University and Wagga Wagga Chamber of Commerce and Industry, NSW, pp. 24–35, <http://www.clw.csiro.au/staff/WilliamsJ/Wealth_from_Water.pdf>

Williams, J. and Goss, K. 2002, 'Our difficult bequest: the collision of biophysical and economic reality, cultural values and public policy', in D. Connell (ed.), *Unchartered Waters*, Murray–Darling Basin Commission, Canberra.

Williams, J. and McKenzie, F. 2008. 'Agriculture' in *10 Commitments: Reshaping the Lucky Country's Environment*. Lindenmayer, D.B., Dovers, S., Hariss Olsen and Morton, S. (eds). CSIRO Publishing, Melbourne.

2. Freshwater Ecosystem Conservation: Principles versus policy

Jamie Pittock, C. Max Finlayson

Introduction

The Ramsar Convention on Wetlands and the Convention on Biological Diversity provide much of the constitutional mandate for the Australian *Water Act 2007* and proposed Murray–Darling Basin Plan, due to be adopted in 2011. The Plan is meant to give effect to these international agreements by reducing excessive water diversions to environmentally sustainable levels, using environmental flows to conserve key ecological assets such as wetlands, manage emerging risks to water supplies and then optimise socioeconomic benefits. The policies proposed in the October 2010 *Guide to the proposed Basin Plan* fail, however, the fundamental test in the Ramsar Convention—namely, maintaining the ecological character of wetlands. The proposed return of 3000–4000 gigalitres per annum on average to the environment is insufficient to conserve more than 75 per cent of the red-gum floodplain forests and designated Ramsar Wetlands of International Importance—or a representative range of wetland types. The threat to wetlands would be exacerbated by the minimal measures proposed to manage emerging risks to water supplies with climate change.

We conclude that the measures proposed in the Guide are insufficient to fulfil Australia's Ramsar Convention obligations, and recommend seven additional measures, including the return of 4000–7600 GL/year of water to conserve Ramsar wetlands. Other required improvements include undertaking a rigorous assessment of ecosystem services and of representation of biota in key ecological assets, protection of remaining unregulated water flows, more robust precautions to adapt to climate change, reoperating water infrastructure and identifying priorities for research to improve the subsequent iteration of the Basin Plan.

> In its dryness, Australia suggests the Planet's future, as the vast human population and the demands of its industries intensify competition for an unchanging quantity of freshwater; in water terms, Australia is a warning, and Chowilla [floodplain forests of the Murray–Darling Basin] is its immediate expression. The Chowilla red gums are part of a vast death event, encompassing hundreds of thousands, perhaps millions of trees...extending six hundred miles. (Leslie 2005:223)

In this chapter, we focus on management steps needed to sustain the freshwater ecosystems and other biodiversity of the Murray–Darling Basin (MDB). Although the Basin is located entirely within Australia, its management holds lessons for the conservation of an increasing number of river basins globally (Falkenmark and Molden 2008; Pearce 2006; Powell 2008). We begin by summarising the key features of a number of conservation treaties that Australia signed and upon which the Federal Government derives a large portion of its constitutional mandate for management of the Basin. The status and trends of freshwater ecosystems of the Basin, and past management responses, are described only briefly as they are covered in existing authoritative publications.

We then focus on the requirements of the *Commonwealth Water Act* (Commonwealth of Australia 2007) to develop a Basin Plan to 'give effect to relevant international agreements' by maintaining and restoring environmental assets through a return to 'environmentally sustainable levels' of extraction of water resources (MDBA 2010a:xii). The ecological and risk-management measures suggested in the first step in meeting this requirement—the *Guide to the proposed Basin Plan* (MDBA 2010a)—are then assessed, and conclusions are drawn on the extent to which the mooted policies would fulfil conservation principles, and international and national legal obligations for freshwater conservation. In particular, we focus on whether the proposed measures would conserve a representative range of wetland types, sustain the ecological character of the wetlands across the Basin, and adequately manage climate change-induced risks. Finally, further measures are recommended to ensure the survival of freshwater ecosystems in the Basin.

The International Context

In this chapter, the freshwater biome—which embraces rivers, floodplains, groundwater-dependent ecosystems, lakes and estuaries—is interchangeably described as: 'inland waters' biodiversity, as defined by the Convention on Biological Diversity (CBD); wetlands, as defined by the Ramsar Convention on Wetlands (Ramsar); and freshwater ecosystems. Ramsar defines wetlands broadly to embrace all freshwater plus coastal ecosystems to 6 m depth (Ramsar 2009a). Globally, freshwater ecosystems are especially threatened, over-exploited and poorly conserved (MEA 2005b; Pittock et al. 2008).

Starting in the 1970s, Australia joined other countries in adopting multilateral environmental agreements and in developing conservation targets and measures to conserve freshwater biodiversity. These actions were taken in response to growing concerns about the state of ecosystems around the world. The CBD commits member states to the conservation and sustainable use of biodiversity, which in the freshwater context has been elaborated through programs of work on protected areas and on the conservation of inland waters' biodiversity (CBD 2004a, 2004b; United Nations 1992). More specifically, the Ramsar Convention on Wetlands commits contracting parties to the wise use (ecologically sustainable development) of all wetlands, and to maintain the ecological character of all designated Ramsar wetlands of international significance (Pittock et al. 2010; Ramsar 2009a); 16 wetlands have been designated as Ramsar sites in the Murray–Darling Basin (DEWHA 2009). A common principle of both the CBD and Ramsar—and an obligation on their members—is the identification, conservation and protection of sites that are representative of the diversity of freshwater ecosystems (CBD 2004b; Ramsar 2008b), such as the different wetland types identified by Ramsar (2009b). Australia is also a member of a number of bilateral and multilateral conventions for the conservation of migratory species—notably for waterbird species, whose populations are declining (Finlayson et al. 2006).

These environmental agreements have weak enforcement provisions in international law, but they are significant in providing a constitutional mandate for the Federal Government to enact legislation to implement Australia's obligations (Pittock et al. 2010). Thus, the Basin Plan and other measures under the *Water Act 2007* are intended to faithfully reflect provisions of the conventions. If they were substantially inconsistent with the conventions their validity may be challenged in the courts (Pittock et al. 2010). Consequently, this assessment considers whether the conservation measures proposed in the Guide give effect to the CBD and Ramsar agreements by fulfilling relevant conservation requirements, including the conservation of adequate and representative examples of freshwater ecosystems, conservation of threatened species, and maintenance of the ecological character of designated Ramsar wetlands. Further, the *Water Act* requires management of risk—in particular, of changes in hydrology due to climate change (Commonwealth of Australia 2007). This assessment looks at the effectiveness of the proposed management of the risk to freshwater ecosystems from climate-induced changes to hydrology.

Freshwater Ecosystems of the Basin

The Murray–Darling Basin covers 1 061 500 sq km (Figure 2.1) and contains the longest river system in Australia. The high-rainfall mountain districts in the east account for 5 per cent of the area and yet contribute more than 50 per cent of the run-off (CSIRO 2008). The MDB has diverse species and ecosystems, nearly 57 000 sq km of wetlands, and 16 wetlands covering 6363 sq km listed as internationally important under Ramsar (DEWHA 2009; Kingsford et al. 2004; MDBA 2010a, 2010b). By distinguishing between 60 000 sq km of floodplains and 25 000 sq km of 'wetlands', the Murray–Darling Basin Authority (MDBA 2010a:59) does not appear to understand the broad definition of a wetland under the Ramsar Convention.

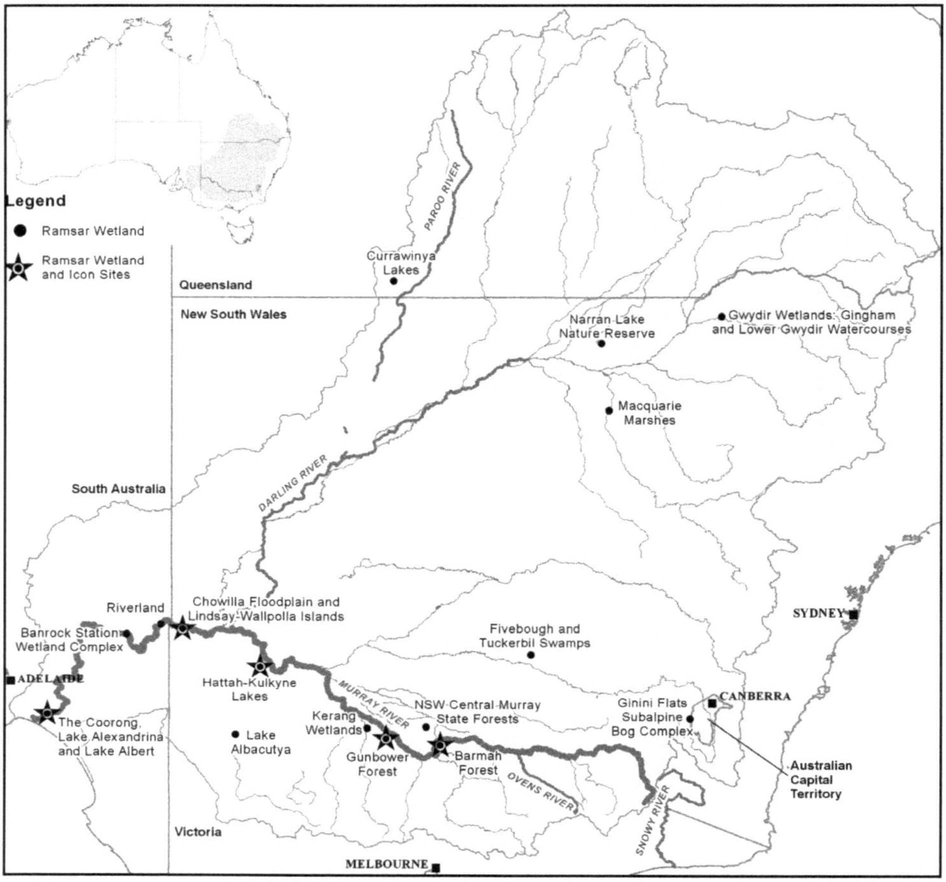

Figure 2.1 The Murray–Darling Basin showing the location of designated Ramsar wetlands

A 2007 audit of the 23 river valleys in the Basin found that only one was in good condition versus five classified as moderate, nine as poor, and eight as very poor (Davies et al. 2010). The condition of freshwater ecosystems of the Basin depends substantially upon adequate water flows of the right quality and at the right time. Flood pulses are particularly important, but their frequency has been greatly reduced such that only one-quarter of the active floodplain in the Basin had been inundated in nine years—a period approaching the tolerance thresholds of some floodplain species (Overton et al. 2009). Importantly, different wetland ecological communities lie on an inundation (or elevation) gradient on the floodplain, and thus a reduction of environmental flows disproportionately impacts on wetlands on the higher reaches of floodplains, such as black-box floodplain forest (NRC 2009; Overton 2010). Further, as river flows are diminished upstream in wetlands or through extractions, over-bank flood frequency diminishes further down the rivers and the ecological impacts increase (CSIRO 2008).

Figure 2.2 Psyche Bend Lagoon on the floodplain of the River Murray near Mildura, Victoria, showing the impact of reduced environmental flows that have led to salinisation, acidification and the death of wetland vegetation

Photo: J. Pittock, 2009

The Basin's freshwater ecosystems are declining primarily as a result of over-extraction of water for consumptive uses—a threat that might now be exacerbated by climate change (Pittock et al. 2010). Expressions of this decline include the death of extensive areas of floodplain forests and their replacement with terrestrial species, toxic cyanobacteria blooms, increasing salinity, oxidation of sulphate soils to form sulphuric acid, declines in waterbird and native fish populations, and expansion of invasive species (Pittock and Finlayson 2011; Pittock et al. 2010). A number of these changes might be difficult if not impossible to reverse, involving such elements as the deaths of trees that are hundreds of years old and changes to the chemistry of sediments (Figure 2.2). For these reasons, the restoration of water flows of adequate volume, quality and timing is essential to sustain ecological assets.

Irrigated land covers only 2 per cent of the Basin yet uses 90 per cent of diverted waters to produce 70 per cent of Australia's irrigated agricultural output—valued at A$7 billion per year (ABS et al. 2009). With increasing concern over the state of the river and its wetlands, the MDBA (2010a:xiv) states: 'The real possibility of environmental failure now threatens the long-term economic and social viability of many industries and the economic, social and cultural strength of many communities.'

Past Efforts to Restore the Basin's Wetlands

The European history of the Basin is marked by a sequence of attempts to reform management (Connell 2007). In recent decades, the failure of previous institutional reform has been exposed by another ecological crisis, sparking the adoption of yet another reform effort. Three recent measures adopted for ecological conservation are worth noting in an analysis of the likely effect of the policies proposed in the Guide. First, under the National Water Initiative (NWI) (COAG 2004), the State governments agreed to prepare water-sharing plans for all rivers, yet New South Wales in 2006 and Victoria in 2007 suspended their plans as the 'drought' from 2002 deepened, highlighting the sovereign risk that bedevils river-conservation measures (NWC 2009).

Second, a number of government programs were established to recover water through efficiency measures because of the perceived benefits for both irrigated agriculture and the environment. These measures proved, however, to be an 'efficiency trap' because they were: slow and unreliable in returning water to the environment, double counting leaking water that returned to the environment, unfairly subsidising inefficient farmers, and risking investing in infrastructure that could become stranded (Grafton and Hussey 2007; Productivity Commission 2010).

Finally, governments prioritised investments in 'environmental works and measures', and engineered structures to enable the use of less water to maintain larger areas of wetlands (Pittock and Lankford 2010). In part this was due to a decision by the governments to reallocate just 500 GL/yr to the River Murray (MDBMC 2003)—a manifestly inadequate volume to sustain major wetlands (Jones et al. 2002)—and the resulting need to make such a small volume deliver greater environmental benefits. There was, however, little consideration of the path dependency this infrastructure establishes, the resulting fragmentation of riverine ecosystems by weirs, levees and channels, or the need for ongoing and high-quality management interventions for this strategy to be successful (Pittock and Finlayson 2011). Indeed, the opportunity cost of this expensive infrastructure has not been considered by governments when the same funds could purchase sufficient water entitlements to make these engineering measures largely redundant (Kingsford et al. 2009; Pittock et al. 2010).

The modest targets adopted by governments in 2003 to conserve iconic wetlands along the River Murray had failed by 2009, with extensive loss of floodplain forests, including in designated Ramsar wetlands (Pittock et al. 2010). The lesson from the unsuccessful attempts to use water-sharing plans, agricultural water-use efficiency and environmental water demand management measures to sustain the health of freshwater ecosystems is that trying to manage the Basin with maximum water-use efficiency leaves no room for error; unplanned events such as poor governance, changes in water use, drought or extreme climate change will derail institutions that lack resilience. We argue that these types of events should be better anticipated and the risks proactively managed. The measures proposed in the *Water Act* and the *Guide to the proposed Basin Plan* were assessed in this context.

The *Water Act*

The *Water Act* established the MDBA with a remit to develop and supervise a framework for the management of the Basin's water in the national interest. The Authority is required to: give effect to relevant international agreements; conserve ecological values and ecosystem services; reduce water extraction to environmentally sustainable levels; and, subject to this, optimise socioeconomic and environmental outcomes (Commonwealth of Australia 2007). The Basin Plan is the primary instrument for implementing the Act.

In October 2010, the Authority published the *Guide to the proposed Basin Plan* for public comment (MDBA 2010a). The MDBA is scheduled to publish a draft plan in early 2011 before submitting a final plan for the minister's consideration in late 2011. We now critique the proposals in the Guide and conclude by suggesting changes required to adequately implement international agreements, conserve ecological values and maintain ecosystem services.

Proposals in the *Guide to the proposed Basin Plan*

The Thresholds for Sustainable Take

The question of an environmentally sustainable level of take is in part a value judgment since the freshwater ecosystems of the Basin evolved with all the available water, and regular extraction of any large amount of water will have a greater or lesser environmental impact. The *Water Act* says that the environmentally sustainable level of take must not compromise: key ecosystem functions, key environmental assets, the productive base and the key environmental outcomes for the water resource (MDBA 2010a:xvii). The CBD and Ramsar have, however, more specific principles that should be used to determine the extent of the freshwater ecosystems to conserve (CBD 2004a, 2004b; Ramsar 2008b, 2009b). These are

- designated wetlands protected areas, such as Ramsar sites
- habitats of threatened biota
- habitats of migratory species
- important breeding habitat for biota such as fish
- comprehensive, adequate and representative examples of each wetland type and the habitat of species dependent on aquatic ecosystems.

The approach adopted by the MDBA (2010a:63) applies five criteria—namely, for: the presence of listed migratory species; natural, rare or unique ecosystems; 'vital' habitat; habitat of listed threatened species or ecosystems; and supporting 'significant' biodiversity. In particular, no effort has been made to identify representative examples of wetland biota for conservation as required by the two conventions. While many wetland types would be conserved through a system of environmental flows that maintain ecological processes, others will not. The wetland types that could be overlooked include those that are: on higher elevations of floodplains (for example, black-box floodplain forests), in the higher parts of catchments and minor tributaries (for example, tableland wetlands), or are disconnected from the rivers (for example, groundwater-dependent ecosystems).

Wise Use of All Wetlands

In addition, the Ramsar Convention (2005:Cl. 11) requires the 'wise use' of all wetlands, which is defined as 'the maintenance of their ecological character, achieved through the implementation of ecosystem approaches, within the context of sustainable development'. 'Ecological character' is defined as: 'the combination of the ecosystem components, processes and benefits/services that characterise the wetland at a given point in time' (Ramsar 2008a:Cl. 15). Consequently, the Australian Government may not decide to change the character of significant areas of wetlands, and the various services that they provide, and adhere to the Ramsar Convention. The proposal not to conserve 25 per cent of the red-gum forests, and, as a result, larger areas of black box and other wetlands that are less frequently watered, amounts to a significant change in wetlands character (Pittock et al. 2010).

Ecosystem Services

The *Water Act 2007* has an objective 'to protect, restore and provide for the ecological values and ecosystem services' (Commonwealth of Australia 2008:s. 3[d][ii]). This is, however, not clearly linked to the 'key ecosystem functions' that the *Water Act* specifies should not be compromised in determining the environmentally sustainable level of take (MDBA 2010a:xvii). The Guide scarcely mentions ecosystem services, but extensively considers ecosystem functions, which are defined as: 'the fundamental physical, chemical and biological processes that support the Basin's environmental assets' (Commonwealth of Australia 2008:59; see also Alluvium 2010).

Curiously, the Act and the MDBA do not draw on international academic literature or the conventions to place these concepts in a broader context. Ecosystem services (called 'ecosystem benefits/services' by Ramsar) are defined as 'the benefits that people receive from ecosystems' (Ramsar 2005:Cl. 11). International environmental institutions are increasingly adopting ecosystem services as a framework for their work, including Ramsar, the CBD and, most recently, the Intergovernmental Science-Policy Platform on Biodiversity and Ecosystem Services (IPBES 2010). Ecosystem services have a broader, defined typology for describing and managing the different regulating, cultural, supporting and provisioning services (MEA 2005a; Ramsar 2005)—an approach that brings together ecological and socioeconomic values. What the MDBA calls ecosystem functions are clearly just a subset of the regulating and supporting ecosystem services. The failure of the Basin Plan process to fully use the ecosystem-services approach risks omitting measures required to maintain key services and misses an opportunity to more cogently assess and explain the links between a healthy ecosystem and the wellbeing of people in the Basin (Figure 2.3).

The lack of consideration of ecosystem services in the Guide highlights the MDBA's evident lack of understanding of the definitions, relevance and application of key concepts of the Ramsar Convention. In particular, the Ramsar definition of wetlands is not applied, ecological character and ecosystem services are cited only in passing, and representativeness is omitted entirely. This is a questionable approach to faithfully implementing the conventions, which the proposed Basin Plan needs to demonstrate if it is to justify its constitutional mandate.

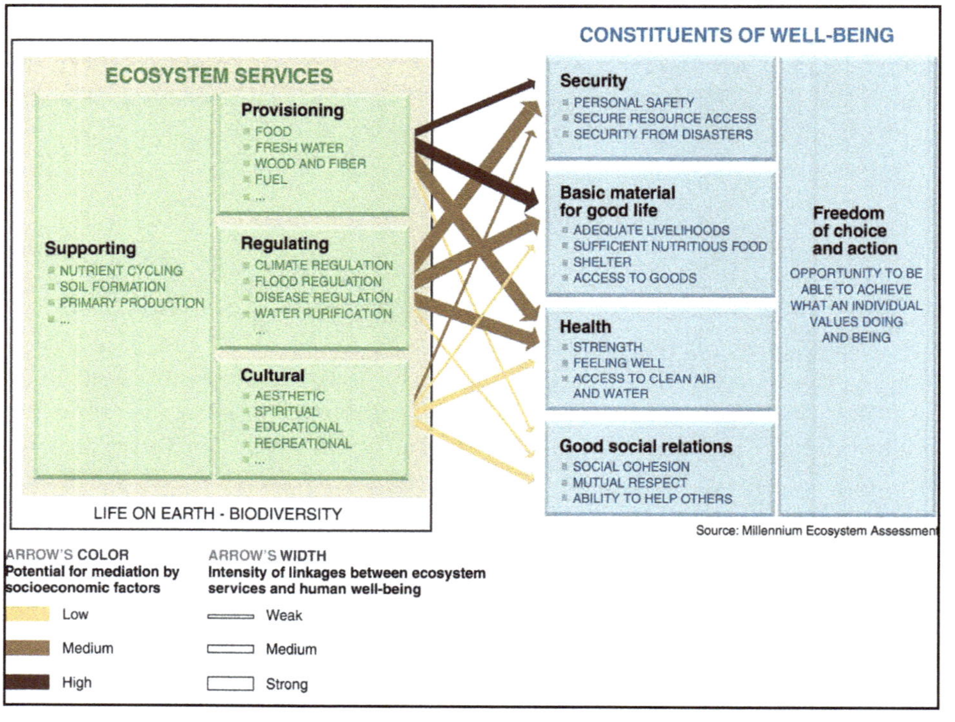

Figure 2.3 The links between biodiversity conservation, ecosystem services and human wellbeing

Source: Millennium Ecosystem Assessment (2005a).

Conservation Targets

The Guide proposes a target for the conservation of 75 per cent of the culturally iconic red-gum floodplain forests in good condition (MDBA 2010a:117). There are a number of issues associated with this choice of target. First, it is not clear why 75 per cent was chosen, and as it countenances the permanent loss of one-quarter of these forests, it is inconsistent with the Ramsar Convention's requirement to maintain the ecological character of wetlands, especially designated Ramsar sites. Second, it is unclear why no targets are considered for other wetland types. Indeed, as black-box floodplain forests are at higher-elevation (less frequently flooded) parts of the floodplains, conserving only a portion of the red-gum forests implies there will not be enough water to conserve many black-box forests. Third, this 75 per cent target amounts to a lower level of ambition for conservation of red-gum forests at Barmah–Millewa and Chowilla, while in The Living Murray program, targets to conserve the entire area had been proposed (MDBMC 2003; Pittock et al. 2010). It appears that not only is the red-gum target an unsupported compromise but is also used to justify focusing on a lower level of reallocation of water for the environment.

Water Reallocation

The Guide outlines hydrological modelling that recommends reallocation of between 3000 and 7600 GL/yr, on average, to the environment (MDBA 2010a:75–6). This is based on an assumption that adequate environmental flows for 18 major wetlands designated as 'indicator assets' will sustain ecological functions and a great many of the 30 000 wetlands throughout the Basin. The MDBA (2010a:74) admits, however, that with an extra 3000 GL/yr, the River Murray would still be ranked as having a 'poor' environmental-flow condition, and the environmental objectives of the *Water Act* are achieved only with an optimistic 'long term return to wetter conditions across the Basin' (p. 75). Further, only a 4000 GL/yr reallocation is likely to achieve the 75 per cent red-gum target (MDBA 2010a:118).

'Giving effect to relevant international agreements' is acknowledged only in relation to a reallocation of 7600 GL/yr. The MDBA then declares, however, that reallocation of more than 4000 GL/yr would have unacceptable socioeconomic impacts, and consequently it countenances reallocation of just 3000–4000 GL/yr on average (MDBA 2010a:100, 107). This contravenes the requirement in the *Water Act* to first give effect to international agreements and determine an environmentally sustainable level of take, and only then optimise the environmental, social and economic outcomes (Commonwealth of Australia 2008).

As a result of the proposed lower levels of water reallocation, the Guide suggests that there could be trade-offs between adequate watering of the Coorong and Lower Lakes Ramsar site versus upstream red-gum floodplain forests, and between watering red-gum versus black-box floodplain forests (MDBA 2010a:126). Neither of these measures would maintain the ecological character of designated Ramsar sites. The MDBA also proposes the politically beguiling, but environmentally risky, option of using engineering measures to apply less water to more efficiently maintain large areas of wetlands (as discussed earlier).

Management of Climate Risks

A further concern in the Guide is the limited consideration given to managing risks of reduced water availability arising from climate change. There are three aspects to this question—namely, whether: a) sufficient reductions in water allocations are considered to account for potential reductions of inflows; b) the adaptation measures adopted adequately spread risk; and c) new, climate change-related inflow interception activities are adequately regulated.

The MDBA has based its consideration of climate-induced hydrological changes on CSIRO modelling for the period 1990–2030 that suggested that surface-water availability could increase by up to 7 per cent in an extreme wet scenario, decrease by as much as 12 per cent with a median forecast, or decline by 24 per cent in an extreme dry scenario (CSIRO 2008). In the Guide, the MDBA (2010a:33–4) then proposes a reduction of water allocations of just 3 per cent based on the median forecast—halved on the basis that part of the impact should already be present, and halved again to reflect the envisaged 10-year (2011–21) life of the Basin Plan (even though Victorian implementation would commence only in 2019 and finish in 2024). This approach is questionable for a number of reasons. First, good risk management considers management for less likely but more catastrophic events. Second, the experience of climate-change impacts on water thus far in Australia has not entailed a linear reduction but rather a 'step-change' (for instance, in south-west Western Australia), and there is every possibility that such a change is under way in south-eastern Australia (CSIRO 2010). In this event, a 3 per cent reduction in water extractions by 2011–24 is likely to be most inadequate.

By default, the Guide proposes managing climate risks almost entirely by water reallocations and environmental flows. Such flows are highly vulnerable to poor government decision making in periods of water scarcity (as discussed earlier). Environmental flows are an important response but far from the only measure that would aid adaptation of freshwater ecosystems to climate change (Palmer et al. 2008; Pittock and Finlayson 2011). Unfortunately, a number of adaptation measures—such as re-establishment of riparian vegetation—lie outside the mandate of the Basin Plan (MDBA 2010a), but others should be considered and would help spread risk by maintaining refuges using approaches that do not require positive management intervention. One key measure would be the reservation of remaining free-flowing rivers and tributary streams from further water-resources development so as to maintain natural flow regimes and other ecological processes (Pittock and Finlayson 2011). Another could be to protect gaining river reaches as refuges from excessive groundwater extraction (CSIRO 2008; Pittock and Finlayson 2011).

Policy responses to climate change are likely to substantially change water use in the Basin, and the Basin Plan needs to ensure all major and emerging inflow-interception activities are adequately regulated. The Guide adopts a status-quo perspective—for instance, modelling only the likely impact of currently planned forestry plantations (MDBA 2010a:51). Yet all sides of politics in Australia have proposed mechanisms to subsidise landholders to undertake afforestation for carbon sequestration (ALP 2010; LPA 2010) without considering the substantial reductions of inflows that could ensue (Herron et al. 2002). Similarly, water demand for climate-related technologies such as bio-char, solar-thermal power plants and carbon capture and storage is likely to change water use in the Basin (Inhaber 2004; Smart and Aspinall 2009).

The MDBA (2010a:34) has also adopted 'a principle of equitable sharing of any reductions in water availability between consumptive and environmental uses' under future climate conditions. There is no doubt that this would be substantially better for the environment than the current water-allocation rules that overwhelmingly favour consumptive users when water is scarce (CSIRO 2008). This is not, however, consistent with the NWI agreement that water entitlement-holders should bear this risk (COAG 2004:s. 48). It also would not meet Australia's obligation under the Ramsar Convention to maintain the ecological character of all designated Ramsar sites while ever it is physically possible to do so, unless the Government declares that it cannot in the 'urgent national interest' (Pittock et al. 2010). In other words, in the event of climate-induced reductions in water availability, Australia is obliged to give priority to maintaining the Ramsar wetlands.

Conclusions

There are major consequences of the MDBA's proposals to manage the Basin with only a minimal reallocation of an extra 3000–4000 GL/yr on average of water for the environment—namely

- large areas of wetlands cannot be sustained, including about one-quarter of the red-gum forests, extensive areas of black-box forest, and trade-offs between maintaining different wetlands would be necessary
- the system would be managed without any substantial ecological capacity to withstand the impacts of unanticipated events, such as a step-change reduction in water inflows due to climate change or changes in inflow-interception activities in the catchment
- increasingly, conservation of key sites would rely on constant intervention through environmental-water demand-management measures when State government authorities in Australia have shown themselves to be incapable

of reliably implementing such freshwater-conservation measures by suspending previous water-sharing agreements (NWC 2009)

- Australia would continue to be in breach of a number of its international obligations under the CBD and Ramsar, such as maintaining the ecological character of all of the designated Ramsar sites.

Compared with the current, deplorable situation, there is no doubt that the freshwater ecosystems of the Basin will be in better condition with a reallocation of water to the environment and with new rules for sharing future water losses between consumptive users and the environment (Pittock et al. 2010). The *Water Act*, however, in requiring the Basin Plan to give effect to relevant international agreements, sets a threshold for better conservation of wetlands that the compromise proposals and trade-offs mooted in the *Guide to the proposed Basin Plan* fail to meet. In particular, omitting measures to conserve a representative range of wetland types and the envisaged loss of ecological character from large areas of Ramsar sites abrogate Australia's international obligations.

The MDBA still has the opportunity to revise its policies to recommend a Basin Plan in 2011 that would meet Ramsar Convention obligations and better conserve freshwater ecosystems in the Basin. We recommend seven key changes

1. identification and valuation of the full range of ecosystem services in the Basin as a basis for better decision making
2. assessment of wetlands against both the diversity of wetland types established by the Ramsar Convention and their nine site-listing criteria, to ensure that an adequate and representative range of wetland biota will be sustained across the Basin in key ecological assets
3. reallocation of sufficient water—in the 4400–7600 GL/yr on average range—to maintain the ecological character of all areas of the 16 designated Ramsar wetlands, as well as key ecological assets that are not located on main rivers
4. identification and protection of remaining free-flowing tributary rivers and of river reaches that receive net inflows from groundwater ('gaining reaches') as additional means of conserving freshwater biota and for climate-change adaptation that are largely independent of management interventions
5. more robust management of climate change-induced risks to water availability, including greater reallocation of water to the environment, implementation of the NWI policy for future reductions in water entitlements to be borne by water entitlement-holders for reductions due to climate change; and provision for contingency measures that would be triggered from 2015 should climate change severely reduce water availability

6. proactive management of existing water infrastructure to ensure that water quality and ecosystem functions are improved, including control of thermal pollution and reinstatement of fish passage
7. identification of additional measures for research by the MDBA in the 2011–20 period, such as conservation of groundwater-dependent ecosystems and climate-change adaptation options, to ensure that the subsequent iteration of the Basin Plan is further improved.

A reform plan that again compromises on allocation of adequate water to sustain the Basin's wetlands will result in ongoing environmental degradation, another crisis in the next, inevitable drought, and demands for further reform. We argue that implementing these stronger environmental measures, based on the Ramsar Convention's provisions, will provide greater certainty for people and nature, and enhance the possibility of maintaining healthy communities across the Basin.

References

Alluvium 2010, *Key ecosystem functions and their environmental water requirements*, Report by Alluvium for Murray–Darling Basin Authoriy, Canberra.

Australian Bureau of Statistics (ABS), Australian Bureau of Agricultural and Resource Economics (ABARE) and Bureau of Rural Sciences (BRS) 2009, *Socioeconomic context for the Murray–Darling Basin—descriptive report*, Murray–Darling Basin Authority, Canberra.

Australian Labor Party (ALP) 2010, *Carbon farming fact sheet*, Australia Labor Party, Canberra.

Commonwealth of Australia 2007, *Water Act 2007*, No. 137 (2007), Commonwealth of Australia, Canberra.

Commonwealth of Australia 2008, 'Water Act 2007', in Attorney-General's Department (ed.), *Act No. 137 as amended*, Commonwealth of Australia, Canberra.

Commonwealth Scientific and Industrial Research Organisation (CSIRO) 2008, *Water availability in the Murray–Darling Basin*, A Report from CSIRO to the Australian Government, Canberra.

Commonwealth Scientific and Industrial Research Organisation (CSIRO) 2010, *Climate Variability and Change in South-Eastern Australia: A synthesis of findings from Phase 1 of the South Eastern Australian Climate Initiative (SEACI)*, CSIRO, Canberra.

Connell, D. 2007, *Water Politics in the Murray–Darling Basin*, The Federation Press, Leichardt, NSW.

Convention on Biological Diversity (CBD) 2004a, *Decision VII/4. Biological Diversity of Inland Waters*, Convention on Biological Diversity, Montreal.

Convention on Biological Diversity (CBD) 2004b, *Decision VII/28. Protected Areas*, Articles 8(a)–(e), Convention on Biological Diversity, Montreal.

Council of Australian Governments 2004, *Intergovernmental Agreement on a National Water Initiative*, Council of Australian Governments, Canberra.

Davies, P. E., Harris, J. H., Hillman, T. J. and Walker, K. F. 2010, 'The sustainable rivers audit: assessing river ecosystem health in the Murray–Darling Basin, Australia', *Marine and Freshwater Research*, vol. 61, pp. 764–77.

Department of Environment, Heritage, Water and the Arts (DEWHA) 2009, *Australian Ramsar Wetlands*, Department of Environment, Heritage, Water and the Arts, Commonwealth of Australia, Canberra.

Falkenmark, M. and Molden, D. 2008, 'Wake up to realities of river basin closure', *International Journal of Water Resources Development*, vol. 24, pp. 201–15.

Finlayson, C. M., Gitay, H., Bellio, M. G., van Dam, R. A. and Taylor, I. 2006, 'Climate variability and change and other pressures on wetlands and waterbirds: impacts and adaptation', in G. C. Boere, C. A. Galbraith and D. A. Stroud (eds), *Waterbirds Around the World*, The Stationery Office, Edinburgh.

Grafton, Q. and Hussey, K. 2007, 'Buying back the living Murray: at what price?', *Australian Journal of Environmental Management*, vol. 14, pp. 74–81.

Herron, N., Davis, R. and Jones, R. 2002, 'The effects of large-scale afforestation and climate change on water allocation in the Macquarie River catchment, NSW, Australia', *Journal of Environmental Management*, vol. 65, pp. 369–81.

Inhaber, H. 2004, 'Water use in renewable and conventional electricity production', *Energy Sources*, vol. 26, pp. 309–22.

Intergovernmental Science-Policy Platform on Biodiversity and Ecosystem Services (IPBES) 2010, *Intergovernmental Science-Policy Platform on Biodiversity and Ecosystem Services*, United Nations Environment Programme, Nairobi.

Jones, G., Hillman, T., Kingsford, R., McMahon, T., Walker, K., Arthington, A., Whittington, J. and Cartwright, S. 2002, *Independent report of the Expert Reference Panel on environmental flows and water quality requirements for the River Murray system*, Cooperative Research Centre for Freshwater Ecology, Canberra.

Kingsford, R. T., Brandis, K., Thomas, R. F., Crighton, P., Knowles, E. and Gale, E. 2004, 'Classifying landform at broad spatial scales: the distribution and conservation of wetlands in New South Wales, Australia', *Marine and Freshwater Research*, vol. 55, pp. 17–31.

Kingsford, R. T., Fairweather, P. G., Geddes, M. C., Lester, R. E., Sammut, J. and Walker, K. F. 2009, *Engineering a Crisis in a Ramsar Wetland: The Coorong, Lower Lakes and Murray Mouth, Australia*, Australian Wetlands and Rivers Centre, University of New South Wales, Sydney.

Leslie, J. 2005, *Deep Water: The epic struggle over dams, displaced people, and the environment*, Farrar, Straus & Giroux, New York.

Liberal Party of Australia (LPA) 2010, *The Coalition's Direct Action Plan: Environment and climate change*, Liberal Party of Australia, Canberra.

Millennium Ecosystem Assessment (MEA) 2005a, *Ecosystems and Human Well-Being: Synthesis*, Island Press, Washington, DC.

Millennium Ecosystem Assessment (MEA) 2005b, *Ecosystems and Human Well-Being: Wetlands and water synthesis*, World Resources Institute, Washington, DC.

Murray–Darling Basin Authority (MDBA) 2010a, *Guide to the proposed Basin Plan. Volume 1: Overview*, Murray–Darling Basin Authority, Canberra.

Murray–Darling Basin Authority (MDBA) 2010b, *Guide to the proposed Basin Plan. Volume 2: Technical background*, Murray–Darling Basin Authority, Canberra.

Murray–Darling Basin Ministerial Council (MDBMC) 2003, *Murray–Darling Basin Ministerial Council communiqué*, 14 November 2003, Murray–Darling Basin Commission, Canberra.

National Water Commission (NWC) 2009, *Australian Water Reform 2009: Second biennial assessment of progress in implementation of the National Water Initiative*, September 2009, National Water Commission, Canberra.

Natural Resources Commission (NRC) 2009, *Riverina bioregion regional forest assessment river red gums and woodland forests*, Final assessment report, NSW Natural Resources Commission, Sydney.

Overton, I. 2010, 'Ecological outcomes of flow regimes in the Murray–Darling Basin', *Distilled*, <http://www.nwc.gov.au/www/html/2750-ecological-outcomes-of-flow-regimes-in-the-murray-darling-basin.asp?intSiteID=1]>

Overton, I. C., Colloff, M. J., Doody, T. M., Henderson, B. and Cuddy, S. M. (eds) 2009, *Ecological Outcomes of Flow Regimes in the Murray–Darling Basin*, CSIRO, Canberra.

Palmer, M. A., Reidy Liermann, C. A., Nilsson, C., Flörke, M., Alcamo, J., Lake, P. S. and Bond, N. 2008, 'Climate change and the world's river basins: anticipating management options', *Frontiers in Ecology and the Environment*, vol. 6.

Pearce, F. 2006, *When the Rivers Run Dry*, Eden Project Books, London.

Pittock, J. and Finlayson, C. M. 2011, 'Australia's Murray–Darling Basin: freshwater ecosystem conservation options in an era of climate change', *Marine and Freshwater Research*, vol. 62, pp. 232–243.

Pittock, J. and Lankford, B. A. 2010, 'Environmental water requirements: demand management in an era of water scarcity', *Journal of Integrative Environmental Sciences*, vol. 7, pp. 75– 93.

Pittock, J., Finlayson, C. M., Gardner, A. and McKay, C. 2010, 'Changing character: the Ramsar Convention on Wetlands and climate change in the Murray–Darling Basin, Australia', *Environment and Planning Law Journal*, vol. 27.

Pittock, J., Hansen, L. J. and Abell, R. 2008, 'Running dry: freshwater biodiversity, protected areas and climate change', *Biodiversity*, vol. 9, pp. 30–8.

Powell, J. L. 2008, *Dead Pool: Lake Powell, global warming and the future of water in the west*, University of California Press, Berkeley and Los Angeles.

Productivity Commission 2010, *Market mechanisms for recovering water in the Murray–Darling Basin*, March, Final report, Productivity Commission, Melbourne.

Ramsar 2005, *Resolution IX.1 Annex A: A conceptual framework for the wise use of wetlands and the maintenance of their ecological character*, Ramsar Convention on Wetlands, Gland, Switzerland.

Ramsar 2008a, *Resolution X.15 Describing the ecological character of wetlands, and data needs and formats for core inventory: harmonized scientific and technical guidance*, Ramsar Convention on Wetlands, Gland, Switzerland.

Ramsar 2008b, *Strategic Framework and Guidelines for the Future Development of the List of Wetlands of International Importance of the Convention on Wetlands (Ramsar, Iran, 1971)*, Ramsar Convention on Wetlands, Gland, Switzerland.

Ramsar 2009a, *Convention on Wetlands of International Importance Especially as Waterfowl Habitat. Ramsar (Iran), 2 February 1971*, UN Treaty Series No. 14583, As amended by the Paris Protocol, 3 December 1982, and Regina Amendments, 28 May 1987, Ramsar Convention on Wetlands, Gland, Switzerland.

Ramsar 2009b, *Information sheet on Ramsar wetlands (RIS)*, Ramsar Convention on Wetlands, Gland, Switzerland.

Smart, A. and Aspinall, A. 2009, *Water and the Electricity Generation Industry. Implications of use*, National Water Commission, Canberra.

United Nations 1992, *Convention on Biological Diversity (with annexes). Concluded at Rio de Janerio on 5 June 1992*, Treaty Series No. 30619, Multilateral, United Nations, New York.

3. Environmental Water: The Benefits of Ecological Goods and Services

Richard H. Norris

Introduction

Water has been viewed primarily as an exploitable commodity for human consumption, either directly or for growing food and fibre. As such, the wider range of benefits that might be derived from the ecological goods and services that are also dependent on water has not been recognised. Stocks of natural capital and flows of ecosystem goods and services are rapidly shrinking, caused by unprecedented rates of land-use change, water scarcity, and changing climate. Thus, there is a need for quantified trade-offs between social, ecological and economic objectives across a landscape that informs initiatives in three areas. First, to increase the equity, effectiveness and/or efficiency of resource use by revealing information on the sources and drivers of change, and the economic and social values generated across the full range of ecological goods and services. Second, to recognise and then improve the management and protection of existing ecosystem goods and services with a view to understanding their relevance and value. Third, to introduce incentive-based mechanisms (such as payments for maintaining ecosystem goods and services) for ecosystem conservation and management. This chapter aims to explore the range of water-dependent ecological goods and services and their importance in sustaining environmental futures in the Murray–Darling Basin.

The Murray–Darling Basin, like all river basins, naturally comprises a set of interconnected physical, chemical and biological elements, which revolve around the flow of water. Thus, the outcomes of activities and management decisions in one place have implications elsewhere in the Basin. This interconnectedness and the limited amount of water in the Basin means that the flow needs of any ecosystem along the rivers cannot be met without compromising the needs of another elsewhere (Roberts et al. 2001). Markets, property/water rights, government structures and social networks might all lead to overuse or misuse of natural resources. Treating ecological goods and services as if they were 'free' when they are not, or not recognising the adverse consequences of their use, will lead to reductions in potential human welfare or increased real costs to meet their loss. Explicit consideration is needed of the ecological implications of social and economic decisions of water use to understand the true costs and benefits (Costanza and Farber 2002). The Basin Plan should provide a framework for feedback to facilitate the management of water use in the context of the human and ecological benefits and costs, and therefore better understanding of their wider 'value'.

An assessment of the ecosystem services provided by the lowland rivers, wetlands and floodplains of the Murray–Darling Basin using the approach adopted by Costanza et al. (1997) estimates their value at between $187 and $302 million per annum (Thoms and Sheldon 2000). While this is small compared with the 2004–05 estimate of agricultural production of $15 billion (Australian Bureau of Statistics: <http://www.abs.gov.au/AUSSTATS/abs@.nsf/mf/4610.0.55.007/>), it does underpin much of that production, and its potential loss is not usually factored into considerations of water use. To provide the ecological goods and services that are fundamental to sustaining the uses to which they are put, the rivers need to be in good condition.

The Basin Plan is a key requirement of the *Water Act 2007* and includes an Environmental Water Plan (EWP) to protect and restore rivers, wetlands and other environmental assets, and to protect biodiversity dependent on the Basin's water holdings in accordance with the EWP.

Under Section 28(2) of the *Water Act 2007*, the EWP must specify

a) the overall environmental objectives for the water-dependent ecosystems of the Murray–Darling Basin;

b) targets by which to measure progress towards achieving the environmental objectives specified in accordance with paragraph (a);

c) an environmental management framework for planned environmental water and held environmental water;

d) the methods to be used to identify environmental assets in the Murray–Darling Basin that will require environmental watering;

e) the principles to be applied, and methods to be used, to determine the priorities for applying environmental water (including applying that water to environmental assets that are identified using the methods specified under paragraph (d));

f) the principles to be applied in environmental watering.

These specifications require that water allocations should be in the context of how ecosystems function and to achieve ecological outcomes. They also require ecological understanding of both the ecological processes and the features being protected and restored, and for relevance these could be related to specific ecological goods and services (see Table 3.1). The Department of the

Environment, Water, Heritage and the Arts (DEWHA 2009) has produced a discussion paper on *A framework for determining Commonwealth environmental water actions*. The paper reiterates an aim of the *Water Act 2007* that ecosystems have sufficient water to perform key ecological functions into the future. Central to this is an Environmental Water Plan to enable the environmental water of all holders and managers to be coordinated in a complementary basin-wide manner. An environmental plan also needs to be in the context of the National Water Initiative (NWI), which defines the environmentally sustainable level of extraction as 'the level of water extraction from a particular system which, if exceeded would compromise key environmental assets, or ecosystem functions and the productive base of the resource' (Commonwealth of Australia 2004). These definitions fit the specifications noted above but are much wider than the current focus on icon sites through the Living Murray and floodplain wetlands noted in the Sustainable Yields project (CSIRO 2008). Also, the NWI definition focuses on limiting extraction to maintain ecological values, rather than how much should be put back—a subtle but important difference in line with the restatement of the title question.

Thus, a primary aim of the Basin Plan is to fulfil the specifications above and guide water trading and recovery to achieve efficiencies and desired ecological outcomes. Stating these ecological outcomes in terms of ecological goods and services would provide a clearer understanding of the basis for their relevance and how they might be valued more widely than just maintaining the environment for its own sake. This chapter aims to elaborate on how various strategies might maintain ecological goods and services supplied by rivers and wetlands. In taking this approach, communities should be better informed as to the potential costs of losing ecological goods and services and of other ways of gaining benefit, rather than just consuming water. Inevitably, there is also a need to consider that ecological damage is caused by more than just water use, and that there is a requirement for assessment and the need for adaptive management.

Water to Achieve Ecological Outcomes

Rivers supply ecological goods and services that are much more than just water (Table 3.1). At present these goods and services are not captured in the market systems that primarily consider only the value of water for human consumption or the production of food and fibre. Defining 'ecological goods and services' is done to convey the important idea that ecosystems are socially valuable and to convey the benefits of nature to human welfare, households, communities and economies (Boyd and Banzhaf 2007; Daly and Cobb 1989). This section considers each of the major functions in Table 3.1 and how they might be addressed by current or proposed management activities likely to come from the Basin Plan.

Regulation Functions

Humans rely on healthy, functioning systems extensively for waste treatment and nutrient regulation (Table 3.1) at significant cost savings. Whenever effluents that are not fully treated are discharged to rivers (for example, from sewage-treatment plants), the riverine ecological processes are being relied on to undertake the final treatment, especially when downstream users depend on the water. The failure of the river ecosystem to supply this service with regard to nutrient regulation resulted in the Darling River algal bloom in 1992, which effectively shut down the entire community at substantial costs way above any treatment costs. The financial advantage of this ecological service would be easily calculated by factoring the additional costs of full treatment. The current scourge of European carp in many lowland rivers is, in part, likely to result from rivers fragmented by water-regulation structures and low flows that together reduce the competitiveness of native species and provide a habitat advantage to alien species. Flow management orientated to supplying human needs without consideration of environmental needs has exacerbated these problems again with significant environmental and economic costs caused directly by the activities of carp and loss of recreational amenity.

Habitat Functions

There has been an emphasis in past management—which is likely to be continued in the Basin Plan—on flow management to make use of refuges (for example, icon sites) to provide some protection against low flows and the current dry conditions. This approach will meet some of the requirements of habitat functions. Icon sites and other floodplain refuges that might be protected will also meet some of the needs of nurseries, although many species breed in areas other than such sites. This approach has not, however, properly considered other ecological goods and services provided by habitat, or been implemented with a view to protect them to maintain other benefits such as biodiversity (to meet the requirements of the *Environment Protection and Biodiversity Conservation Act*), production and dispersal, particularly with regard to connectivity throughout the Basin (Table 3.1).

Table 3.1 Ecologically dependent functions, goods and services of rivers.

Functions	Ecosystem processes and components	Goods and services
Regulation functions	*Maintenance of essential ecological processes and life-support systems*	
Waste treatment	Role of vegetation and biota in removal or breakdown of discharges to rivers	Pollution control Reduction in full treatment costs
Nutrient regulation	Role of biota in storage and recycling of nutrients (N&P)	Maintenance of water quality Reduction of algal blooms
Biological control	Population control through trophic relationships	Balanced native populations Control of pest numbers (eg., European carp)
Habitat functions	*Providing habitat for native plants and animals*	
Refuges	Suitable living space for native plants and animals	Maintenance of biodiversity Sources for re-colonisation. Minimum population support
Nurseries	Suitable reproduction habitat	Maintenance of population numbers Natural recruitment
Complexity	Variety of niches to support complex communities	Resilient food webs Diverse ecosystem structure supporting long-term stability
Vertical structure	Floodplain inundation and riparian growth	Vertical habitat, especially in arid zones Connected zones throughout catchments
Connectedness	Migration and dispersal throughout catchments	Catchment-wide maintenance of ecological communities via channels and riparian corridors
Production functions	*Provision of natural resources*	
Genetic resources	Genetic material, evolution and adaptation flexibility in native plants and animals	Adaptation to changed conditions because of use or climate change Chemical models and tools Test and assay organisms
Recreation	Sport fishing, aquarium plants	Populations with sufficient production for harvesting
Food	Commercial fishing and aquaculture	Harvestable populations Source material for aquaculture
Raw materials	Conversion of solar energy into biomass for human construction and other uses	Specialist riparian species—eg. river red gum

Functions	Ecosystem processes and components	Goods and services
Information functions	*Providing opportunities for education and cognitive development*	
Aesthetic value	Attractive landscapes	Enjoyment of scenery
Recreation	Variety in riverine landscapes	Travel and ecotourism Outdoor sports
Culture	Traditional people's values and significance	Understanding the place and its value for long-term human habitation
Art	Natural features with artistic value	Nature as motive in books, film, painting, folklore, national symbols, advertising, and so on
History	Variety of features with value	Historical development of the country via rivers
Science and education	Variety in nature with scientific and educational value	Use of natural systems for education Use of nature for scientific research

Source: Adapted from de Groot et al. 2002.

Production Functions

Healthy ecological resources also provide for fishing, boating and passive recreation that all benefit from proper environmental flow management (Table 3.1). Fish-population recruitment, healthy riparian zones and the maintenance of genetic diversity are dependent on adequate flows. Genetic diversity will be fundamental to river ecosystems being able to adapt to changed conditions including climate change.

Information Functions

Rivers also have considerable traditional, historical, cultural and educational values. They have been central to Aboriginal inhabitation, avenues for exploration and features of early inland development. The natural thirst for knowledge and understanding of the world around us is also provided for by natural systems. Thus, ecotourism, travel and understanding of our cultural heritage and history are fundamentally linked to healthy river systems. In places, the rivers have been so damaged that these ecological goods and services have been all but lost.

Current conservation assessment for rivers is based on the principles of selecting areas that are comprehensive, adequate and representative (CAR). While useful, this approach neglects other important features of condition, vulnerability, irreplaceability and complementarity (see Linke et al. 2007). Complementarity involves selecting successive adjacent sections of rivers to achieve the highest levels of protection. Irreplaceability represents the likelihood that an area will be required as part of a conservation system that achieves all conservation targets (Linke et al. 2008). In conservation planning for rivers it is important to recognise their inter-connectedness, so that the upstream catchments of selected reaches are also protected, and vulnerability also recognises threatening processes such as flow regulation (Cullen 2003). This latter approach applied to Victoria indicated that quite small catchment areas (<20 per cent; Linke et al. 2007, 2008) might achieve desired conservation goals, although the authors hasten to point out that their study should not be seen as an attempt to direct conservation policy. Thus, if decisions on flows are to be made that adequately protect biodiversity, a broader approach to planning is needed that also includes how successive river sections are chosen and linked and their irreplaceability, condition and vulnerability.

Generally, desired ecological outcomes from environmental watering are stated only in general terms and from the position of the water available, rather than how much water is needed to achieve the outcomes. The science behind the development of the Living Murray Initiative (<http://thelivingmurray.mdbc.gov.au/>) initially made strong arguments that 1500 gigalitres of water allocated for the environment would have a moderate chance of achieving ecological outcomes. As a first step, 500 GL was agreed but with a second step now being $3 billion of government purchases that would be likely to retrieve a further 1200–1500 GL, albeit with varying levels of security. Supplying water to achieve specific ecological outcomes should, however, involve an integrated approach that begins with ecology, rather than economics and availability. Water purchased from the Riverina cannot be supplied to the Darling and water bought from the northern rivers is unlikely to be of benefit to the Murray mouth. The ecological outcomes, and the ecological goods and services that they supply, integrated across whole catchments need to be decided first and the water recovery and delivery decided in response to the desired outcomes. In places, this might mean that sufficient water will never be available to achieve the desired outcomes and this should be clearly acknowledged. In other places, more expensive water might need to be purchased because it is the only source to achieve the desired outcome.

Desired ecological outcomes can be determined from the general to specific and from short to long term. Some outcomes will be basin wide, such as connectedness and fish-species distribution; some might be highly targeted such as endangered species and refuge sites. It is important to decide on short-term outcomes (days to weeks) such as carbon and nutrient processing and flushing algal blooms and sediment. Defining and targeting these things will demonstrate the immediate benefits of environmental watering, which is also important in maintaining support for the allocations. Medium-term ecological outcomes (months to one year) might include providing water for habitat maintenance, channel-forming flows and specific flows for recruitment events. Long-term outcomes might include overall ecosystem condition, maintenance of population numbers of endangered or threatened species and maintenance of biodiversity including distribution ranges and survival. Clearly, the watering regimes (timing and volumes) will be different for each of these sets of outcomes, although there might be some overlap.

Assets and Condition

Considerable ecological information is available on the rivers of the Murray–Darling Basin. The *Snapshot of the Murray–Darling Basin River Condition* (Norris et al. 2001a) assessed the overall condition of rivers in the Basin, and in particular the Murray River, using information available at the time. The Sustainable Rivers Audit is a major program now run by the Murray–Darling Basin Authority (MDBA) in conjunction with the Basin States, which collects data to assess the condition of all rivers in the Basin using an approach with multiple ecological indicators similar to the snapshot. The Australian Water Resources 2005 assessment (Norris et al. 2006) summarised and mapped environmental assets that had some legal protection, national parks, designated rivers, wetlands of national significance and Ramsar wetlands. The recent Sustainable Yields project (CSIRO 2008) showed that historical water availability was greatest in the southern part of the Basin. Therefore, ecological outcomes might need to be viewed both basin wide and regionally.

The poor ecological condition—and thus loss of ecological goods and services—of rivers in the Murray–Darling Basin is clearly demonstrated by all of these studies. The snapshot concluded that 40 per cent of the river length assessed had biota that was significantly impaired and more than 95 per cent of the river length assessed had degraded environmental condition. Catchment disturbance and changes to nutrient and suspended sediment loads were the greatest contributors. Data on flows, however, were not available for much of the Basin, although more than half the river reaches assessed had modified hydrology, with the greatest changes immediately downstream of dams and in lowland reaches used for irrigation supply—for example, the River Murray, the Murrumbidgee, Wimmera/Avon, Loddon and Darling rivers (Norris et al. 2001a). Nationally, the results were similar for the intensive land-use zone, with almost one-quarter having lost at least 20 per cent of the kinds of aquatic biota expected and more than 85 per cent of the river length assessed as significantly modified in terms of catchment disturbance, hydrological disturbance, habitat condition and nutrient and suspended sediment loads (Norris et al. 2001b, 2007). Multiple causes of damage were noted, with unseasonal flooding of wetlands, loss of connection with the floodplain, habitat simplification, water quality and bank erosion all being assessed as significant issues. Thus, managing flows is only one part of sustainable use.

The Sustainable Rivers Audit provided an assessment of rivers in the Basin for 2004–07 (MDBC 2008). Fish condition was considered to be very poor in most river valleys in the Basin and worst in the more heavily developed areas of the south and east. The assessment confirmed the well-known decline of native fish in the Basin, which was also compounded by alien species that rivalled, outnumbered or outweighed native fish in nine of the 23 valleys of the Basin. Although not as severely damaged, the macro-invertebrate assessment showed a similar trend, with the highest ranking being moderate and falling to very poor. Therefore, it is clear that the ecological goods and services represented by these assessments have been severely degraded. Assessment of their loss, however, hardly figures in arguments about potential loss of food and fibre production through reduced water availability.

Legislated protection, such as national parks, is one way of preserving Australia's natural capital. The Australian Water Resources assessment done by the National Water Commission (NWC 2005) showed more than 3500 wetlands nationally that had some sort of legislative protection (Ramsar Convention, National Significance and State legislation), but only 48 rivers (43 in Victoria and a few more that have since been protected in Queensland). While the number of wetlands seems large, it is actually a very small proportion of the tens or hundreds of thousands that exist nationally, or even in the Basin alone. National parks and other reserves are not designed to protect rivers or wetlands and their boundaries do not coincide with catchments. Additionally, there is considerable development in national parks that harms rivers (for example, ski resorts and the Snowy Mountains Hydro-Electric Scheme in the Kosciuszko National Park). It was concluded that Australia's legislative commitments to the National Strategy for Conservation of Australia's Biological Diversity 1996 and the *Environmental Protection of Biodiversity Conservation (EPBC) Act 1999* are unlikely to be well served in most of Australia under the current level of conservation protection that is afforded to rivers and associated wetlands (Norris et al. 2006).

In summary, damage to the ecological condition of Australia's rivers (including floodplain wetlands) is extensive and has been demonstrated in several reports. Rivers are beset by multiple stressors including changes to their catchments, water quality, habitat, riparian vegetation and hydrology, and these are likely to act synergistically meaning that addressing only one aspect is unlikely to achieve the expected benefits. There are few rivers that have explicit legislative protection, and current protected areas such as national parks are generally not designed to protect rivers and their biota. Thus, many of the ecological goods and services listed above are likely to be degraded or lost with little assessment of the consequences relative to the use of water for consumption.

Climate Change and Environmental Water

Climate-change predictions so far deal mostly with likely changes in the volumes of water available, with predictions that water availability might fall by 9–11 per cent in the north of the Murray–Darling Basin and 13 per cent in the south (CSIRO 2008). Little is known, however, about the concomitant changes in water quality (Whitehead et al. 2009) and ecological outcomes. Climate-change projections indicate that Australia will see changes in the magnitude and frequency of extreme events (Francis and Hengeveld 1998; Jackson et al. 2001), particularly droughts, floods and fires. Although droughts, floods and fire are the major disturbance events in Australia, their infrequency means they have generally been studied and managed in isolation. Thus, individually there is some understanding of the water-quality and ecological responses to drought, flood and fire (that is, it is known that disturbance plays a major role in structuring stream ecological communities and stream water quality), but they are rarely considered in combination. Nor have they been considered with regard to the management and supply of ecological goods and services of the rivers.

An increase in the frequency of extreme events might result in an increase in the frequency with which ecosystem thresholds would naturally be exceeded and consequently the capacity for ecosystem recovery might be severely impeded (Murdoch et al. 2000). Under such circumstances, simply managing to avoid exceeding thresholds will no longer be sufficient to protect ecosystem function, but other factors such as the period of, and time between, exceeding events must also be considered. Thus, the way in which environmental watering is planned and managed might need to fundamentally change. In part, this underpins the current flow-management predicament, especially for the lower Murray River.

If the frequency of disturbance events increases, it becomes likely that they will act in combination to cause changes to water quality and ecosystem processes. Such a combination of effects might lead to ecological responses outside current ranges. Thus, there is a need to consider disturbance events not individually, but in combination. The potential impact of changes in the frequency of events requires that these are also considered over the long term. Such forecasting of future conditions has only recently attracted the attention of the international research community (Batterbee et al. 2008; Kundzewicz et al. 2007) and has yet to be considered by the natural-resource management community, particularly in relation to environmental watering.

Need for Assessment to Determine Effectiveness and Adaptive Management

Statement of desired ecological outcomes of environmental watering effectively provides hypotheses that can be tested to determine whether management actions have been effective. This is a necessary step that subsequently leads to adaptive management, but that has too often been neglected. The term 'adaptive management' has been used for more than 30 years to describe an approach to natural-resource management (Holling 1978). It is the application of experimentation to the design and implementation of natural-resource and environmental management, which relies on learning from experience (Walters and Holling 1990). Adaptive management is done to learn and better manage ecosystems (Holling 1978; Lee 1999). The same search for scientific understanding is shared by scientists and those wanting to make better management decisions (Norton 2005). Yet it has been said that adaptive management has been more influential as an idea than as a practical means of gaining insight into ecosystem behaviour (Lee 1999).

Features of adaptive management critical for success are

1. scientists independent of stakeholder groups leads to efficiencies through single agreed projects, rather than each group commissioning their own and each consequently being viewed suspiciously by the other parties
2. clear statement of desired ecological outcomes of flow management and assessment to test the effectiveness of the decisions made to achieve them
3. the translation of ecological outcomes to flow volumes and the release regime
4. robust study design critical for unforeseen events
5. implementation of the feedback loop and flexibility.

Independent Scientists

Scientists provide crucial input in structuring the questions, developing the models for testing and in formulating appropriately scaled experiments that can test and improve system understanding and provide alternative options for the future under changing conditions (Hughes et al. 2007). Testability, objectivity, and impartiality are the criteria used by science to evaluate the reliability of a scientific finding (Christie 2008). The usual give and take of criticism that science relies on (Christie 2008) might also be conducive to the flow of information and open communication among the parties needed for an effective adaptive approach to environmental management (Norton 2005). These characteristics of scientific objectivity and independence from the stakeholders are critical for efficiency and success.

Clear Statement of Desired Ecological Outcomes and Assessment to Test the Effectiveness of Decisions to Achieve Them

Adaptive management is 'place sensitive' in both a physical sense and the need to take into account the local environmental values (Norton 2005). The desired ecological outcomes of the management actions will be motivated by the environmental values held for any given place. Adaptive management requires the parties involved to have agreed on the desired ecological outcomes of the adaptive approach. According to Lee (1999), there was little evidence that the adaptive-management approach was being used this way. The lack of agreed questions and failure to clearly state the desired ecological outcomes would be two reasons adaptive management would fail to provide a practical means of gaining insight into ecosystem behaviour. Some iconic examples of adaptive management have declared great success—for example, the experimental releases of large volumes of water to the Colorado River from the Glen Canyon Dam (Walters et al. 2000). These experimental releases enabled better scientific understanding of sediment dynamics and of how water temperature and introduced pests influence the recruitment of endangered native fish (Hughes et al. 2007).

The Translation of Ecological Outcomes to Flow Volumes and the Release Regime

> In principle, the scientific approach leads to reliable determination of causes; in practice, that means being able to learn over time how management does and does not affect outcomes. (Lee 1999)

Environmental watering guidelines should specify precise flow volumes, and include specification of drought and other special-purpose flow rules. The transformation of ecological outcomes into flow volumes, release regimes, and published rules will have the effect of removing environmental flows as a source of great uncertainty from those with responsibility for release and those who see it as a potential loss (Purves et al. 2009).

Robust Study Design Critical for Unforeseen Events

Adaptive management for environmental flows requires a shift from measuring change (for example, more fish were found) to 'measuring and understanding change' (for example, the fish community responded because of the changed flow regime) (Souchon et al. 2008). Making that shift requires acknowledgment that the distinction between survey and experimental study designs is in the replication. Surveys (without sample replication) are generally used only to make correlative assessments, whereas field experiments that employ replication and keep particular factors constant while others vary can be used to investigate causation (Souchon et al. 2008). While trial and error are part of the adaptive-management process, nothing can be learned from the trial or the error without an adequate study design to learn something about the ecosystem's processes and structure (Lee 1999), and thus build capacity to apply that knowledge and achieve desired ecological outcomes in a dynamic environment. Thus, evidence-based decisions for setting environmental flows require rigorous data collection and study designs that test hypotheses, answer specific questions and provide for extrapolation to similar systems or systems undergoing change (Souchon et al. 2008).

Feedback Loop

There is no 'one' or right solution to most environmental problems; rather there is a choice of better or worse solutions (Norton 2005). Further, the resolution might be only temporary until things change—for example, with drought, fire, climate change, or changing demands of competing values. Adaptive management requires flexibility to take advantage of the inherent feedback loop between science and management so that decisions can be modified based on new information (Souchon et al. 2008). This flexibility and willingness to confront uncertainty provides the capacity to deal with moving targets presented by environmental change (Hughes et al. 2007). A static approach to water-resource management that ignores the likelihood of unforeseen dynamics is doomed to failure (Hughes et al. 2007); the many NSW water-sharing plans currently under implementation in Australia might prove to be such examples (<http://www.dwe.nsw.gov.au/water/plans.shtml>). The ability to reassess and make changes while a project is ongoing adds to the likelihood of success (Souchon et al. 2008)

Approach to Managing Environmental Watering

The approach suggested here includes consideration of the ecological goods and services supplied by the Murray–Darling Basin in addition to water as a commodity. This opens the way for communities to consider the wider set of values and the benefits derived from them, rather than the current narrow, exploitative view that tends to dominate the debate over water use. The steps taken could include the following.

1. Determine and map ecological assets—places, features, species, and communities. Cullen (2003) lists four reasons to maintain the biodiversity of aquatic ecosystems. First, national (Commonwealth of Australia 1992, 1999) and international (UNCED 1992) obligations on biodiversity conservation have to be met. Second, protected areas provide a reference condition for comparative assessment of impacted regions (Bailey et al. 2004; Reynoldson et al. 1997); and third, act as re-colonisation pools for taxa extinct in other parts of a catchment. Fourth, aquatic species bear intrinsic values (Angermeier 2000; Sarkar 2002) and often provide irreplaceable ecosystem services

(Cullen 2003). Ecological processes such as final treatment of effluents that have been discharged not fully treated should be regarded as assets, and the full range should be defined (Table 3.1).

2. Determine the desired ecological outcomes in relation to the range of recognised ecological goods and services, including protection of those defined in (1) above, and various processes such as management of algal blooms, maintenance of habitat, channel-forming flows, and flushing of nutrients and organic matter.
3. Prioritise the desired ecological outcomes both basin wide and regionally as suggested by the Sustainable Yields project.
4. Estimate how much water would be needed to achieve the desired outcomes. This important step would be a marked departure from much current practice whereby decisions are made on how to distribute available water, rather than how much is needed to achieve particular outcomes.
5. Based on the preceding estimates of how much water is needed, determine which of the ecological outcomes and where they are that could be achieved with the available water.
6. What would be the risks associated with achieving, or failing to achieve, the desired ecological outcomes?
7. What would be the indicators of environmental watering outcomes in the short (weeks–months), medium (months to a year) and long (several years) term?
8. Model where can water be bought or traded to give the best return in volumes for the environment for the dollars spent. This would be part of another, related project and would need iteration dependent on where water was needed to achieve priority ecological outcomes.
9. Model where will the water retrieved for the environment actually be stored, or let run? Understanding this question will direct how much water and where it can be allocated for the environment. Answering this question will require hydrological modelling based on economic-modelling scenarios.

Water Recovery

Water that becomes available through recovery and trading will determine what can be achieved through appropriate allocation. If desired ecological outcomes are also framed in the context of ecological goods and services, however, there is also a need to determine the volumes and delivery regimes to achieve specific desired ecological outcomes and the water that might be needed to do this. Such an approach is philosophically different from the former in that desired outcomes are determined first and then the watering regime is estimated subsequently. Thus, the questions being asked would be what do we want to achieve and how much water will it take, rather than what changes might be possible with whatever water becomes available? The former would be realistic and targeted, and the latter just making do with whatever becomes available without clear direction or incentive for modifying volumes and delivery.

Economic modelling has indicated that significant volumes of water might be reclaimed for the environment for only small changes in net economic returns on production. To maximise environmental benefits of reclaimed water, the relationships between trading/buyback and environmental outcomes need to be understood. Ecological modelling has not been combined with economic modelling to prioritise trading/buyback so that both economic and ecological outcomes can be modified.

It is also possible that the highest-priority ecological outcomes might be in places that do not coincide with the cheapest or largest volumes of water that can be delivered. Thus, it might be necessary to prioritise the ecological outcomes and direct water-reclamation buyback, trading and infrastructure investment to places that might be less economic but more ecologically beneficial.

Conclusion

1. The National Water Initiative defines the environmentally sustainable level of extraction as 'the level of water extraction from a particular system which, if exceeded would compromise key environmental assets, or ecosystem functions and the productive base of the resource' (Commonwealth of Australia 2004).
2. The full range of ecological goods and services that is supplied by rivers needs to be recognised before they can be properly valued and managed. Some of these might be difficult to value economically but they provide important components of our culture and quality of life. Recognition of these values will broaden the debate over water use and clarify its true costs as well as suggest alternative economic benefits that might be derived from healthy rivers.
3. There is ample evidence to show that Australia's rivers are damaged, and recognition of this has led to several pieces of legislation: the NWI and government water recovery.
4. Few rivers are explicitly protected and conservation planning for rivers has been based largely on defining areas that meet the criteria of comprehensive,

adequate and representative. There is also an urgent need to consider complementarity, condition, irreplaceability and vulnerability across whole catchments. Arguments for conservation might be strengthened if they are made in the context of the ecological goods and services that will be maintained or enhanced.

5. Climate change will almost certainly require adaptation of current views on environmental water and its management to achieve environmental outcomes.
6. Managing only environmental water is unlikely to achieve the maximum desired benefits if other factors damaging rivers are not also coincidentally addressed.
7. Desired ecological outcomes need to be determined first; then these should be translated into the volumes and regime of water needed to achieve them, rather than the reverse. The desired outcomes need to cover the short, medium and long term.
8. Assessment and adaptive management are necessary for achieving the best ecological outcomes most efficiently in the long term.

Bibliography

Andersen, H. E., Kronvang, B., Larsen, S. E., Hoffmann, C. C., Jensen, T. S. and Rasmussen, E. K. 2006, 'Climate-change impacts on hydrology and nutrients in a Danish lowland river basin', *Science of the Total Environment*, vol. 365, pp. 223–37.

Angermeier, P. L. 2000, 'Regional frameworks and candidate metrics for assessing biotic integrity in mid-Atlantic highland streams', *Transactions of the American Fisheries Society*, vol. 129, pp. 962–81.

Bailey, R. C., Norris, R. H. and Reynoldson, T. B. 2004, *Bioassessment of Freshwater Ecosystems. Why and how*, Kluwer, New York.

Batterbee, R. W., Kernan, M., Livingstone, D., Nickus, U., Verdonschot, P., Hering, D., Moss, B., Wright, R., Evans, C., Johnson, R., Maltby, E., Linstead, L. and Skeffington, R. 2008, *Freshwater Ecosystem Response to Climate Change: The Euro-limpacs project*, John Wiley & Sons, Chichester, UK.

Boulton, A. J. and Lake, P. S. 1992, 'The ecology of two intermittent streams in Victoria, Australia. III. Temporal changes in faunal composition', *Freshwater Biology*, vol. 27, pp. 123–38.

Boyd, J. and Banzhaf, S. 2007, 'What are ecosystem services? The need for standardized environmental accounting units', *Ecological Economics*, vol. 63, pp. 616–26.

Bunn, S. E. and Hues, J. M. 1997, 'Dispersal and recruitment in streams: evidence from genetic studies', *Journal of the North American Benthological Society*, vol. 16, pp. 338–46.

Burger, G. 2002, 'Selected precipitation scenarios across Europe', *Journal of Hydrology*, vol. 262, pp. 99–110.

Caruso, B. S. 2002, 'Temporal and spatial patterns of extreme low flows and effects on stream ecosystems in Otago, New Zealand', *Journal of Hydrology*, vol. 257, pp. 115–33.

Chang, H. J. 2004, 'Water quality impacts of climate and land use changes in southeastern Pennsylvania', *Professional Geographer*, vol. 56, pp. 240–57.

Christie, E. (2008). *Finding solutions for environmental conflicts: power and negotiation*. Cheltenham, UK: Edward Elgar.

Commonwealth of Australia 2004, *Intergovernmental Agreement on a National Water Initiative*, Commonwealth of Australia, Canberra.

Commonwealth of Australia 1992 *Endangered Species Protection Act 1991, No. 194 of 1992*, http://www.austlii.edu.au/au/legis/cth/num_act/espa1992282/sch1.html

Commonwealth of Australia 1999 *Environment Protection and Biodiversity Conservation Act 1999* http://www.environment.gov.au/epbc/

Commonwealth Scientific and Industrial Research Organisation (CSIRO) 2008, *Water availability in the Murray–Darling Basin*, A report to the Australian Government from the CSIRO Murray–Darling Basin Sustainable Yields Project, CSIRO, Melbourne.

Costanza, R. and Farber, S. 2002, 'Introduction to the special issue on the dynamics and value of ecosystem services: integrating economic and ecological perspectives', *Ecological Economics*, vol. 41, pp. 367–73.

Costanza, R., d'Arge, R., de Groot, R., Farber, S., Grasso, M., Hannon, B., Naeem, S., Limburg, K., Paruelo, J., O'Neill, R. V., Raskin, R., Sutton, P. and van den Belt, M. 1997, 'The value of the world's ecosystem services and natural capital', *Nature*, vol. 387, pp. 253–60.

Council of Australian Governments (COAG) 1994, *Council of Australian Governments' Communiqué*, 25 February 1994, Hobart, <http://www.coag.gov.au/coag_meeting_outcomes/1994-02-25/index.cfm>

Council of Australian Governments (COAG) 2004, *Council of Australian Governments' Communiqué*, 25 June 2004, Canberra, <http://ncp.ncc.gov.au/docs/Council%20of%20Australian%20Governments%20Meeting%20-%2025%20June%202004.pdf>

Cullen, P. 2003, 'Challenges to the conservation of Australian freshwater biodiversity: an epilogue', *Aquatic Ecosystem Health & Management*, vol. 6, pp. 97–101.

Daly, H. E. and Cobb, J. B. 1989, *For the Common Good*, Beacon Press, Boston.

de Groot, R.S., Wilson, M.A. and Bpoumans, R.M.J. (2002). A typology for the classification, description and valuation of ecosystem functions, goods and services. *Ecological Economics*, 41, 393-408.

Department of the Environment, Water, Heritage and the Arts (DEWHA) 2009, *A framework for determining Commonwealth environmental water actions*, Discussion Paper, Department of the Environment, Water, Heritage and the Arts, Canberra, <http://www.environment.gov.au/water/policy-programs/cewh/pubs/cehw-framework-discussion-paper.pdf>

Everard, M. 1996, 'The importance of periodic droughts for maintaining diversity in the freshwater environment', *Freshwater Forum*, vol. 7, pp. 33–50.

Francis, D. and Hengeveld, H. 1998, *Extreme Weather and Climate Change*, Canadian Ministry of Supply and Services, Ottawa.

Fu, G. B., Charles, S. P. and Chiew, F. H. S. 2007, 'A two-parameter climate elasticity of streamflow index to assess climate change effects on annual streamflow', *Water Resources Research*, vol. 43, p. 12.

Holling, C. S. 1978, *Adaptive Environmental Assessment and Management*, John Wiley & Sons, New York.

Hughes, J. M. 2007, 'Constraints on recovery: using molecular methods to study connectivity of aquatic biota in rivers and streams', *Freshwater Biology*, 52, pp. 616–31.

Hughes, T. P., Gunderson, L. H., Folke, C., Baird, A. H., Bellwood, D., Berkes, F., Crona, B., Helfgott, A., Leslie, H., Norberg, J., Nystrom, M., Olsson, P., Osterblom, H., Scheffer, M., Schuttenberg, H., Steneck, R. S., Tengoe, M., Troll, M., Walker, B., Wilson, J. and Worm, B. 2007, 'Adaptive management of the Great Barrier Reef and the Grand Canyon world heritage areas', *Ambio*, vol. 36, pp. 586–92.

Jackson, R. B., Carpenter, S. R., Dahm, C. N., McKnight, D. M., Naiman, R. J., Postel, S. L. and Running, S. W. 2001, 'Water in a changing world', *Ecological Applications*, vol. 11, pp. 1027–45.

Junk, W. J., Bayley, P. B. and Sparks, R. E. 1989, 'The flood pulse concept in river–floodplain systems', *Special Publication of the Canadian Journal of Fisheries and Aquatic Sciences*, vol. 106, pp. 110–27.

Kundzewicz, Z. W., Mata, L. J., Arnell, N. W., Doll, P., Kabat, P., Jimenez, B., Miller, K. A., Oki, T., Sen, Z. and Shiklomanov, I. A. 2007, 'Freshwater resources and their management', Contribution of Working Group II to the *Fourth Assessment Report of the Intergovernmental Panel on Climate Change*, Cambridge University Press, UK.

Lake, P. S. and Barmuta, L. A. 1986, 'Stream benthic communities: persistent presumptions and current speculations', in P. De Deckker and W. D. Williams (eds), *Limnology in Australia*, CSIRO, Melbourne.

Lake, P. S., Bond, N. and Reich, P. 2007, 'Linking ecological theory with stream restoration', *Freshwater Biology*, vol. 52, pp. 597–615.

Lane, S. N., Tayefi, V., Reid, S. C., Yu, D. and Hardy, R. J. 2007, 'Interactions between sediment delivery, channel change, climate change and flood risk in a temperate upland environment', *Earth Surface Processes and Landforms*, vol. 32, pp. 429–46.

Lee, K. N. 1999, 'Appraising adaptive management', *Conservation Ecology*, vol. 3, p. 3.

Linke, S., Norris, R. H. and Pressey, R. L. 2008, 'Irreplaceability of river networks: towards catchment-based conservation planning', *Journal of Applied Ecology*, vol. 45, pp. 1486–95.

Linke, S., Pressey, R. L., Bailey, R. C. and Norris, R. H. 2007, 'Management options for river conservation planning: condition and conservation re-visited', *Freshwater Biology*, vol. 52, pp. 918–38.

Murray–Darling Basin Commission (MDBC) 2008, *Murray–Darling Basin rivers: ecosystem health check, 2004–2007*, Summary report based on the Independent Sustainable Rivers Audit Group, *SRA Report 1: A report on the ecological health of rivers in the Murray–Darling Basin 2004–2007*, Submitted to the Ministerial Council in May 2008, Murray–Darling Basin Commission, Canberra, <http://www.mdbc.gov.au/__data/page/2260/34373_MDBC_SRA_Report_Web.pdf>

Murdoch, P.S., Baron, J.S. and Miller T.L. (2000). Potential effects of climte change on surface-water quality in North America. *Journal of the American Water Resources Association*. 36, 347-366

National Water Commission (NWC) 2006, *Australian Water Resources 2005. A baseline assessment of the water resources for the National Water Initiative, Level 1 assessment*, National Water Commission, Canberra, <http://www.nwc.gov.au/resources/documents/awr-2005-level1-report-text-pub-1006.pdf>

Neff, R., Chang, H. J., Knight, C. G., Najjar, R. G., Yarnal, B. and Walker, H. A. 2000, 'Impact of climate variation and change on mid-Atlantic region hydrology and water resources', *Climate Research*, vol. 14, pp. 207–18.

Norris, R. H., Linke, S., Prosser, I., Young, W. J., Liston, P., Bauer, N., Sloane, N., Dyer, F. and Thoms, M. 2007, 'Very-broad-scale assessment of human impacts on river condition', *Freshwater Biology*, vol. 52, pp. 959–76.

Norris, R. H., Liston, P., Davies, N., Coysh, J., Dyer, F., Linke, S., Prosser, I. and Young, B. 2001a, *Snapshot of the Murray–Darling Basin River Condition*, Murray–Darling Basin Commission, Canberra.

Norris, R. H., Prosser, I., Young, B., Liston, P., Bauer, N., Davies, N., Dyer, F., Linke, S. and Thoms, M. 2001b, *The assessment of river condition (ARC). An audit of the ecological condition of Australian rivers*, May 2001, Report submitted to the National Land and Water Resources Audit Office, Canberra.

Norris, R, Dyer, F, Hairsine, P, Kennard, M, Linke, S, Read, A, Robinson, W, Ryan, C, Wilkinson, S & Williams, S 2006, *'Australian Water Resources 2005. Assessment of River and Wetland Health'*, National Water Commission, Canberra

Norton, B. G. 2005, *Sustainability: A philosophy of adaptive ecosystem management*, University of Chicago Press, Ill.

Purves, T., Lindner, D., Salma, S. and Larkings, K. 2009, *Water Planning 2009 Review of Planning Variables for Water Supply and Demand Assessment: A review of the changes in water resources modelling assumptions*, Infrastructure Development Branch Water Division, Actew Corporation, Canberra, <http://www.actew.com.au/publications/WaterPlanning2009Review_WaterSupplyandDemandAssessment.pdf>

Reynoldson, T. B., Norris, R. H., Resh, V. H., Day, K. E. and Rosenberg, D. M. 1997, 'The reference condition: a comparison of multimetric and multivariate approaches to assess water-quality impairment using benthic macroinvertebrates', *Journal of the North American Benthological Society*, vol. 16, pp. 833–52.

Roberts, J., Bourman, B., Close, A., Erdmann, B., Hillman, T., Jones, G., Ganf, G., Geddes, M., Koehn, J., Newman, B., Paton, D., Suter, P., and Thoms, M. 2001, *A scientific panel assessment of modeled flow and technical options for the River Murray*. A report to the Murray-Darling Basin Ministerial Council, Canberra

Sarkar, S. 2002, 'Defining biodiversity: assessing biodiversity', *The Monist*, vol. 85, pp. 131–55.

Souchon, Y., Sabaton, C., Deibel, R., Reiser, D., Kershner, J., Gard, M., Katopodis, C., Leonard, P., Poff, N. L., Miller, W. J. and Lamb, B. L. 2008, 'Detecting biological responses to flow management: missed opportunities; future directions', *River Research and Applications*, vol. 24, pp. 506–18.

Thoms, M.C., Sheldon, F., 2000. Lowland rivers: an Australian introduction. *Regulated Rivers: Research and Management* 16, 375–383.

Thorp, J. H. and Delong, M. D. 1994, 'The riverine productivity model: an heuristic view of carbon sources and organic processing in large river ecosystems', *Oikos*, vol. 70, pp. 35–60.

United Nations Conference on Environment and Development (UNCED) 1992, *Convention on Biological Diversity*, United Nations Conference on Environment and Development, Rio de Janeiro.

Vannote, R. L., Minshall, G. W., Cummins, K. W., Sedell, J. and Cushing, C. E. 1980, 'The river continuum concept', *Canadian Journal of Fisheries and Aquatic Sciences*, vol. 37, pp. 130–7.

Walters, C. J. and Green, R. 1997, 'Valuation of experimental management options for ecological systems', *The Journal of Wildlife Management*, vol. 61, pp. 987–1006.

Walters, C.J. and Holling, C.S. (1990). Large-scale management experiments and learning by doing. *Ecology*, 71, 2060-2068.

Walters, C., Korman, J., Stevens, L. E. and Gold, B. 2000, 'Ecosystem modeling for evaluation of adaptive management policies in the Grand Canyon', *Conservation Ecology*, vol. 4, p. 1.

Whitehead, P. G. and Hornberger, G. M. 1984, 'Modeling algal behaviour in the River Thames', *Water Research*, vol. 18, pp. 945–53.

Whitehead, P. G., Wilby, R. L., Batterbee, R. W., Kernan, M. and Wade, A. J. 2009, 'A review of the potential impacts of climate change on surface water quality', *Hydrological Sciences*, vol. 54, pp. 101–23.

Wilby, R. L., Dalgleish, H. Y. and Foster, I. D. L. 1997, 'The impact of weather patterns on historic and contemporary catchment sediment yields', *Earth Surface Processes and Landforms*, vol. 22, pp. 353–63.

4. Climate Change and its Impacts: Current understanding, future directions

Nick Schofield

Introduction

Global climate change is well documented through warming of the atmosphere and oceans, sea-level rise and changes in the cryosphere (the portions of the Earth's surface where water is in solid form) over the past few decades. Climate change is also occurring across the Murray–Darling Basin (MDB), as is evidenced by increasing temperatures. There is strong evidence that changes in greenhouse gas concentrations due to human activity are the dominant cause of the global warming that has taken place over the past half-century. Global warming is, in turn, causing changes to the whole *climate system*—the highly complex interaction between the atmosphere, oceans, water cycle, ice, snow and frozen ground, land surface and living organisms. There will be environmental, economic and social 'impacts' resulting from all these changes.

The MDB is one of Australia's largest drainage divisions, covering approximately one-seventh of the continent. It incorporates Australia's three longest rivers (the Murray, Darling and Murrumbidgee rivers) and contains more than 30000 wetlands, including 16 internationally significant wetlands that provide habitat for migratory birds. The MDB is also very important for rural communities and Australia's economy, with three million Australians inside and outside the Basin directly dependent on its water. About 85 per cent of all irrigation in Australia takes place in the MDB, which supports an agricultural industry worth more than $9 billion per annum.

The impact of climate change on the natural resources, industries and communities of the Basin is, arguably, the region's most pressing issue. In response to this, the Murray–Darling Basin Authority (MDBA) has recently funded a series of scientific reviews and syntheses, as well as more fundamental research, to begin to comprehend the effects and develop policy and management responses. The Authority has also taken climate change into account in its draft Basin Plan. This chapter broadly describes the observations of climate change in the MDB, future climate projections, recent advances in our understanding, incorporation of climate change in basin planning and future directions.

Human-induced climate change is a complex and contentious issue of global, regional and local significance. The risks and ramifications of climate change are large in terms of both impacts and measures to mitigate the causes. Consequently, the topic has been one of fierce debate, with science playing a central role in attempting to describe the causes, impacts and solutions. The far-reaching nature of the issue has meant that the science itself is frequently challenged.

This chapter attempts to summarise the current state of knowledge of climate change in the MDB and suggests future directions for research and investigation. In doing so, the chapter draws on the substantial body of international and national research and recent projects funded by the MDBA.

Definition of Climate Change

It is important for any chapter discussing climate change to be clear on definitions, scope and perspective. The Intergovernmental Panel on Climate Change (IPCC 2007:943) defines climate change as 'a change in the state of the climate that can be identified (for example, by using statistical tests) by changes in the mean and/or variability of its properties, and that persists for an extended period, typically decades or longer'. This definition is adopted here.

In terms of scope, climate change is taken to mean changes to the *climate system*—the highly complex interaction between the atmosphere, oceans, water cycle, ice, snow and frozen ground, land surface and living organisms. This is an important concept as it recognises the complexity, interactions, and chaotic nature of the system under study.

A simpler concept also used here is 'climate change and its impacts'. This conceptualisation essentially attempts to separate 'cause' and 'effect', with the change in climate variables (temperature, rainfall, and so on) causing impacts (bushfires, floods, and so on). This approach recognises 'feedbacks' from impacts to 'causes' that might amplify or reduce climate change.

The former *climate system* approach is an important scientific framework, whilst *climate change and impacts* can be a useful way of addressing the economic, policy and social issues.

The Science of Climate Change

The science of climate change has recently been well summarised by the Royal Society (2010 p13). Their general conclusion is:

> There is strong evidence that changes in greenhouse gas concentrations due to human activity are the dominant cause of the global warming that has taken place over the last half century. This warming trend is expected to continue as are changes in precipitation over the long term in many regions. Further and more rapid increases in sea level are likely which will have profound implications for coastal communities and ecosystems.

The Royal Society goes on to define the climate science that is agreed on, that has wide consensus but is still debated, and topics that are not well understood. The last aspects relevant to the MDB are:

- the net effect of changes in the carbon cycle in all current models is to increase warming—by an amount that varies considerably between models because of uncertainties in how to represent the relevant processes
- projections of climate change are sensitive to the details of the representation of clouds in models and the influence of particles on the properties of clouds; these are poorly understood
- the ability of the current generation of models to simulate some aspects of regional climate change is limited; there is little confidence in specific projections of future regional climate change, except at continental scales.
- the future strength of the uptake of carbon dioxide by land and oceans is very poorly understood
- observations are not yet good enough to quantify, with confidence, some aspects of the evolution of either climate forcing or climate change, or for helping to place tight bounds on climate sensitivity

In the MDB context, it could further be added that

- Global Climate Model (GCM) forecasts for precipitation are distributed almost equally between positive and negative change—hence, all that can be said reliably is that 'it will be warmer and wetter OR warmer and drier'
- the regional climate modes such as the El Niño–Southern Oscillation (ENSO), Indian Ocean Dipole (IOD), Southern Annular Mode (SAM), Pacific Decadal Oscillation (PDO), Inter-Decadal Pacific Oscillation (IPO) and the Sub-Tropical Ridge (STR) are known to be important 'drivers' of MDB climate over a range of spatial and temporal scales; the interactions between these modes and with global climate change are not well understood.

Observations of Climate Change in the MDB

This section tabulates some observations of changes to the *climate system* in the MDB.

Table 4.1 Summary of observations of climate change in the MDB

Aspect of the climate system	Observed change
Annual mean temperature	A general upward trend in temperature is observed since the 1950s, with 2009 being the hottest year (Bureau of Meteorology: <www.bom.gov.au>; Figure 4.1). The decadal mean temperatures in the MDB (2000–09) show an increase of 0.53°C over the previous decade (1990–99). The spatial trend in mean annual temperature since 1950 shows the whole of the Basin increasing in temperature, with slightly higher rates towards the north-west. Warming has occurred in all seasons, however, the strongest warming has occurred in spring (about 0.9°C) and the weakest in summer (about 0.4°C) (CSIRO 2010).
Precipitation	The annual rainfall anomaly across the MDB shows no clear temporal trend at the Basin scale. 2010 saw the end to a long drought with the highest rainfall on record at 808 mm. When viewed spatially (Figure 4.2), however, there is a trend of increased drying in the south-eastern corner of the Basin. The spatial trend in heavy rain days for the Basin for the period 1970–2009 shows a declining trend across the MDB. Seasonal temporal rainfall trends for the MDB show little long-term temporal trends except for autumn, which shows a marked downward trend over the past two decades.
Pan evaporation	Annual pan evaporation temporal trends for the MDB show no clear trend; however, the spatial trends show broadly increasing pan evaporation in the southern Basin and decreasing in the northern Basin.
Stream flow	Precipitation in the MDB from 1997-2009 was significantly lower than the long-term average, and was accompanied by a 40 per cent reduction in stream flow in the southern MDB, where the majority of run-off is generated. During the period 2000–07, average annual inflow was 4150 GL/yr. In 2006–07, the 12-month stream flow reached a historical low of 770 GL/yr-1 to March 2007 (Cai and Cowan 2008; Figure 4.3). The average stream flow between 1998 and 2008 was 5700 GL—substantially lower than the long-term average of 11600 GL/yr between 1892 and 1997 (Kiem and Verdon-Kidd 2010). Despite these observations of unprecedented low stream flow in the Basin, the attribution of the reduction to climate change is highly complicated. Stream flow is heavily influenced by land-management practices, such as irrigated agriculture, forestry and water-management regimes determining the levels of diversions and water use from streams. Another confounding factor lies in the emerging understanding of the impact temperature change has on stream flow in the MDB. Determining plausible physical mechanisms underlying empirical temperature–stream flow relationships is subject to ongoing research (Yu et al. 2010). The record high rainfall of 2010 and resulting floods in 2010 and 2011 has re-directed attention to the potential effects of climate change on La Niña events (with added ocean warming), and cyclone, storm and rainfall intensities.

Aspect of the climate system	Observed change
Flooding/tropical cyclones	The flood seasonality for the north-east of the Basin shows predominantly summer/autumn flood events deriving from tropical troughs/lows (31 per cent) and remnants from tropical cyclones (12 per cent) (Grootemaat 2008). Trends in tropical cyclone activity in the Australian region (south of Equator; 105–160°E) show that the total number of cyclones has decreased in recent decades; however, the number of stronger cyclones (minimum central pressure less than 970 hPa) has not declined.
Drought	The relative frequency of El Niño to La Niña events within a 15-year moving window during the past 600 years shows when more frequent El Niño events occur there are more likely to be periods of drought in Australia (Verdon-Kidd and Kiem 2010). This reconstruction indicates that more severe and prolonged periods of drought might have occurred in the Australian landscape in the past than what has been experienced since European settlement. Verdon-Kidd and Kiem (2009) conducted an analysis examining the climatic drivers and rainfall characteristics of the Federation (1895–1903), World War II (1935–45), and 'Big Dry' (1997–2009) droughts in Australia. These three droughts were found to vary in terms of their: primary climatic drivers (ENSO, IPO, IOD, SAM); severity (in terms of reduction of rainfall experienced from normal); spatial footprint; seasonality of rainfall reductions; and daily rainfall characteristics such as intensity and number of rain days.
Extreme temperatures and heatwaves	The annual maximum and minimum temperatures for the MDB over the period 1910–2009 show an upward trend and approximate increase of 1°C (Bureau of Meteorology: <www.bom.gov.au>). The spatial trends in maximum and minimum daily temperatures (1970–2009) show strongly rising values across the MDB. The incidence of warm spells across the MDB has been increasing since 1910. The incidence of heatwaves (the number of worst annual three-day heatwave indices) is increasing across eastern Australia (Deo et al. 2006).
Bushfires	Since the 1950s, the climate has changed in ways that are likely to increase fire frequency and intensity in the MDB (CSIRO and BOM 2007): the average maximum temperature has warmed; south-eastern Australia has become drier; droughts have become hotter (Nichols 2004); and the number of extremely hot days (maximum temperature >40°C) has risen. Although the relationship between climate and fire is confounded by factors such as arson and fire management, it is clear that hotter and drier years have greater fire risk.

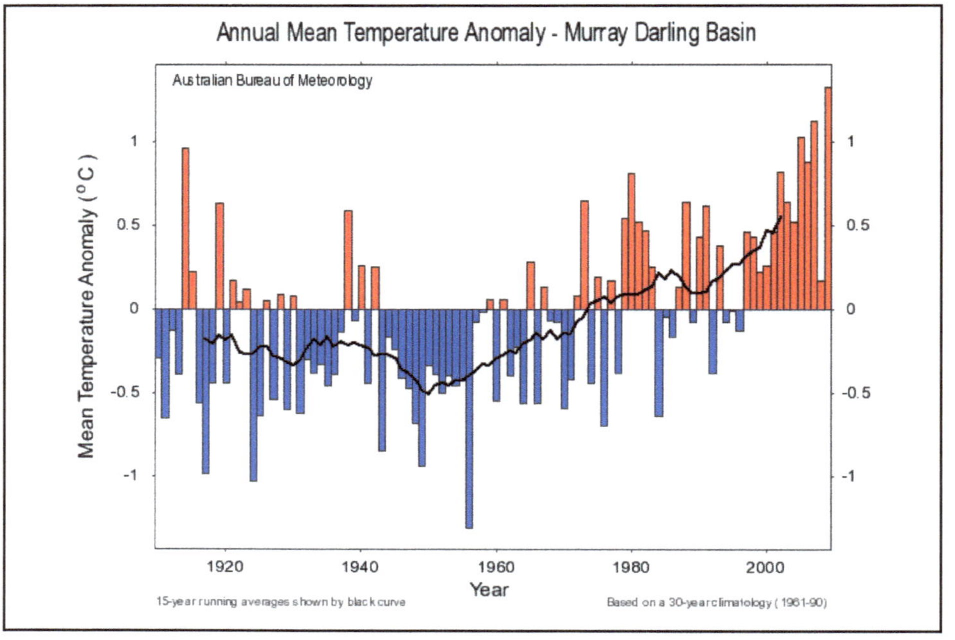

Figure 4.1 Annual mean temperature anomaly for the Murray–Darling Basin, 1910–2010, with 15-year running average (black line)

Source: Bureau of Meteorology: <http://www.bom.gov.au> accessed 20 March 2011

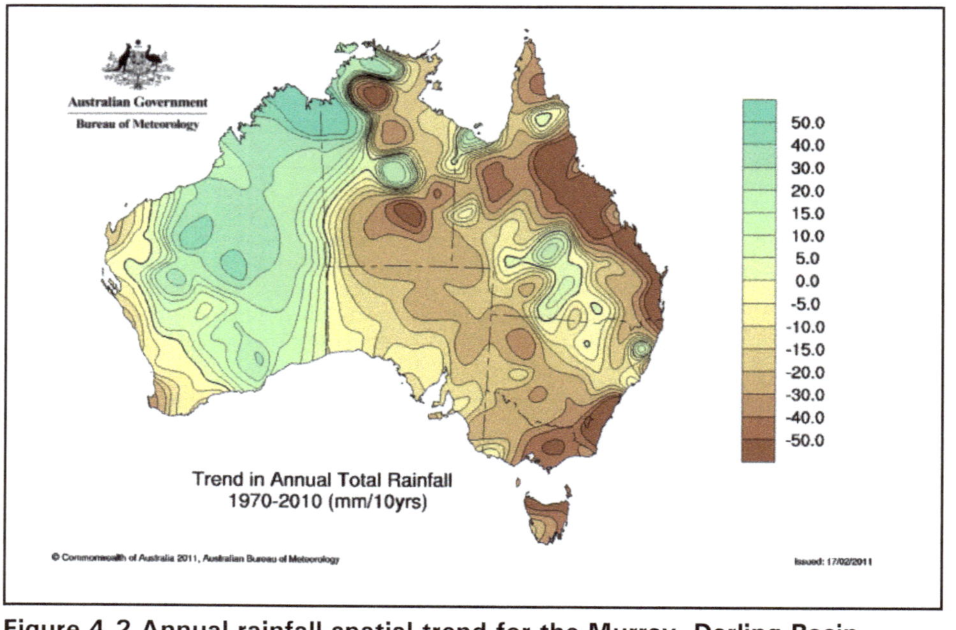

Figure 4.2 Annual rainfall spatial trend for the Murray–Darling Basin, 1970–2010

Source: Bureau of Meteorology: <http://www.bom.gov.au> accessed 20 March 2011

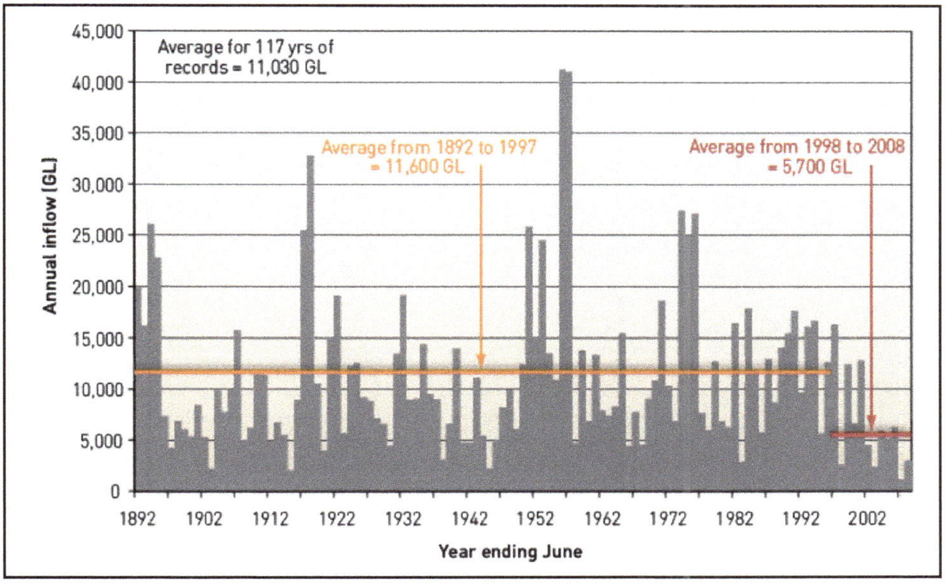

Figure 4.3 Record of annual inflows to the MDB 1892-2008

Source: Kiem and Verdon-Kidd (2010).

Future Climate Projections

Projections of future climate change in the MDB are summarised in Table 4.2.

Table 4.2 Summary of future climate-change projections in the MDB

Aspect of the climate system	Forecast changes in the MDB
Annual mean temperature	Increase in maximum surface temperatures across the MDB of 1–2°C by 2030 and up to 7°C by 2100; and increase in average surface temperatures across the MDB, particularly in the northern Basin, of 1–2°C by 2030 and up to 7°C by 2100. In winter, warming is projected to be as low as 0.5°C for the far south (CSIRO and BOM 2007).
Precipitation	The CSIRO's best estimate indicates that the future mean annual rainfall in the MDB in 2030 relative to 1990 will be lower by about 2 per cent in the north, and 5 per cent in the south (CSIRO 2008). Averaged across the Basin, the extreme estimates range from a 13 per cent decrease to an 8 per cent increase in mean annual rainfall (CSIRO 2008). In the southernmost MDB, the extremes range from a decrease in mean annual rainfall of up to 20 per cent to little change in mean annual rainfall (CSIRO 2008). In contrast to the CSIRO results, Sun et al. (in press) found almost zero change in MDB precipitation between 1970-99 to 2070-99, based on using an ensemble of 39 IPCC AR4 climate model runs for the A1B emissions scenario.

Evapotranspiration	Wet area evapotranspiration in the MDB has been projected for the years 2030 and 2100 based on a prediction from the CSIRO-Mk3.5, under SRES marker scenario A1F1, under a high rate of global warming. Predictions for 2030 show an increase in evapotranspiration in the range of 75–100 mm annually in the far north-east, 25–50 mm in the north-west, and 50–75 mm through the central and south-eastern parts of the Basin. By 2100, projections indicate a change of >175 mm annually for the entire Basin, with the exception of the far north-west, which is predicted to see increases between 125 mm and 150 mm annually (Figure 4.4).
Stream flow	The CSIRO Sustainable Yields project estimated changes in run-off in the MDB ranging from −33 per cent under a dry extreme scenario, and −9 per cent under a median scenario to a +16 per cent change under the wet extreme scenario (CSIRO 2008).
Flooding/tropical cyclones	It is likely that flooding in the northern part of the Basin will increase as monsoons are projected to be enhanced through climate change, and the northern Basin could become increasingly affected by tropical weather patterns (Grootemaat 2008). Abbs et al. (2006) indicate that tropical cyclone frequency could decrease by 9 per cent in 2070, but increases of 60 per cent (2030) and 100 per cent (2070) in the intensity of extreme tropical storms are possible.
Drought	Hennessy et al. (2008) project more frequent, longer, and more intense droughts in the MDB: • by 2010–40, exceptionally hot years are likely to affect about 65 per cent of the MDB and occur every 1.6 years on average • by 2010–40, little change is likely in the frequency and areal extent of exceptionally low rainfall years • by 2030, exceptionally low soil moisture years are likely to affect about 7 per cent of the MDB, and occur about once every 13 years on average.
Extreme temperatures and heatwaves	Global warming is projected to be associated with an increase in the frequency of hot days and warm nights. Daily maximum temperature data from six climate models were used to generate the ratio of the change in maximum to mean temperature. The ratio was more than 1 for the southernmost part of the MDB (CSIRO and BOM 2007).
Aspect of the climate system	**Forecast changes in the MDB**
Bushfires	Climate-change projections produced by the CSIRO show an overall increase in accumulated fire risk for Australia (CSIRO and BOM 2007). The combined frequencies of days with very high and extreme Forest Fire Danger Index (FFDI) ratings are likely to increase 4–25 per cent by 2020 and 15–70 per cent by 2050 (Lucas et al. 2007).

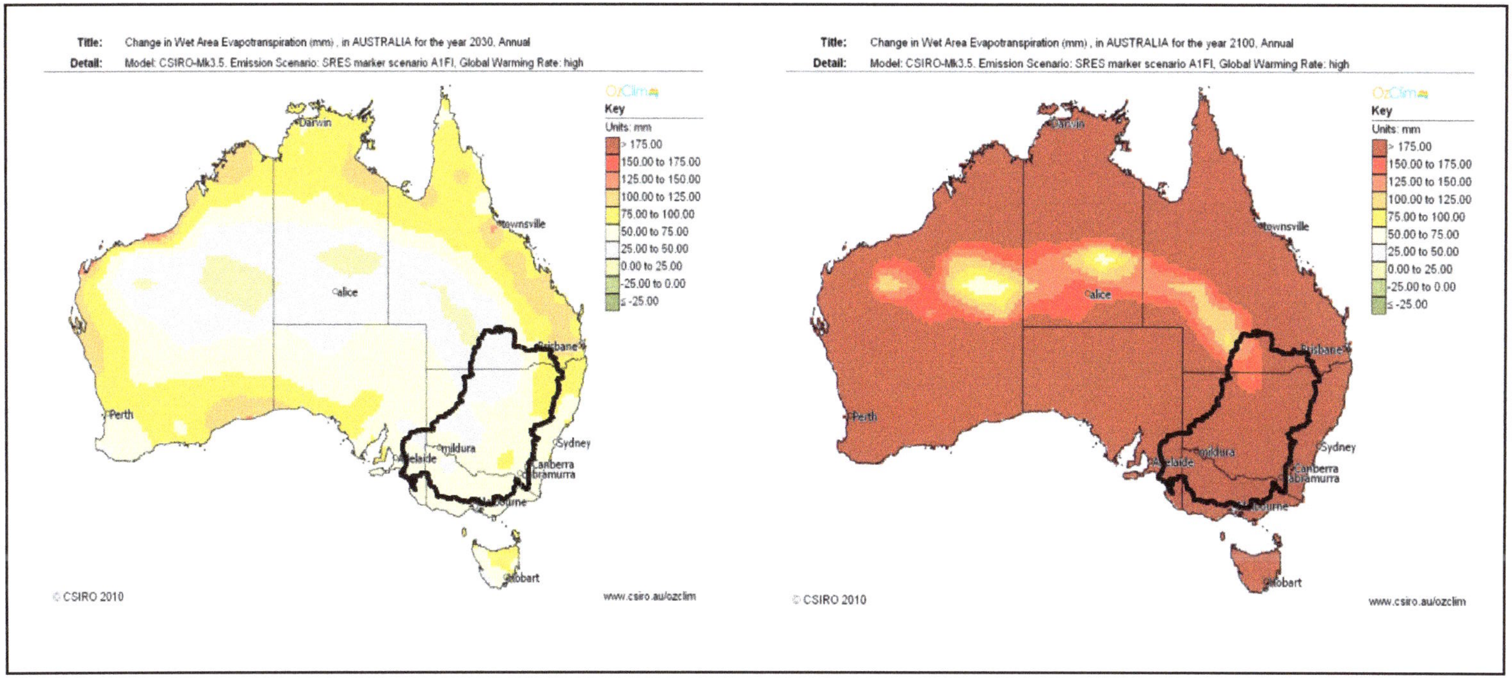

Figure 4.4 Wet area evapotranspiration (mm) in the MDB, 2030 and 2100

Note: Based on a prediction of the CSIRO-Mk3.5, under SRES marker scenario A1FI, under a high rate of global warming.

Source: CSIRO (2010).

Recent Advances in Understanding

Recent syntheses and investigations funded by the MDBA are summarised in Table 4.3. This work is as yet unpublished and summaries are presented with author and MDBA permission.

Table 4.3 Recent synthesis studies funded by the MDBA

Author(s)	Project title	Summary
CSIRO/BOM	South East Australia Climate Initiative	Links between the observed autumn rainfall decline in the MDB and a strengthening of the subtropical ridge (STR) have been made.
Anthony Kiem and Danielle Verdon-Kidd	Review of current understanding of Murray–Darling Basin climate patterns and causal processes	Dry conditions in autumn across the southern MDB are most likely if an El Niño event occurs in combination with a positive SAM. It was also found that, unlike the majority of eastern Australia, in the southern MDB, a La Niña event is not necessarily always associated with above-average rainfall. In fact, La Niña events occurring in conjunction with a positive SAM phase are often as dry as an El Niño event for the southern MDB.
Ailie Gallant and David Karoly	Climate patterns and causal processes	The MDB is broadly discussed in two regions: the north, where tropical influences dominate, and the south, where mid-latitude processes are most important. The ENSO is the most important driver of inter-annual variability across the MDB. Indian Ocean sea-surface temperatures (SSTs), including the IOD and the SAM, are regionally important during some seasons. The PDO is an important regulator of decadal climate variations in the MDB.
Roger Stone	Comprehensive review into the core patterns, droughts, rainfall systems and associated causal processes relevant to the Murray–Darling Basin at a range of associated time scales	ENSO has been and remains a major driver of year-to-year rainfall variability over the MDB, with strong impacts during the winter, spring and summer. Impacts in summer are especially relevant for northern regions of the MDB through affecting inflow into the MDB system via northern river systems influenced by tropical and extra-tropical systems. Additionally, the IOD, especially when considered in conjunction with ENSO, can influence rainfall variability over the MDB.
Jason P. Evans, Andy J. Pitman and Faye T. Cruz	Scientific review of the atmospheric and land-surface dynamics of the Murray–Darling Basin	While the climate of the MDB is dominated by large-scale processes, the nature of the landscape, the vegetation, soil moisture, fire, irrigation and orography interact with the large-scale forcing. Some of these terrestrial processes are locally important but regionally are likely to be insignificant. Others, through spatial aggregation of the small-scale processes, might lead to amplification or moderation of the larger-scale forcing.

Michael Roderick & GrahamFarquhar	Water availability and evapotranspiration in the Murray–Darling Basin: a look at the past and a glimpse into the future	The Budyko framework is used to calculate catchment-scale evaporation and run-off as a function of two climatic factors—precipitation (P) and evaporative demand (E)—and a third parameter (n) that encodes the catchment properties. The effect of GCM-projected changes in P and E on run-off is estimated (other factors remaining constant).
Peter Gell and collaborators	Palaeo-climate studies relevant to natural-resource management in the Murray–Darling Basin	Studies relating to past climate change and variability across south-eastern Australia have used speleothems, tree rings, river channels and terraces, dune systems and lake sediments. Available scientific evidence reveals that the Murray–Darling Basin has, over the past few hundred years, been subjected to extended inter-decadal variability, known as flood and drought-dominated phases, and year-to-year ENSO variability.
Peter Helman	Droughts in the Murray–Darling Basin since European settlement	In general, MDB droughts occur during positive IPO phases, with severe drought years linked to negative SOI or IOD. These climatic indices, while providing indicative correlations, do not correlate with all events. The drought history of the Basin shows that for more than one-third of the 221 years since European settlement part or all of the Basin has been in drought. Two long drought periods from 1795 to 1830, extending over 35 years, and 1980–2009, extending over 29 years, show a complete alteration to the water cycle.
Tim Low	Climate Change, Weeds and Pests in the Murray–Darling Basin	The evidence suggests that most of the weeds assessed are likely to benefit overall from climate change. Reasons for this include longer growing seasons, less frost, more droughts reducing competition, and higher temperatures along with carbon dioxide fertilisation increasing plant growth rates in situations where water is not limiting and competition from native plants is minimal. Climate change should lower populations of exotic fish because there will be less habitat for them to occupy.
Wendy Proctor, Karin Hosking, Thomas Carpenter, Mark Howden, Mark Stafford Smith and Trevor Booth	Future research needs for climate-change adaptation in the Murray–Darling Basin	Knowledge about adaptation is quite limited. There is a need for adaptation thinking to embrace the need for transformative, rather than incremental, adaptation in the region, which requires a major program to help the policy and management communities to envision different futures. It is likely that institutional change will be required as part of the process of responding to this challenge.
Bonnie Bonneville, Julie Morris, Emma Collins, David Dettrick and Annie Sanderson	Impacts of Climate Change on the Murray–Darling Basin's Water Quality	Impacts of climate-related extreme events (prolonged droughts, bushfires, heatwaves, flooding, and dust storms) on water-quality parameters (dissolved oxygen, nutrients, turbidity, salinity, water temperature, pH and BOD) were assessed across the MDB. Resulting risks to beneficial uses were rated for each event type for different regions of the Basin.

Fran Sheldon, Nick Bond, Nick Marsh, Stephen Balcombe, Samantha Capon, Wade Hadwen and Mark Kennard	Impacts of Climate on the Aquatic Ecosystems of the Murray–Darling Basin	Three climate-change drivers for inland aquatic ecosystems were reviewed: increasing temperature (air and water), changing patterns of precipitation (rainfall and run-off), and increased UVB-radiation. Impacts were reviewed for a range of fish, riparian and floodplain vegetation species.
Quentin Grafton, Angella Duvnjak, Chris Miller, Paul Ryan, Fiona Verity, Mavis Zutshi, Jared Dent, Qiang Jiang, Michael Ward, & William Nikolakis	Potential Water Quantity and Quality Impacts in the Murray–Darling Basin from Communities and Industries Responding to Climate Change	Impacts of wet and dry future climate scenarios on irrigated agriculture, aspects of dryland agriculture, aspects of forestry and tourism, basin communities, and Indigenous peoples were assessed though modelling and community workshops.
Greg Holland, Keith Collett, Nicole Caruso and Bonnie Bonneville	Risk of climate-change impacts on salinity dynamics and mobilisation processes in the Murray–Darling Basin	A quantitative and qualitative assessment of the impact of wet and dry climate scenarios to 2050 on salinity loads and concentrations is made for broad regional landscapes as well as the Darling tributaries and Murray River. A risk assessment is conducted and some options for management presented.
Tim McVicar, Randall Donohue, Anthony O'Grady and Lingtao Li	The effects of climate change on plant physiological and catchment eco-hydrological processes in the high-rainfall catchments of the Murray–Darling Basin: a scoping study	This project explores the sensitivity of run-off in the context of climate change across the MDB to changes in five key eco-hydrological parameters: annual precipitation, annual potential evaporation, average storm depth, catchment-average rooting depth, and atmospheric carbon dioxide concentration. The sensitivities were analysed for five MDB water-resource yield zones: the extremely high yield zone (EHYZ), the very high yield zone (VHYZ), the southern high yield zone (sHYZ), the northern high yield zone (nHYZ), and the whole Murray–Darling Basin.
Peter Gehrke	Afforestation risks to water resources in the Murray–Darling Basin	The risks to MDB water resources of afforestation under climate change were assessed. Catchment water yields were analysed for climate change alone and climate change with afforestation. Recent work on the adaptive capacity of forests to climate change and effects of industry plantation projections were also taken into account.
Leon Bren, Jeya Jeyasingham, Stuart Davey, Patrick Lane, Richard Benyon and Ian Ferguson	Impacts of native forest management practices in silvicultural systems on catchment water yield in the MDB	The past 13 years in Victoria were viewed as similar to the more extreme levels of reduced rainfall estimated to potentially occur due to future climate change. Using this, estimates were made of the reduction in rainfall that might be expected and the impact of this on stream flow should logging cease in the higher-rainfall, forested catchments of the MBD.

Basin Planning and Climate Change

The *Guide to the proposed Basin Plan* (MDBA 2010a) was prepared on the basis of modelling that shows surface-water availability declining by 10 per cent between 1990 and 2030. The story for groundwater is somewhat different, with the MDBA's modelling for the same period suggesting that groundwater recharge will remain at about historical levels. The Authority also noted that while 'there is an increasing likelihood that climate change is part of the reason for the recent drought, it is not yet possible to distinguish this component from the naturally high climatic variability experienced in the Basin'.

The MDBA's modelling is based on the entire historical record (1895–2009), which incorporates data from the first half of the period 1990–2030. As a consequence, the 10 per cent reduction due to climate change is already partially accounted for in existing hydrological modelling used for basin planning. With the Basin Plan set for review by 2021, the Authority considered it 'only appropriate to incorporate a percentage of the remaining change not already in the modelling' in the Basin Plan. The Authority determined that a 3 per cent reduction in surface-water availability was appropriate and factored this reduction into calculations for the sustainable diversion limits (SDLs). The implication here is that the 3 per cent reduction will be revisited when the Basin Plan is reviewed in 2021. The arrival of the MDBA at a 3 per cent reduction in the availability of surface water is not made clear in the Guide, but is understood to be a proportional estimate. This 3 per cent is applied across the Basin without attempting to take into account local variations. It should be noted that as water-resource plans will allocate water on an annual basis, reductions in water availability due to climate change will be accounted for and shared between the environment and other users.

The MDBA has relied upon the Sustainable Yields methodology and models for determining the impacts of climate change on rainfall, run-off and other climate-driven variables (MDBA 2010b). Volume 2 of the Guide quotes the CSIRO Sustainable Yields project, which found there would be an 11 per cent reduction in surface-water availability in 2030 under the median climate scenario (MDBA 2010b). This methodology uses high, medium and low global greenhouse gas emission scenarios based on the IPCC's *Fourth Assessment Report* and the work of CSIRO and the Bureau of Meteorology (BOM). The changes in temperature under these scenarios are multiplied by the estimated changes in rainfall and other variables per degree of warming that have been calculated from 15 global climate models from the *Fourth Assessment Report*. The results are 45 sets of 'seasonal scaling' factors that are applied to the historical record to arrive at a range of predicted outcomes of climate change, including changes to rainfall and run-off (Chiew et al. 2008).

The MDBA (2010b) considers itself to have factored climate change into the Basin Plan in three ways

1. the 3 per cent reduction in surface-water availability by 2021
2. requiring flexibility in water-resource plans to operate across a range of climatic conditions
3. the sharing of climate-change effects evenly between the environment and water users.

Work on the final Basin Plan is ongoing and revisions are expected in the way that climate-change impacts are determined in the final Basin Plan (due in 2012).

Concluding Remarks and Future Directions

The constraints of word length and unpublished material mean that justice has not been done to this topic and the growing body of knowledge it encompasses. It is clear, however, that climate change is now an integral part of the environmental, economic and social futures of the MDB. This chapter at best gives an introduction to this complex sphere and some of the emerging ideas and analyses. A more comprehensive synthesis is under way.

In this final section, some broad priorities for future investigation are proposed. Many more specific needs have been identified but are too numerous to report here.

Perhaps the overriding constraint in describing and assessing future climate change and its impacts is the ability of GCMs to provide accurate climate forecasts at the scale of the MDB. This issue has been highlighted by the Royal Society (2010), which states that 'there is little confidence in specific projections of future regional climate change'. GCMs also do not adequately take into account regional modes of climate variability (prominent in the Australian context), land surface–atmosphere coupling and some feedback processes in the carbon cycle. This further limits their capacity to provide realistic forecasts at the MDB scale. This issue of forecast quality is critical to assessment of future impacts of climate change in the MDB as climate inputs drive most impact assessments.

To address this quandary, modellers have developed future climate scenarios that capture both a range of future greenhouse gas emission scenarios and a range of forecasts by the different GCMs. Such scenarios are useful for asking 'what if' questions, and perhaps in putting some reasonable bounds on future possibilities.

Despite the scenario approach, it is proposed that uncertainty in climate-change forecasting needs to be addressed more explicitly and communicated more effectively. There are many sources of uncertainty in models to be considered, including the: key processes selected and omitted; different representations/codifications of processes; selection of model parameter values; representation of interactions between processes; inclusion of important feedback mechanisms; treatment of processes not fully understood; effects of different model structures; impacts of using different antecedent conditions; error cascades in down-scaling; selection of future emission scenarios. These and other factors can create a wide range in the uncertainty of forecasts. Current methods to account for uncertainty in climate projections are also limited—for example, the use of statistics applied to ensembles of projections based on intrinsically different models, or assumptions relating to commonality in forecasts from different models equating to accuracy of forecasts. Hence, greater effort to characterise and communicate uncertainty is proposed.

Most users of climate information seek forecasts at a particular spatial scale of relevance. This scale is often small in relation to the forecast capability of GCMs. Down-scaling techniques have been used to assist in translating GCM forecasts to local scales, but it should be noted that these techniques do not improve the quality of the GCM forecasts used.

The natural variability of the Australian climate—and that of the MDB—is amongst the highest in the world. This is due in part to a number of regional-scale climate 'modes' that affect the climate over weeks, months, seasons, years and decades. The modes include: ENSO, IOD, SAM, IPO/PDO and STR, amongst others. Each of these modes affects climate outcomes over different spatial and temporal scales and some interact with each other. There are also some indications that global warming is influencing some of these modes. Research on these aspects of climate variability and interactions with climate change is progressing, but much more effort is required in this domain.

Landscape type, vegetation, soil moisture, fire, irrigation and orography interact with the large-scale climate forcing in complex ways. Whilst these are primarily second-order effects, more research is required on the role of each aspect in climate outcomes.

Warming has been occurring in the MDB since the 1950s (Figure 4.1) and numerous climate trends are evident in climate observations. It is proposed here that more emphasis should be placed on observations of climate change in the instrumental and palaeo-climate records as a means of understanding and assessing impacts of past and future climate change, particularly given the limitations of future GCM forecasts and their inability to account for some current observed trends such as autumn rainfall decline.

The scientific evidence for global warming and ongoing changes in the climate system is very strong, and the potential short, medium and long-term consequences range from substantial to catastrophic. Climate change and climate variability are arguably the most important drivers of future change to the natural resources of the MDB and might have significant impacts on industries and communities. Clear, well-argued priorities for research to meet the adaptation needs of all sectors of the community should be agreed to and financially supported.

Acknowledgments

The author is grateful for comments on an earlier manuscript by Jason Alexandra, Dr Mathew Maliel, Dr Anthony Kiem, Steve Purbrick and Annie Sanderson.

Bibliography

Abbs, D. J., Aryal, S., Campbell, E., McGregor, J., Nguyen, K., Palmer, M., Rafter, T., Watterson, I. and Bates, B. 2006, *Projections of extreme rainfall and cyclones*, A report to the Australian Greenhouse Office, Canberra.

Bureau of Transport Economics (BTE) 2001, *Economic Costs of Natural Disasters in Australia*, Bureau of Transport Economics, Canberra, <www.bte.gov.au/docs/r103/contents.htm>

Cai, W. and Cowan, T. 2008, 'Evidence of impacts from rising temperature on inflows to the Murray–Darling Basin', *Geophysical Research Letters*, vol. 35, no. L07701, <doi:10.1029/2008GL033390>

Chiew, F. H. S., Teng, J., Kirono, D., Frost, A. J., Bathols, J. M., Vaze, J., Viney, N. R., Young, W. J., Hennessy, K. J. and Cai, W. J. 2008, *Climate data for hydrologic scenario modelling across the Murray–Darling Basin*, A report to the Australian Government from the CSIRO Murray–Darling Basin Sustainable Yields Project, CSIRO, Canberra.

Commonwealth Scientific and Industrial Research Organisation (CSIRO) 2006, *Climate change impacts on Australia and the benefits of early action to reduce global greenhouse gas emissions*, Prepared by B. L. Preston and R. N. Jones for the Australian Business Roundtable on Climate Change.

Commonwealth Scientific and Industrial Research Organisation (CSIRO) 2008, *Water availability in the Murray Darling Basin*, Summary report to the Australian Government by CSIRO Murray–Darling Basin Sustainable Yields Project, Canberra.

Commonwealth Scientific and Industrial Research Organisation (CSIRO) 2010, OzCLIM Online Prediction Generator, CSIRO, Canberra, <http://www.csiro.au/ozclim/home.do>

Commonwealth Scientific and Industrial Research Organisation (CSIRO) and Bureau of Meteorology (BOM) 2007, *Climate change in Australia*, Technical Report, CSIRO and Bureau of Meteorology, Canberra, <www.climatechangeinaustralia.gov.au>

Commonwealth Scientific and Industrial Research Organisation (CSIRO) and Bureau of Meteorology (BOM) 2009, *Climate Science Update 2009*, CSIRO and Bureau of Meteorology, Canberra, <http://www.climatechangeinaustralia.gov.au/documents/resources/ClimateScienceUpdate2009_2.pdf>

Deo, R. C., McApline, C. A., Syktus, J., McCowan, H. A. and Phino, S. 2006, *On Australian Heat Waves: Time series analysis of extreme temperature events in Australia, 1950–2005*, University of Queensland, St Lucia.

Grootemaat, G. D. 2008, The relationship of flooding in Australian dryland rivers to synoptic weather patterns, El Nino southern oscillation, sea surface temperature and rainfall distribution, PhD Thesis, School of Earth and Environmental Sciences, University of Wollongong, NSW.

Hennessy, K., Fawcett, R., Kirono, D., Mpelasoka, F., Jones, D., Bathols, J., Whetton, P., Stafford Smith, M., Howden, M., Mitchell, C. and Plummer, N. 2008, *An Assessment of the Impact of Climate Change on the Nature and Frequency of Exceptional Climatic Events*, Commonwealth of Australia, Canberra, <http://www.appslabs.com.au/salinity.htm>

Intergovernmental Panel on Climate Change (IPCC) 2007, 'Climate change (2007). The physical science basis', Contribution of Working Group I, S. Solomon, D. Qin, M. Manning, Z. Chen, M. Marquis, K. B. Averyt, M. Tignor and H. L. Miller (eds), *Fourth Assessment Report of the Intergovernmental Panel on Climate Change*, Cambridge University Press, UK and New York.

Kiem, A. S. and Verdon-Kidd, D. C. 2010, Presentation to Murray–Darling Basin Authority NCCARF Workshop, 4 March 2010.

Lucas, C., Hennessy, K., Mills, G. and Bathols, J. 2007, *Bushfire weather in southeast Australia: recent trends and projected climate change impacts*, Consultancy report prepared for the Climate Institute of Australia by the Bushfire Cooperative Research Centre and CSIRO, Melbourne.

Murray–Darling Basin Authority (MDBA) 2010a, *Guide to the proposed Basin Plan. Volume 1: Overview*, Murray–Darling Basin Authority, Canberra.

Murray–Darling Basin Authority (MDBA) 2010b, *Guide to the proposed Basin Plan. Volume 2: Technical background*, Murray–Darling Basin Authority, Canberra.

Nicholls, N. 2004, 'The changing nature of Australian droughts', *Climatic Change*, vol. 63, pp. 323–36.

Royal Society 2010, *Climate Change: A summary of the science*, September 2010, Royal Society, London.

Sun, F., Roderick, Lim, W. H. and Farquhar, G.D. (in press). 'Hydro-climatic projections for the Murray-Darling Basin based on an ensemble derived from IPCC AR4 climate models', *Water Resources Research*.

Verdon-Kidd, D. C. and Kiem, A. S. 2009, 'Nature and causes of protracted droughts in southeast Australia: comparison between the federation, WWII and big dry droughts', *Geophysical Research Letters*, vol. 36, no. L22707, <doi:10.1029/2009GL041067>

Verdon-Kidd, D. and Kiem, A. 2010, 'Quantifying drought risk in a nonstationary climate', *Journal of Hydrometeorology*, August, pp. 1019–31, <doi:10.1175/2010JHM1215.1>

Yu, J., Fu, G., Cai, W. and Cowan, T. 2010, 'Impacts of precipitation and temperature changes on annual streamflow in the Murray–Darling Basin', *Water International*, vol. 35, no. 3, pp. 313–23.

Part II. Communities

5. Basin Bookends, the Community Perspective[1]

Leith Boully, Karlene Maywald

Introduction

There is a collective understanding across the Murray–Darling Basin (MDB) that water has been over-allocated, and, generally speaking, most people know in their hearts and minds that this must be addressed. Responding to calls from people across the nation (including in the MDB), the Australian Parliament passed the *Water Act 2007* (Cwlth) with the support of the opposition and minor parties. With this level of consensus, it is reasonable to expect that implementation of the provisions of the Act would be well received. The Act has fundamentally redefined the water-policy priorities for the Murray–Darling Basin. It requires the preparation of a Basin Plan, which is the legal instrument with which the Australian Government has the capacity to return the system to a sustainable balance.

It is well understood by communities across the MDB that having a healthy environment is fundamental to economic prosperity and social harmony. So why has there been such an emotional outcry from individuals and communities to the release of the *Guide to the proposed Basin Plan* (MDBA 2010), which recommends the return of at least 3–4000 gigalitres of water to the environment? Could it be that the process and communication styles have so threatened people's core values and sense of identity and purpose that the only responses available are fight or flight?

This chapter seeks to put into perspective the communities' responses to the release of the Guide. It will demonstrate that process and communication are central to engaging individuals and communities in and achieving significant change. It will seek to make evident the need for a balanced environmental, economic and social response in shaping the final Basin Plan. Successful implementation of the Basin Plan is unquestionably one of the most important reforms facing our nation. The Australian commitment to a 'fair go for all'

1 The authors collectively have extensive experience working with regional communities in the Murray–Darling Basin. The opinions expressed in this chapter regarding the community are the views of the authors and have been drawn from these experiences.

demands that we do it in a way that respects all of the individuals involved. We must get the balance right so the health of our rivers can underpin the future prosperity of our regions for generations to come.

What Does Water Mean to People in the MDB?

Water is Life—We cannot survive without it

There is nothing more important to Australians than water. The past 10 years of severe drought have severely impacted on the agricultural sector, but interestingly, the period has also highlighted the vulnerability of the nation's major-city water supplies. City and regional communities agree—they want to feel confident that governments are doing what is necessary to ensure greater certainty of water supply. There is also agreement that the environment needs its share.

Our nation prospers from it

Communities and industries have emerged from and grown on the back of development of the water resources of the Murray–Darling Basin. State Governments have actively promoted development based on the prosperity offered by irrigation, and communities have responded. According to the Guide, 'The Murray Darling Basin's agriculture produces $15 billion worth of produce annually. It contains around 65% of Australia's irrigated land area and about 40% of Australia's farms' (MDBA 2010 p13). The Basin is economically significant in terms of Australia's agricultural production.

There currently are historical access and rights to it

Irrigators own legally issued access entitlements to water. They have invested heavily in infrastructure to support their businesses based on their legal access to water. State and federal governments have invested heavily in:

1. river-regulation infrastructure—for example, dams, locks and weirs
2. irrigation-delivery infrastructure—for example, extensive channel systems, pumps, pipelines, and so on, to support the development of irrigation.

Each basin community has a different settlement story and each State has a different development story. Successive governments over-allocated water-access rights, and successive communities prospered and developed on the basis of the access rights legally issued.

Water Entitlement-Holders want to provide for their families from it

Irrigators are people, running businesses, and providing for their families and/ or shareholders. Many families are attracted to irrigation towns where business opportunities and jobs provide the opportunity to bring lifestyle and livelihood together.

Local Communities and the Nation Thrive from the Prosperity It Generates

Communities have thrived and grown on the back of successful irrigation development. Communities have also struggled to cope when irrigation production has not performed well. Irrigators are risk takers. Irrigation businesses are subject to many variables: commodity prices, the Australian dollar, trade barriers, water availability, oversupply, under-supply, market failure, and the weather. Generations of irrigators have survived the cycles through extraordinary resilience and adaptively managing the changing economic circumstances. They are now overexposed as a result of previous government policy, which over-allocated water-access entitlements. Their future is uncertain.

People want to camp, catch fish and yabbies, and love the environment it creates

'The Murray Darling Basin consists of 23 major river valleys and covers one million km^2 across four states and the ACT...Twenty of the 23 major river valleys of the Basin are in poor to very poor ecological condition' (MDBA 2010 p13). A do-nothing scenario will lead to further degradation and possibly irreversible decline in the health of the system. This is not the future that basin communities want for future generations.

So why is getting the balance right so hard?

No matter where you stand in the Basin, communities look upstream with envy and downstream with disdain. All people see the problem through the lens of their own circumstances and limited knowledge: 'Of course it should be fixed, but why is it my fault? Why should I wear the cost of fixing it? Governments stuffed it up, so Governments should fix it.' This is a normal human response to a threat to personal security.

Process

> The thing that we have to understand is that values and ethics are not formed overnight. They are deep seated. So persuasion and manipulation towards solutions that do not include community values and ethics are a waste of time and of money. (Nancarrow 2006)

Criticism of process and at least a perceived lack of communication have been the dominant issues raised by commentators from across the MDB regarding the Guide to the draft Basin Plan. Linked closely to this has been a call for amendments to the *Water Act* to allow for the 'proper' consideration of social and economic issues and robust community-engagement processes. The definition of 'proper', however, changes depending on the perspective of the individual.

'The truth is that irrespective of whether you live in Adelaide or Griffith, Dirranbandi or Waikerie, water is not only technical it is social—how it is used or abused is about people and their values, needs and aspirations' (Boully 2010, p 1).

It is not possible to 'fix the problem' without tackling the needs of the people affected by the proposed changes.

A complicating factor is that the values, needs and aspirations of people will differ greatly from community to community, region to region and State to State.

It does not help that the Murray–Darling Basin is littered with historical prejudices and spin. Every community, industry sector, local government and State government is guilty of pointing the finger and playing the blame game.

South Australians blame greedy upstream States and rice and cotton growers, while upstream they think South Australians are just a bunch of whingers and cannot understand why the barrages are not opened to flood the Lower Lakes with sea water.

Right across the Basin there are cries of 'It's not my fault'; 'It has to be fixed but why should I bear the brunt of the cost'; 'The numbers can't be right'; 'The environment is important, but so is my family/community'; 'I've done my bit—look how efficient I am.'

Decades of entrenched, ill-informed prejudices fuelled by local political point scoring have tended to polarise the debate.

Add to this a deep cynicism of government and a media that will always prefer controversial opinion to boring facts and it is easy to see why decision makers have shied away from making the tough decisions.

We can draw on recent experience to look for alternative processes. The development of the concept of a 'healthy working river' as a means to reconcile multiple values and objectives in the Living Murray Initiative (Murray–Darling Basin Commission: <http://thelivingmurray2.mdbc.gov.au/>) was a critical contribution to the success of what was the first step taken by the Murray–Darling Basin Ministerial Council to improve the health of the MDB.

A healthy, working river is defined as a river that is managed to provide a sustainable compromise, agreed to by the community, between the condition of the river and the level of human use (Cooperative Research Centre for Freshwater Ecology 2003). Implicit in this definition is the recognition that a complex socio-biophysical system exists and that rigorous processes must be provided through which the community and governments can, in an informed manner, negotiate a range of trade-offs or optimise objectives.

This language engaged the hearts and minds of basin communities so that the process and subsequent decisions were warmly embraced.

The National Water Initiative (NWI) (Commonwealth of Australia 2004), which has widespread stakeholder support, recognised that settling the trade-offs between competing outcomes for water systems will involve judgments informed by the best-available science, socioeconomic analysis and community input. Open and inclusive discussion of the range of values held by different interests must be incorporated into the process. The values held collectively in relation to water, rivers and groundwater systems can, if properly explored, lead to the development of an ethical framework through which to negotiate objectives and management arrangements (Boully 2007).

The *Guide to the proposed Basin Plan* has, for the most part, been designed by scientists, engineers and other technical experts, with little input from planners and social scientists. The NWI has established the theoretical opportunity for communities and governments to develop sophisticated, values-based, ethical adaptive-management arrangements, while it appears that the *Water Act* has restricted this philosophical approach.

The process adopted by the Murray–Darling Basin Authority (MDBA) to date has followed the approach that governments have traditionally employed in relation to water-resource planning—that is, an adversarial and expert decision-making approach to water allocation, management and planning processes. This has been described by Vanderbyl and Boully (2004), who suggest that such approaches tend to

- rely mainly on the input of scientific experts, scientific organisations, lobbyists or other people who might be strongly aligned with particular 'sides' or a specific point of view

- place emphasis on the importance of research, technical reports or other information generated for a specific purpose and without reference to, or integration with, other work that might be relevant
- involve decisions to strategically either withhold or release information (rather than to openly pool it)
- keep 'grassroots' stakeholders at arm's length from committees, expert panels or other administrative or scientific bodies involved in the decision-informing stages of the processes
- take no account of the professional and personal damage done to individuals on all 'sides' through the highly emotive and destructive approaches people feel forced to adopt in order to have their view heard.

Every decision has impacts that are felt in both positive and negative ways across river reaches and communities. Every refinement has a new set of impacts. What is fair to people with one set of values will be seen as unfair to others. Only rigorous processes that take account of these issues can reduce the potential for conflict, or manage it well, by ensuring that the decisions are defensible on a broad range of grounds (Vanderbyl and Boully 2004).

Only a sophisticated and well-resourced adaptive-management philosophy, framework and practice will meet the multiple and changing values and objectives of the mature water economy. Vanderbyl and Boully (2004) conclude that the primary objective of reform must be the development of rigorous, adaptive, inclusive, informed and fair processes for managing complex water-resource systems.

The voices of individuals and communities across the MDB suggest that these processes are absent.

Communication

It is said that 'communication is the response that you get'.

Senior officials in the MDBA point out that they have acted in accordance with the *Water Act*, which was passed unanimously by both houses of the Commonwealth Parliament. They are communicating the 'facts' about the state of the rivers of the MDB and the rational solution to the challenge of over-allocation. Why, then, is this not getting traction and support?

Olson (2009), in his book *Don't Be Such A Scientist: Talking substance in an age of style*, tells us that there are 'layers' of communication, rather like a pyramid. At the top of the pyramid is the 'mind', which is arguably where most scientists and policy makers spend most of their time. They communicate learnedly with each other in a careful, heavily footnoted style. The next layer down is the 'heart': the locus of love. The third layer is the 'gut': the locus of fear.

The MDBA's interpretation of the *Water Act* has caused it to focus on the science first and the community second. The Authority has tried to communicate to people who are not 'just like them' through engaging with the mind in a logical and rational manner. They have failed to engage the hearts by neglecting the social and economic matters that are as fundamental to an individual's survival as the water itself. People have received the communication as a blow to the gut. They are now fearful for their basic survival and their very identities are threatened. There is a very real risk that the extreme polarisation of the debate will result in decisions being made that will be detrimental to the Basin environment, its communities and individuals.

Creating the Environment for Change

Bringing about sustainable change requires that people believe there is a need for change, are committed to it, know how to go about it and have sufficient resources to do so.

In order to create the environment for change given the current circumstances, a rethink of the processes being employed is required. This rethink should take into account Ingram and Schneider's (1998) proposition that water is a fundamental social resource, and collective decision making about its distribution and protection is fundamental to building a sense of community. This sense of community is important for building resilience and to assist difficult decision making that affects community, rather than individual, interests.

In relation to rivers and groundwater systems, the behaviour of communities is strongly influenced by how they value water. Different beliefs and values will lead people to define water in terms of how it can provide: economic opportunity; recreational opportunity; cultural and spiritual needs; basic drinking needs; places of great beauty; abundance of wildlife; and other ecosystem services. Whilst all communities have deeply entrenched beliefs and value systems, they are not necessarily the same across the Basin. The impacts of the Basin Plan will be felt differently in each community depending on their socioeconomic status and their capacity to afford to manage the necessary changes. Programs to deal with the impacts will need to be flexible enough to respond to the specific needs of each community.

Adopting a bottom-up approach to enable communities to determine what the future might look like for them in an environment with less water would deliver better results than a top-heavy, 'one-size-fits-all' approach. Transition time frames will also be important as large volumes of water are transferred from economic use to the environment. Providing support for communities to play a part in determining what programs will best suit their local needs will empower them to think about how they can influence their own future. Communities cannot yet see what the future looks like in an environment with less water. They need help to participate in designing a future that will sustain them and their children and grandchildren.

There is a need for a new process at the regional scale that can take account not only of economic values and interests in water, but also—more explicitly—the social and cultural values that underpin them. These processes will enable policy makers and planners to explore the full range of strategies that can be used to respond to the diversity of values and motivations within the community and the attitudes and behaviours that result (Boully 2007).

Key features that should be evident in the process include

- formal and permanent engagement arrangements focused on improving water-resource management and meeting community aspirations
- a focus on joint discovery and mutual-gains decision making in realistic time frames
- consideration of a broad range of economic, social and environmental interests
- processes to identify and manage conflict
- commitment to building the capacity of those in the community who have trouble participating due to lack of knowledge or for social or economic reasons
- innovation in the provision of environmental flow
- certainty that rights are secure and that the market will be used to adjust levels of consumptive use
- establishment of multiple objectives that reflect multiple values that are measurable.

With the appropriate process in place, the grassroots players will be able to be proactive in designing a future rather than devoting that energy to trying to destroy a top-down process.

Who Are the Players Today?

Retirees

The average age of our farmers is fifty-eight and many sons and daughters are not likely to take over the family farm. This means that for some irrigators the answer will be to exit irrigation and retire from the land. Selling water at market rates might be attractive, however, the capital will most likely go to retire debt and perhaps fund resettlement. It will not necessarily stay in the community.

In addition, the land will be rendered unproductive in many districts and irrigation-delivery infrastructure might be stranded, resulting in increasing costs to the stayers.

Stayers

For those who stay, the cost will be significant. Despite the short-term policy of allowing irrigation-infrastructure operators to charge modest exit fees to cover the cost of asset rationalisation, it will not mitigate the very significant issue of stranded assets. Irrigation fixed costs will be a significant impediment to the profitability of the stayers. In addition, the stayers will be subjected to the increasingly conservative lending criteria of the banking sector due to their perceptions of the risks inherent in and consequential to the reform process.

Others

The non-irrigator population of the small and medium-sized towns in the MDB who are reliant mostly on agriculture for their sustainability will face significant challenges. As economic output declines, jobs are lost and services downsized or removed. While this is the way it has always been, there has never been a policy shock of the scale of that proposed by the *Guide to the proposed Basin Plan* and it is difficult to imagine any complementary or replacement industry emerging to mitigate the impact. That being the case, it is only fair that these communities are provided with the processes and resources to plan the adjustment.

Finding the Way Forward

The release of *The Guide to the proposed Basin Plan* is the first glimpse the community has had of the work undertaken by the independent MDBA in accordance with the *Water Act*. Public meetings have been held throughout the Basin as the first step in the consultation process and irrigation communities have turned out in droves to express their opposition to the quantum of water proposed for return to the environment. It is not surprising that the initial reaction from irrigation communities is one of anger and opposition. Communities are frightened by what the future will hold for them and they lack confidence in the decision-making process.

What has become evident very early is that a process to deal concurrently with the social and economic issues is missing. The Minister for Regional Development has subsequently announced that a Parliamentary Inquiry will be held into the social and economic issues of the proposed Basin Plan. The original vision for minimising the impact of the necessary reform included significant investment in infrastructure-upgrade programs to save water and water-purchasing programs. Water-purchase programs compensate an entitlement-holder for giving up water, while investment in infrastructure provides support for those staying in the irrigation industry, saving water through efficiency gains and creating employment. While there has been criticism of the way in which both approaches have been implemented, there is no doubt that the withdrawal of investment from either program would create yet more conflict.

For some time community advocates have been desperately trying to impress on the Australian Government that while individual irrigators might be treated fairly in a financial sense, their communities have no safety net to rely on. What is missing is a program that will support the development of a new vision for the future and deal with the broader-reaching social and economic impacts.

Knowing the real impact of the proposed Plan on regional communities is important, but it is equally important to determine what needs to be done about it. Those best placed to do this are the communities themselves. Acknowledging the fact that water reform is a social process and, based on successful experiences in South Australia and Queensland, we would argue that in addition to the existing process there are three critical steps required to re-engage communities in designing their future in an environment where less water is available for consumptive use.

Recommit to sustainability

Recommit to the concept of achieving healthy, working rivers in the MDB. The MDBA should support and resource State governments to engage their communities in water-resource planning processes including designing stepped approaches to achieving the environmental outcomes articulated in the *Guide to the proposed Basin Plan* and taking into account the social and economic consequences of doing so. Application of local knowledge will reveal much more sophisticated and efficient means to provide water for key ecological assets and functions than can be achieved from an office in Canberra.

The evidence that this can be done is provided by Vanderbyl and Boully (2004, p 1), who summarise a very successful and timely process undertaken in 2002 in the Lower Balonne, Queensland, as follows:

> Water sharing between users and between users and the environment has been a focus for the Lower Balonne community for a long time. Conflict between users within the community and between community and government has been significant for over a decade. Both government

and the community have been able to rise to the challenge, put the past behind them and develop an approach to sharing water resources between competing values.

Through the use of the best available hydrologic data and ecological science community and government have together acknowledged the risks to specific ecological assets, and to the economic viability of the region if the 'business as usual' scenario is adhered to. With the ultimate goal being the establishment of 'social harmony' the process of learning, debate and negotiation has successfully dealt with equity between most users and between users and the environment.

Government has taken the unprecedented step of providing unrestricted access to all information and adequately resourcing the process of supporting a community reference group. This group has been able to interact directly with an independent Scientific Review Panel and in partnership with government subsequently develop through a process of 'joint discovery', innovative solutions to the challenges involved in managing a large, ephemeral floodplain river system with multiple values.

Support for regional communities

The Australian and State governments should jointly fund regional communities to develop adjustment prospectuses and provide investment capital through a future fund.

An example of a regional development program designed to assist communities to adjust is the SA Government-funded 'Riverland Futures' program.

The Riverland is a SA irrigation-based community that has been hard hit by the drought.

The Riverland Futures Task Force was established to drive the program. The Task Force was made up of local government mayors and CEOs and leaders from the local health, education and industry sectors, primary industries and regional development officers.

The SA Government provided funding to prepare a Riverland Prospectus to explore development opportunities and guide investment in the region. It sets out the region's goals and aspirations, its strengths and weaknesses and the opportunities to diversify the economic base.

The SA Government has also committed $20 million as seed funding to support projects that meet investment criteria.

Invest

The Australian Government should significantly increase investment in irrigation research and development with the goal of bringing about transformational change to both practices and production. Australian agriculture has a proud history of generating growth in productivity and innovation in environmental practice. Sustained high levels of investment in this professional and innovative sector would be a mark of respect for those who produce some of the best food and fibre in the world.

We would not suggest that these are the only ways to retrieve what has become a sad and divisive argument about the most precious of our resources, people and water, but our practical and pragmatic community experience leaves us in no doubt that significant initiatives in these three areas would provide hope to those who have lost it and produce sustainable outcomes for the environment and communities. Even though there will be many who would say that this will take too long, we are in no doubt that continuing the current adversarial approach will take even longer and have fewer beneficial outcomes.

References

Boully, L. 2007, 'Water resource management: multiple values demand multiple objectives', *Farm Policy Journal*, vol. 4, no. 13 (August Quarter).

Boully, L. 2010, The water dance, Presentation to the Australian Academy of Science, Canberra.

Commonwealth of Australia 2004, *Intergovernmental Agreement on a National Water Initiative*, Commonwealth of Australia, Canberra.

Cooperative Research Centre for Freshwater Ecology 2003, *Ecological assessment of environmental flow reference points for the River Murray system*, Interim report prepared by the Scientific Reference Panel for the Living Murray Initiative, Murray–Darling Basin Commission, Canberra.

Ingram, H. and Schneider, A. 1998. 'Science, democracy and water policy', *Water Resources Update*, vol. 113, no. 21-28. < http://www.ucowr.org/updates/113/index.html>

Murray–Darling Basin Authority (MDBA) 2010, *Guide to the proposed Basin Plan*, Murray–Darling Basin Authority, Canberra.

Nancarrow, B. 2006, Social sciences—ethics, attitudes and trust in water management, Proceedings of the Australian Academy of Sciences' High Flyers Think Tank Conference on Innovative Technical Solutions for Water Management in Australia, Adelaide, 30 October, <http://www.science.org.au/events/thinktank2006/nancarrow.htm>

Olson, R. 2009, *Don't Be Such a Scientist: Talking substance in an age of style*, Island Press, Washington, DC.

Vanderbyl, T. and Boully, L. 2004, Rising to the challenge—developing a community-based solution to water sharing in the Lower Balonne in Queensland, Paper presented to the Irrigation Australia Conference, Adelaide, 11–13 May.

6. The Media and the Guide to the Basin Plan

Åsa Wahlquist

The current over-allocated state of the Murray–Darling Basin, and the consequent run-down in its environmental health, is due to more than a century of parochial decisions—decisions typically taken with no regard for downstream users, the Basin as a whole, or the health of the environment. The *Water Act 2007* aims to redress the balance, to restore the environment and to bring about a huge cultural change in managing the waters of the Basin. The Act received bipartisan political support when it was passed by the Federal Government.

Under the Act, the Murray–Darling Basin Authority (MDBA) must prepare a Basin Plan, reallocating the water in the Basin to provide environmental watering, in accordance with international environmental agreements (Commonwealth of Australia 2007).

The process outlined in the Act is essentially a top-down approach: while the MDBA must consult with the basin States, Basin Officials Committee and Basin Community Committee in preparing the plan, no specific mention is made of public consultation.

This stands in stark contrast with the European Union's approach through the 1998 Aarhus Convention. The United Nations Economic Commission for Europe describes it as a 'new kind of environmental agreement', which links environmental rights and human rights. It acknowledges the obligations to future generations, states that 'sustainable development can be achieved only through the involvement of all stakeholders', commits governments to accountability and 'focuses on interactions between the public and public authorities in a democratic context' (UNECE 1998).

Any process of cultural change and of communication involves the media. But reporting on the Basin Plan—and indeed on the ongoing story of managing the waters of the Murray–Darling Basin—is an overwhelming challenge. It is a complex story, but the media is increasingly dominated by the 24-hour news cycle. It is moving towards shorter and shorter radio and television grabs, towards news items brief enough to be read on mobile devices.

The Basin Plan was a long time in gestation. Although several interviews were granted in the early stages, the MDBA remained silent in the months leading up to the release of the *Guide to the proposed Basin Plan* (MDBA 2010a).

The Authority chose to break its silence on the release of the Guide. It set up a 'lock-up' for journalists and interested parties, similar to that used by the Treasury on the release of the annual budget. Journalists—stripped of their phones and communication devices—were allowed to read the report several hours before the release time: 4 pm on Friday, 8 October 2010.

There was a lot of reading and not much time. The Guide is 260 pages long, supported by 21 documents and the Basin Plan Knowledge and Information Directory of about 1000 reports that identifies information and knowledge, including scientific and socioeconomic, that underpins the development of the proposed Basin Plan (MDBA 2010a).

Just before the release, veteran journalist and media commentator Margaret Simons made an unprecedented plea on the web site Crikey. In a piece headlined 'It's the story of a generation, shape up, media', Simons (2010) wrote: 'Today we have one of the biggest stories in the nation's history. And it isn't even on the front page of most of today's newspapers, nor is it heading many news bulletins. I can hardly believe it.'

Simons, who has lived on and written about the Murray River, pointed out:

> The plan, two years in the making, is the result of the first exercise ever in asking the vital question: what is sustainable use of this nation's major river system? It has been an immense research undertaking. And we knew, or should have known, that the answer would be 'something very different to what we are doing now'. But how do we deal with that—with the human suffering, the wholesale changes to land use, the unequal distribution of punishment for 200 years of mismanagement and ignorance and political stuff-ups and lack of will? This is a story about fairness...This is a story about how we exist in this nation. It is about history and the future. It is a landmark moment in our nation's history. (Simons 2010)

Simons ended with a plea to her colleagues: 'Get across the issue. Do better when the plan is released. This is one story that is not spectator sport. This is that rare thing—a story that really matters, and where our reporting can make a real difference' (Simons 2010).

The day after the release of the Guide, the banner headlines on the front page of Sydney's biggest-selling Saturday newspaper (Roy Morgan 2010), the *Daily Telegraph*, read 'Knockout blow', subtitled 'Huge power bills, now soaring food prices' (Rolfe 2010).

The article asserted that the Murray–Darling Basin Plan would slash water allocations and 'inflate the price of vegetables, fruit and even clothing'. Inside articles were titled 'Blood from stone, river study puts environment ahead of food output' (Farr and Townsend 2010), and 'Stopping the flow would finish Glen' (Townsend 2010) (Glen referred to Griffith rice grower Glen Andreazza, who earlier in the year celebrated a record rice crop of 11 tonnes to the hectare).

The national broadsheet, *The Australian*, titled its front-page story 'Reality hits in Murray blueprint'. The article began: 'Restrictions on water use along Australia's biggest river system could wipe out 16 per cent of the irrigated agriculture industry—worth up to $1 billion a year—and have severe social and economic impacts on rural communities in the nation's foodbowl' (Franklin and Karvelas 2010).

The 'reality hits' was presumably written without irony, though it followed several days of front-page stories in that broadsheet detailing mounting farmer and basin community alarm. In remaining silent until the Guide was released, the MDBA relinquished the public platform to those who had most to fear from changes in water allocation. It effectively allowed irrigator groups and river towns to set the agenda, and in the following months the Authority was unable to regain control of the agenda.

On the day of the Guide's release, *The Australian* ran a page-one story titled '"Huge cost" in returning water to Murray'. It stated that 'tens of thousands of jobs' would be lost in rural communities (Wilson and Schliebs 2010). Two days before the release of the Guide, the front-page story—headlined 'Farmers fume over Murray–Darling cuts'—stated between 3000 and 4000 gigalitres 'would be taken away from irrigators and added to water already quarantined for environmental flows' (Owen and Schliebs 2010).

On the day after the release, *The Australian* ran nine stories about the Basin Plan over two pages. They included 'Plan will "save river, kill towns"' (Vasek and Wilson 2010), 'Rice, cotton farmers hardest hit' (Karvelas 2010), and 'Farmers say jobs will flow out with water' (Akerman 2010). All up, nearly 5000 words were printed in that edition on the topic. But just more than 600 words were devoted to the aim of the Plan: to restore the health of the Murray–Darling system. It was left to the environment editor, Graham Lloyd, in a comment piece, to point out: 'Environmentally, the starting point of yesterday's report by the Murray–Darling Basin Authority that the river mouth remain open at least 90 per cent of the times is a good one.' He went on to opine: 'It is unfortunate that a long history of failure by state governments in over-allocating water entitlements has pitted the interests of irrigators against those of the health of the river basin' (Lloyd 2010).

The *Sydney Morning Herald*'s Saturday page-one story was titled 'Selling the farm to save the rivers' (Arup 2010). Inside, it ran a spread with six stories and a large map of the Basin on which, incredibly, every single catchment was misnamed (SMH 2010a).

The paper counterpointed 'Redgum skeletons are stark, eerie sentinels' (Jopson 2010)—a plea from a farmer for more environmental water—with a comment piece by long-time rural writer Paul Myers, 'You can't pay to save the environment if rains fail', which argued the problem was a decade of record low rainfall. He asserted that dams 'would eliminate the claimed need for irrigation water cuts and make more water available to grow food' (Myers 2010).

It was left to the *Sydney Morning Herald*'s editorial to lay out the case for the Basin Plan: 'We can see that if the government proceeds with the plan—as it should—it will be tough going…The problem is no longer that an entire river basin is dying. The problem has suddenly become the government's plan to revive it' (SMH 2010b).

Australia's largest-selling paper, Melbourne's *Herald Sun*, posted an article titled 'Farmers pay high price to save Murray–Darling Basin', which ran fourteenth on its web page after articles about a cat needing a home and cricketer Shane Warne being 'king of the tweets' (Harvey 2010).

Melbourne's *The Age* took a different tack, with its lead article on the Plan titled 'Murray–Darling set for minimum water return', though the subtitle was the by now predictable 'Farmers hit under Murray plan' (Ker and Arup 2010).

It was left to South Australia's *The Advertiser* to remind readers that the Basin Plan aims to restore the degraded river system. *The Advertiser* has conducted a long campaign to restore the Lower Lakes and save the Murray River (see <www.adelaidenow.com.au/news/in-depth/save-the-murray>). Its Saturday-morning headline read 'River rescued but at a cost'. The article, predictably, went on to lament 'our irrigators are facing cuts even greater than those faced by irrigators and other rural industries upstream' (Kelton 2010).

The *Australian Financial Review* ran three news stories the day after the Guide was released, leading off with a page-three story: 'Water cuts to cost 800 jobs, $1bn at farm gate' (Morris 2010b). It also ran a comment piece—'Debate heats up in fluid environment'—in which journalist Sophie Morris argued the post-election climate had 'recast the debate, elevating regional issues but also giving the Greens a bigger voice'.

She wrote: 'And if there are significant reforms…it remains to be seen whether they will eventually be enforced' (Morris 2010a).

The headlines of just that one day are deeply revealing. They reflect the different State and essentially city-based interests of the metropolitan media. They largely emphasised the personal case story of the irrigator facing cuts, the town facing job losses, and the overall forecast economic losses. They pointed to the political debate that threatened to overwhelm a reasoned assessment of the Basin Plan.

While television news items are necessarily brief, the Australian Broadcasting Corporation (ABC) provided the most thorough coverage. Just two hours after the Guide was released, ABC Radio National's *The National Interest* program tackled the Plan. ABC TV's *Lateline* dealt with the topic the same day.

Both *The National Interest* and *Lateline* went live to air. That gave the commentators, and in the case of *Lateline*, the Water Minister, Tony Burke, the opportunity to make their cases unedited.

On *The National Interest*, Australian Farm Institute Executive Director, Mick Keogh, explained it was important to understand that the Basin Plan was drafted in accordance with the requirements of the *Water Act 2007*, which was legislated by the Howard Government, and passed by both sides of politics:

> The legislation tells it what it can and can't do. That is where the problem lies...They have to restore the basin to what is called an environmentally sustainable level of take. That means they have to first and foremost make some judgement about how much extra water is needed to restore the environment, and then after they have done that, they are allowed to look at what the best way is to use the rest of the water to maximise social and economic outcomes. There is no doubt that legislation gives them a very narrow focus. (Keogh in ABC 2010a)

On *Lateline*, Water Minister, Tony Burke, stressed that the Guide to the proposed Plan was just that: 'This is not the Basin plan. It is a Guide to a draft. This is just the beginning of a major round of consultations' (ABC 2010b).

The Guide was originally scheduled for released in July. It was postponed until August, but that date fell within the election campaign, and the MDBA decided to further postpone the release until after the election. Minister Burke rebuffed the suggestion politics had a role in the timing of the release of the Guide: 'They [the MDBA] are truly an independent authority and any minister who tried to treat them as anything other than that would be in for a pretty big shock' (ABC 2010b).

Two days later, on *Australian Agenda* on Sky News, Minister Burke again stressed the point:

> People are referring to it as "the basin plan". It's not. It's a guide to a draft document. It won't be until the end of next year that we actually have a basin plan that has to be signed off by me and then needs to maintain the confidence of each house of Parliament, because either house of Parliament can vote it down once it's put in place...
>
> The last thing I'm going to do is start giving instructions to an independent authority from the sidelines. This authority was given its independence by the Howard Government and given it for good reasons. The Murray–Darling Basin has been plagued by being managed as though it were different river systems that all followed state boundaries. That's part of how we got to the problems that we're in now. (Burke 2010)

The ABC also put together a web site that not only pulled together most of the ABC's coverage of the Plan, it also provided essential background (<www.abc.net.au/rural/murraydarling/>).

Some media outlets, such as the ABC, had long covered the Murray–Darling. Others, such as *The Advertiser* and *The Australian*, had actually conducted campaigns to save the Murray.[1] In its editorial on Monday, 11 October 2010, *The Australian* called the Guide 'a landmark in water policy'. It stated: '*The Australian* feels strongly about the issue. In 2001, we launched a Saving the Murray campaign...it helped kick off the process that led to this report' (The Australian 2010).

The editorial noted that the Government could draw on 'a history of bipartisan support for reclaiming water but navigating the interests of South Australia, Victoria, NSW and Queensland will not be easy'. It pointed out that more work was needed on the economic impact, the job losses and the cost of the buyback. The editorial went on to state: 'There is no disputing the damage inflicted on the system, which has been under extreme stress from drought as well as water over-allocation.' It concluded: 'The authority's report is an important step in developing the sustainable solution that has eluded politicians on both sides for decades' (The Australian 2010).

News reports might be the first draft of history, but their obsession with the immediate, with the breaking news story, is too frequently at the expense of history. Most journalists are non-specialists grappling with unfamiliar topics, pressed for time, striving to render the complex into a news story with a 500-word limit.

---

1 Disclaimer: the author participated in *The Australian*'s 'Saving the Murray' campaign as that paper's rural writer.

The Guide is 260 pages long. It provided a helpful and comprehensive history of the development of the Murray–Darling Basin, a wealth of information on the Basin itself, the role of the MDBA and the objectives of the proposed Plan.

The lock-up before the release of the Guide not only gave journalists time to read it before reporting on it, it also enabled key players to address journalists, and answer their questions.

As one unnamed journalist said plaintively at the Guide release press conference with MDBA Chairman, Mike Taylor: 'we don't have the 300 pages to write about it in our newspapers or television stations…can we make the comments as simple as [possible]…because it is getting quite complex' (Media Monitors 2010).

The MDBA was already having to play catch-up when the Guide was released. The irrigators and river towns had already set the agenda, and the Authority was forced to respond to their claims.

There were a number of strong, newsworthy points that the MDBA either failed to make or failed to interest the media in.

In the same week that the Australian media was celebrating records set at the Commonwealth Games, it overlooked a major home-grown first.

Chairman, Mike Taylor, told journalists

> to plan a basin in the way that is being proposed is really a world first… So it is very much an extraordinary thing, not only in an Australian context, but in an international context.
>
> I think it is clearly the forerunner…of how nations around the world are going to need to deal with water resources. It [water] is the scarcest resource that confronts the world, and communities, going into this century. (Media Monitors 2010)

The huge amount of work involved in preparing the Basin Plan was also overlooked. The Guide stated: 'The task of assessing Basin-wide and catchment specific environmental water requirements has never before been undertaken in the Murray–Darling Basin' (MDBA 2010a:58). Nor had key environmental assets been identified or prioritised, the water needs of key ecosystem functions had not been considered at the Basin scale, and 'very little work had been done to define the productive base and identify key environmental outcomes at the Basin scale. In short, this is largely new territory for the Basin' (MDBA 2010a).

Another key point overlooked by most journalists was how prescriptive the legislation governing the process is. MDBA Chairman, Mike Taylor, told the gathered journalists that the *Water Act*

> sets down a prescription by which the Basin Plan must be established. It requires the Authority to, first of all, identify key ecosystems and key environmental assets, to determine the water requirement that is necessary to both maintain and restore those assets and to not compromise them.
>
> In doing so, the Act asks the Authority to identify the amount of water that's needed for those assets, and consequently to reduce the amount of water that is available for existing human and irrigation purposes. So this Act leads to some very major changes. (Media Monitors 2010).

Section 20 of the *Water Act 2007* spells out the purpose of the Basin Plan. That purpose, it states, is 'to provide for the integrated management of the Basin water resources' by providing for 'the establishment and enforcement of environmentally sustainable limits on the quantities of surface water and ground water that may be taken from the Basin water resources (including by interception activities)'. That use, it states, must be in a way that 'optimises economic, social and environmental outcomes' and enables water 'to reach its most productive use through the development of an efficient water trading regime across the Murray–Darling Basin' (*Water Act 2007*, <www.austlii.edu.au/au/legis/cth/consol_act/wa200783/index.html>).

Taylor told the journalists: 'Interestingly, the Act itself does not deal with… agriculture, food or fibre production. The parliament very much focussed on laying down this process by how we would readdress the environmental issues.' He added that the Act also required the MDBA use 'the best available science, and failing that the precautionary principle'.

He went on to stress the importance of addressing the social and economic aspects of the proposed Plan, through a long process of consultation:

> We look forward to working very, very closely with communities, environmental groups, industry groups, local governments, state governments, [the] federal government, and other parties in making sure not only is the data, the modelling, the information which we've based this on fully tested, but also that the conclusions are properly evaluated. (Media Monitors 2010)

Also largely overlooked was the fact the MDBA actually opted for the lower end of the recommended water return, and the consequences of this decision for some catchments in the Basin.

As Taylor told the journalists: 'The reason we drew the line at 4000 [GL was that the] social and economic impacts are going to be very significant.'

An additional 3000 GL per year would mean the Border and Lachlan Rivers would retain a moderate rating for environmental health outcomes; the Murrumbidgee, Campaspe, Goulburn–Broken, Barwon–Darling and Wimmera–Avoca rivers would move from poor to moderate; while the Condamine–Balonne, Gwydir, Loddon, Lower Darling and Murray rivers regions would retain their 'poor' ranking.

An additional 4000 GL/yr would raise the Condamine–Balonne, Loddon and Murray rivers to moderate, while the Gwydir and the Lower Darling rivers would remain poor (MDBA 2010a:74, 112).

The MDBA stated that 3000 GL was the minimum 'required to achieve the environmental objects of the *Water Act*'. But it did point out: 'This level of reduction has a high dependence on a long-term return to wetter climatic conditions across the Basin.'

This could be a doomed hope, with the MDBA stating earlier in the Guide: 'the latest climate change modelling suggests that, under a median 2030 prediction, conditions are likely to be around 10% drier than past experience' (MDBA 2010a:36).

In fact it is only with a long-term average increase of 7600 GL/yr that 'the environmental targets are all met and all catchments improve from their existing status to good flow levels' (MDBA 2010a:74).

But the MDBA (2010a:82) considered the impact of taking 7600 GL from irrigation too severe: 'the Authority has judged that in order to optimise social, economic and environmental outcomes, as it is obliged to do under the *Water Act 2007*, it can only consider Basin-wide reductions of between 3000 and 4000 GL/yr (reductions of 22–29% of current diversion limits).' As a result, it considered three scenarios—an increase in the water available to the environment of: 3000 GL/yr, 3500 GL/yr and 4000 GL/yr.

By 30 September 2010—a week before the Guide was released—the Federal Government's water buyback had purchased 920 GL, or between nearly one-third and one-quarter of the proposed targets (Department of Sustainability, Environment, Water, Population and Communities 2008). Despite Taylor mentioning the buyback figure several times, this progress was also largely overlooked in the press coverage.

A number of media reports, however, raised the spectre of threats to food security, but these fears do not stand up to scrutiny. In March 2010, the Productivity Commission prepared a report on the market mechanisms for recovering water. It briefly addressed the question of food security: 'At present, Australia exports around 60 per cent of all of its agricultural output in addition to providing the large majority of the food eaten by Australians' (Productivity Commission 2010:140).

The Commission stated that 'food production will continue to be influenced by price signals, including those resulting from a decline in water availability'. It concluded that Australia's food security was very unlikely to be significantly influenced by the buybacks then under way.

Since 2006, water allocations in the Murray–Darling Basin have been slashed due to drought. Former Water Minister Penny Wong, in June 2009, told the Victorian Farmers' Federation: 'In the last three years the yield on Basin water entitlements has averaged 2,500 GL per year compared to a long term average of almost 8,000 GL' (Wong 2009). In other words, despite more severe reductions than those proposed under the Basin Plan, Australia's food security was never in doubt.

News focuses on the immediate, the short term. In the months before the Guide was released, the drought that had gripped the Basin since 2002 eased. The media ran a multitude of stories about farmers rejoicing in the best season for years.

In the first week of September—the month before the Guide was released—1090 GL flowed into the Murray River, which was more than had entered for the entire water year of 2006–07 (MDBA 2010c).

That month was the wettest September since 1993 along the Murray River, and the inflow into the Murray for September totalled 2920 GL—well above the long-term September average of 1605 GL (MDBA 2010d).

The day before the release of the Guide, Water Minister, Tony Burke, flew over the River Murray mouth. The mouth had been kept open by dredging for the past four years, but the good rains that had fallen upstream were at last sufficient for the river to flow naturally to the sea (Wilson and Schliebs 2010).

In the short term of the news cycle, the drought had receded and with it the immediate crisis in the Basin. It was left to the Guide to spell out the dire straits the Basin was still in:

> The environment has not had sufficient water for decades. This has led to serious environmental decline in many parts of the Basin. The real possibility of environmental failure now threatens the long-term economic and social viability of many industries and the...strength of many communities.
>
> If the focus does not swing back towards considering water required for the environment, then the nation risks irretrievably damaging the attributes of the Basin that enable it to be so productive. (MDBA 2010a:xiv)

This fact—the most salient of all: that the very survival of the communities relied on a healthy river system—was the most overlooked of all in the media's coverage of the proposed Basin Plan.

Margaret Simons said her impression of the coverage was that some journalists realised for the first time the importance of the Basin, 'but there was a scramble to "go bush" without any realisation of the complexity and variability of the picture in different regions. For example, how Griffith varies from Renmark, and Forbes from St George.'

She also noted that the only way many journalists could understand the story was in terms of its implications for politics in Canberra: 'That is, whether or not it would be bad for Gillard, whether or not she would have "won" the election if the Murray–Darling Report had been released during the campaign, and so on and so forth.'

Simons noted there were some

> honourable exceptions and notable efforts, but on the whole, the media seemed unable to cope with the scale of the story, or its historical context and future implications. I didn't see a single outlet give it the scale of coverage I think it deserves—which is front page news consistently over a long period, supported by major analytical effort, journalistic depth, local presence and diversity of opinion. (Margaret Simons, Personal communication, 13 October 2010)

As the members of the MDBA conducted often boisterous meetings about the Guide, the Basin was finally receiving drought-ending rains. The rains began in July, then pelted down in November and December, running off the saturated catchments into the rivers.

According to the National Climate Centre (2011), spring 2010 was the wettest on record in the Basin and December 2010 was the second-wettest December on record. The resulting inflows into the Murray set a new December record of 2976 GL—well above the long-term average inflow of 420 GL (MDBA 2011). There were floods in south-east Queensland, in the Murrumbidgee, Lachlan and Castlereagh river catchments in New South Wales and in large parts of northern and western Victoria.

By early December, the River Murray was rising, with 48 000 ML flowing out the Murray mouth every day, and dredging ceased for the first time in more than eight years (Todd 2010).

On 7 December, MDBA Chairman, Mike Taylor, surprised observers when he announced his resignation. In a statement issued by the MDBA, Taylor said:

> Balancing the requirements of the *Water Act 2007* against the potential social and economic impact on communities will be a significant challenge...the Authority has sought, and obtained, further confirmation that it cannot compromise the minimum level of water required to restore the system's environment on social or economic grounds. (MDBA 2010b)

'A sustainable plan for the Basin would require far more than a decision by the Authority on how much water should be transferred from human uses to the environment', he said in his resignation statement, adding: 'I believe it is time for the Government to reconsider the next phase.'

His resignation was widely interpreted by the media as not only challenging the Government's interpretation of the *Water Act*, but threatening the progress of the Basin Plan.

Adelaide's *The Advertiser* reported that Taylor's departure increased

> the likelihood the authority will shift to a more pro-irrigation stance at the expense of the Lower Lakes and downstream users.
>
> Mr Taylor's departure was a direct response to Government pressure to force the authority to take so-called 'social' and 'economic' factors into account, which the Authority believes it does not have the power to consider under the Water Act.
>
> The resignation was merely the latest setback in the political minefield that is basin reform and suggests that the needs of downstream users and the environment will continue to come a poor second to politically powerful irrigators higher in the basin. (Kenny 2010)

The Prime Minister, Julia Gillard, reportedly held firm: 'I can say to you very clearly that the election commitment I made is that the Government will do what is necessary to implement the Murray–Darling Basin plan' (Kelly and Massola 2010).

She reiterated that the Government's position was to optimise the environmental, social and economic areas: 'That is the aim of these reforms—to ensure that we've got a healthy river, we've got food production and we've got viable regional communities...The government will continue to see these reforms with optimisation across these three areas' (Kelly and Massola 2010).

But Mick Keogh, Executive Director of the Australian Farm Institute, told *The Australian* that the MDBA had always felt constrained by the Act because of the environmental requirements, 'in that it put a very high level of focus on environmental requirements and then only allowed consideration of socio-economic factors, subject to having met those environmental requirements or standard' (Kelly and Massola 2010).

Reporter Alexandra Kirk stated on ABC Radio's *PM* program: 'Today Mr Taylor tendered his resignation stating the Authority had sought and obtained further confirmation that it cannot compromise the minimum level of water needed to restore the river system's environment' (Kirk 2010). Heavy rains and flooding through December and into January prompted increased calls for the Basin Plan to be dropped, or at least delayed. National Farmers' Federation President, Jock Laurie, said the strong flows in the river 'buys the government time to sit back and make sure they get this right' (Wilson 2011).

But Prime Minister, Julia Gillard, remained firm:

> Rather than just wait until the next drought hits the Murray–Darling, now is the time to get it right for the future. So we will continue in 2011 to pursue our reforms through the Murray–Darling Basin Authority... We've got to get this right, we've got to make sure that we've got a healthy river, viable communities, as well as food production along the Murray.' (Wilson 2011)

At the end of January—three days before Taylor's resignation took effect—the Government appointed Craig Knowles as the new MDBA Chairman. Knowles is a former member of the NSW State Labor Government. He held the Natural Resource ministry between 2003 and 2005, and during that period was a member of the Murray–Darling Basin Commission Ministerial Council. He helped negotiate the National Water Initiative (NWI) and introduced Water Sharing Plans covering 80 per cent of water extracted in New South Wales (NSW Office of Water n.d.). According to the NSW Office of Water, the plans establish rules 'for sharing water between the environmental needs of the river or aquifer and water users, and also between different types of water users such as town supply, rural domestic supply, stock watering, industry and irrigation'.

Knowles' appointment was reported as being both a Labor-mates deal and a good choice—often in the same story (Coorey 2011; Kruger 2011).

Sydney's *Daily Telegraph* reported:

> The ex-Labor minister's appointment as chairman of the Murray–Darling Basin Authority sparked immediate cries of 'jobs for the boys' from the federal and state oppositions. NSW irrigators applauded the appointment but said Mr Knowles faced an 'impossible task' trying to boost water levels in the Murray–Darling without big changes to the Water Act. (Lewis 2011)

Knowles was quick to distance himself from Taylor's interpretation of the *Water Act*:

> I just disagree with Mike Taylor, let's be frank about that. I've delivered on a lot of environmental legislation over the years—in forestry, in natural resource management, native vegetation, water, and all of those international agreements, all of those environmental imperatives are found in every piece of legislation. The Water Act is no different. I am very comfortable that the scope of the legislation, the objectives of the legislation talk about optimising the economic, social and environmental outcomes as plain as day. (Kruger 2011)

In the same report, the Chief Executive of the National Irrigators' Council, Danny O'Brien, wryly pointed out that 'it's not exactly a job you'd give to a mate'. Meanwhile, the South Australians worried a former NSW politician would overlook that State's special needs, with *The Advertiser* reporting the appointment was greeted with scepticism.

Knowles told *The Advertiser*: 'South Australia has much to be concerned about… [But] I'm very comfortable knowing that South Australia has got particular needs and particular imperatives and they are very, very front and centre' (Martin 2011).

There was widespread agreement that Knowles came to the job at a difficult time. His task was to take on a plan that had been very poorly communicated, and that lacked both strong community and political support.

The Guide outlined a Basin Plan for audacious change. The need to cut water allocations in favour of the environment had been obvious for decades—the difficulty had been finding the political will. Although the States—which were responsible for the over-allocation—had made incremental cuts, it was left to the Federal Government, faced with an unprecedented drought in the Basin, to legislate for the Basin Plan.

It was always going to be very difficult to carry everyone with the changes, especially those who were to bear the brunt of the cuts.

The media played several roles. It reported on the release of the Guide. It reported some of the details of the Guide, but it largely concentrated on those people who thought they would be disadvantaged by it. There was far less reporting on those who saw benefits in the Plan.

But the media also played a pivotal role in shaping the debate.

In the year leading up to the release of the Guide, the irrigation community and the river basin towns grew increasingly anxious about the Basin Plan. The Aarhus Convention (UNECE 1998) states: 'sustainable development can be achieved only through the involvement of all stakeholders.' But instead of using the lead-up as a time to involve all stakeholders—to forge, in the words of the Aarhus Convention, 'a new kind of agreement'—the MDBA largely went into lock-down. At the point when communities were desperate for information from the Authority, journalists seeking to speak to key players at the Authority could get nothing other than 'no comment'—if that.

The irrigators and the towns that depended on them proved far more savvy in their use of the media than the MDBA. They took full advantage of the Authority's silence.

In the week leading up to the release of the Guide, the media ran prominent stories highlighting irrigators' fears, but the MDBA refused to respond until the official release of the Guide, at 4 pm on a Friday—late in the news day and the news week, and an extremely difficult time to organise coverage for big weekend editions of the newspapers and for early evening television news bulletins.

The result was those opposed to the Plan—or at least fearful of what the consequences of the Plan might be—set the agenda. Over the following days, only a careful reader of the newspapers—with the exception of the Adelaide press—would have learned why the Plan had been formulated. The battle was lost before the Guide was even released.

The points the MDBA needed to make—the necessity to make the Murray–Darling Basin environmentally sustainable and to do that only through purchases of water from voluntary sellers, with perhaps one-third of that water already in hand—were unlikely to be heard, even if they had been made with the strength and clarity they required (which they were not).

The initial reporting of a story—whether it is fair or unfair, accurate or erroneous—usually frames the debate. Thus, the irrigators set the agenda, and the story developed a momentum that could not be halted or redirected by a quiet, reasonable speech from the Chairman about the commendable objectives of the Guide.

The media also thrives on stories about conflict. As the MDBA held often-heated meetings across the Basin, the Authority was cast as the opposition, instead of a partner in a democratic debate.

The average news story is less than 500 words long, and usually contains just one new idea. There is little room for detailed history or nuance. There has also been a decline in the number of specialist rural reporters who could have provided that history, and brought to bear a deeper understanding of the Plan on their reporting.

Griffith was arguably one of the towns most impacted by the Plan. To hold the second community session there instead of in a South Australian centre such as Murray Bridge, which largely welcomed the Plan, was a poor choice. The burning of the Guide outside the Griffith meeting set the tone for the coverage of the series of public meetings about the Plan.

Whether the MDBA under the chairmanship of Craig Knowles can rescue the Plan remains to be seen. His main challenge is to get the communication of the Plan back on track, to correct the misconceptions and begin to engage the communities he must carry with him if the Plan is to succeed. He will also need to restore the reputation of the MDBA as a body open to genuine dialogue with basin stakeholders.

Knowles will need strong support from the highest level of government if he is to succeed. The Prime Minister has paid lip-service to the Plan. But imagine the difference it would have made if, on the day of the release of the Guide, she had argued it was a world-leading plan for a better Australia. If she had outlined a glowing vision of a much healthier Murray–Darling Basin, and pointed out all water acquisition would be voluntary and promised no-one and no community would be left behind, the outcome would arguably have been different.

Knowles has a successful track record in pushing through difficult changes in water management with the Water Sharing Plans in New South Wales. Dealing with water at the Commonwealth level, where the only real leverage is the environment, could prove a great deal more challenging, as Mike Taylor found.

If yet another attempt is going to be made to restore the Basin to environmental health, there will have to be a far more concerted effort to engage with the community, to genuinely include them in the process, and to keep talking to them.

This includes communicating with journalists. This means more than being available to journalists, though this would be a good start. And it does not mean bombarding them with piles of information, as useful as this might appear to be.

It means preparing information in a clear, brief and accessible manner, assuming the journalist knows nothing at all about the subject—which, when it comes to rural and science matters, is increasingly likely to be the case.

It also means having a clear time line, providing information so reporters have ample time before their deadlines to come to grips with the subject, to talk to a number of people and line up an appropriate picture story and even write a comment piece. This cannot be done in one and half hours late on Friday afternoon.

A successful launch of the Guide would have meant making the case that the MDBA was rescuing the Basin from environmental catastrophe. But the irrigators got in first, and the overwhelming story the media ran became the catastrophe about to be inflicted on them by the Authority.

References

Akerman, P. 2010, 'Farmers say jobs will flow out with water', *The Weekend Australian*, 9–10 October 2010, p. 7.

Arup, T. 2010, 'Selling the farm to save the rivers', *Sydney Morning Herald*, 9–10 October 2010, p. 1.

Australian Broadcasting Corporation (ABC) 2010a, 'Basin politics', *The National Interest*, ABC Radio National, 8 October 2010, <www.abc.net.au/rn/nationalinterest/stories/2010/3033392.htm>

Australian Broadcasting Corporation (ABC) 2010b, 'Burke defends timing of plan', *Lateline*, ABC Television, 8 October 2010, <www.abc.net.au/news/video/2010/10/08/3033742.htm>

Burke, A. 2010, Population and communities, Australian agenda, Transcript, Minister for Sustainability, Environment, and Water, 10 October 2010, Parliament House, Canberra.

Commonwealth of Australia 2007, *Water Act 2007*, No. 137 (2007), Commonwealth Consolidated Acts, www.austlii.edu.au/au/legis/cth/consol_act/wa200783/

Coorey, P. 2011, 'A cosy number for a party mate or the right man for the job?', *Sydney Morning Herald*, 29 January 2011, <www.smh.com.au/environment/water-issues/a-cosy-number-for-a-party-mate-or-the-right-man-for-the-job-20110128-1a8dv.html>

Department of Sustainability, Environment, Water, Population and Communities 2008, *Progress of Water Recovery under Restoring the Balance in the Murray–Darling Basin Program*, September, Department of Sustainability, Environment, Water, Population and Communities, Australian Government, Canberra, <www.environment.gov.au/water/policy-programs/entitlement-purchasing/2008-09.html>

Farr, M. and Townsend, S. 2010, 'Blood from stone, river study puts environment ahead of food output', *Daily Telegraph*, 9 October 2010, p. 3.

Franklin, M. and Karvelas, P. 2010, 'Reality hits in Murray blueprint', *The Weekend Australian*, 9–10 October 2010, p. 1.

Harvey, M. 2010, 'Farmers pay high price to save Murray–Darling basin', *Herald Sun*, 9 October 2010, <www.heraldsun.com.au/news/victoria/farmers-pay-high-price-to-save-murray-darling-basin/story-e6frf7kx-1225936174055>

Jopson, D. 2010, 'Redgum skeletons are stark, eerie sentinels', *Sydney Morning Herald*, 9–10 October 2010, p. 4.

Karvelas, P. 2010, 'Rice, cotton farmers hardest hit', *The Weekend Australian*, 9–10 October 2010, p. 6.

Kelly, J. and Massola, J. 2010, 'Basin authority Mike Taylor's resignation won't halt Murray–Darling water reforms, says PM', *The Australian*, 7 December 2010, <www.theaustralian.com.au/news/murray-darling-basin-boss-mike-taylor-resigns/story-e6frg6n6-1225966814825>

Kelton, G. 2010, 'River rescued but at a cost', *The Advertiser*, 9 October 2010, <www.adelaidenow.com.au/news/in-depth/river-rescued-but-at-a-cost/story-e6frebju-1225936203940>

Kenny, M. 2010, 'Murray–Darling Basin Authority head Mike Taylor quits', *The Advertiser*, 7 December 2010, <www.adelaidenow.com.au/news/in-depth/murray-darling-basin-authority-head-mike-taylor-quits/story-e6frebju-1225966827897>

Ker, P. and Arup, T. 2010, 'Murray–Darling set for minimum water return', *The Age*, 9 October 2010, <www.theage.com.au/victoria/murray-darling-set-for-minimum-water-return/20101009-16c23.html>

Kirk, A. 2010, 'PM says water reform on track, despite resignation', *PM*, ABC Radio, 7 December 2010, <www.abc.net.au/worldtoday/content/2010/s3086763.htm>

Kruger, P. 2011, 'Former Labor MP new Murray–Darling Basin Authority boss', *PM*, ABC Radio, 28 January 2011, <www.abc.net.au/pm/content/2011/s3124477.htm?site=rural&microsite=murraydarling&section=latest&date=(none)>

Lewis, S. 2011, 'Craig Knowles is ALP's old man river', *Daily Telegraph*, 29 January 2011, <www.dailytelegraph.com.au/news/nsw-act/craig-knowles-is-alps-old-man-river/story-e6freuzi-1225996432267>

Lloyd, G. 2010, 'Plans must stay true to use of scarce resources', *The Weekend Australian*, 9–10 October 2010, p. 6.

Martin, S. 2011, 'Cronyism cries greet river boss', *The Advertiser*, 29 January 2011, <www.adelaidenow.com.au/news/south-australia/former-labor-mp-takes-murray-role/story-e6frea83-1225996035376>

Media Monitors 2010, Questions regarding the Murray–Darling Basin Plan, Transcript of press conference, 8 October 2010, Media Monitors, Canberra, <http://thebasinplan.mdba.gov.au/>

Morris, S. 2010a, 'Debate heats up in a fluid environment', *Weekend Australian Financial Review*, 9–10 October 2010, p. 9.

Morris, S. 2010b, 'Water cuts to cost 800 jobs, $1bn at farm gate', *Weekend Australian Financial Review*, 9–10 October 2010, p. 3.

Murray–Darling Basin Authority (MDBA) 2010a, *Guide to the proposed Basin Plan*, Murray–Darling Basin Authority, Canberra, <http://thebasinplan.mdba.gov.au>

Murray–Darling Basin Authority (MDBA) 2010b, *Plan for the Murray–Darling Basin—Role of authority chair*, 7 December 2010, Murray–Darling Basin Authority, Canberra, <www.mdba.gov.au/files/Media-release-Role-of-Authority-Chair_0.pdf>

Murray–Darling Basin Authority (MDBA) 2010c, *River Murray weekly report for the week ending 8 September 2010*, Murray–Darling Basin Authority, Canberra, <www.mdba.gov.au/files/weeklyreports/WR100917-River-Murray-Operations-Weekly-Report-08-September-2010.pdf>

Murray–Darling Basin Authority (MDBA) 2010d, *River Murray weekly report for the week ending 6 October 2010*, Murray–Darling Basin Authority, Canberra, <www.mdba.gov.au/files/weeklyreports/WR101007-River-Murray-Operations-Weekly-Report-06-October-2010.pdf>

Murray–Darling Basin Authority (MDBA) 2011, *Weekly report for the week ending 5 January 2011*, Murray–Darling Basin Authority, Canberra, <www.mdba.gov.au/water/river_info/weekly_report_archive>

Myers, P. 2010, 'You can't pay to save the environment if rains fail', *Sydney Morning Herald*, 9–10 October 2010, p. 4.

National Climate Centre 2011, *Special Climate Statement 24*, Updated 25 January 2011, National Climate Centre, Bureau of Meteorology, Melbourne, <www.bom.gov.au/climate/current/statements/scs24b.pdf>

NSW Office of Water n.d., Water sharing plans, Office of Water, Government of New South Wales, Sydney, <www.water.nsw.gov.au/Water-management/Water-sharing-plans/Water-sharing/default.aspx>

Owen, M. and Schliebs, M. 2010, 'Farmers fume over Murray–Darling cuts', *The Australian*, 7 October 2010, p. 1.

Productivity Commission 2010, *Market Mechanisms for Recovering Water in the Murray–Darling Basin*, March 2010, Productivity Commission, Melbourne, <www.pc.gov.au/projects/study/water-recovery/report>, p. 140.

Rolfe, J. 2010, 'Knockout blow', *Daily Telegraph*, 9 October 2010, p. 1.

Roy Morgan 2010, Readership estimates for Australia for the 12 months to June 2010, Press release, 20 October 2010, Roy Morgan, <www.roymorgan.com/news/press-releases/2010/1145/>

Simons, M. 2010, 'It's the story of a generation, shape up, media', *Crikey*, 8 October 2010, <www.crikey.com.au/2010/10/08/simons-its-the-story-of-a-generation-shape-up-media/11/10/2010>

Sydney Morning Herald (SMH) 2010a, 'A river runs through it', *Sydney Morning Herald*, 9–10 October 2010, p. 1.

Sydney Morning Herald (SMH) 2010b, 'Murray–Darling plan sails on inland sea of trouble', Editorial, *Sydney Morning Herald*, 9–10 October 2010, p. 10.

The Australian 2010, 'The Murray–Darling is a crucial test for Labor', Editorial, *The Australian*, 11 October 2010, p. 15.

Todd, A. 2010, 'Christmas camping closed by rising River Murray', *The Advertiser*, 6 December 2010, <http://www.adelaidenow.com.au/news/south-australia/river-murray-water-levels-rising/story-e6frea83-1225966640770>

Townsend, S. 2010, 'Stopping the flow would finish Glen', *Daily Telegraph*, 9 October 2010, p. 3.

United Nations Economic Commission of Europe (UNECE) 1998, *Convention on Access to Information, Public Participation in Decision-Making and Access to Justice in Environmental Matters* [Aarhus Convention], United Nations Economic Commission of Europe, Geneva, </www.unece.org/env/pp/>

Vasek, L. and Wilson, L. 2010, 'Plan will "save river, kill towns"', *The Weekend Australian*, 9–10 October 2010, p. 7.

Wilson, L. 2011, 'Gillard to push ahead with Murray–Darling Basin plan despite floods', *The Australian*, 6 January 2011, <www.theaustralian.com.au/national-affairs/gillard-to-push-ahead-with-murray-darling-basin-plan-despite-floods/story-fn59niix-1225983012023>

Wilson, L. and Schliebs, M. 2010, '"Huge cost" in returning water to Murray–Darling river system', *The Australian*, 8 October 2010, p. 1.

Wong, P. 2009, Speech by Minister for Climate Change, Energy Efficiency and Water, Senator Penny Wong, to Victorian Farmers' Federation Annual General Meeting, 11 June 2009, <www.climatechange.gov.au/en/minister/previous/wong/2009/major-speeches/June/sp20090611.aspx>

7. Rethinking Community in the Face of Natural Resources Management Challenges

Martin Mulligan

Introduction

As with natural disasters in the past, the devastating floods that inundated local communities in southern Queensland, northern New South Wales and Victoria in December 2010 and January 2011 brought out some inspirational examples of communities standing together to face an unexpected crisis. Something similar happened in the wake of the Black Saturday bushfires in Victoria, yet the dispersal of the survivors and the passage of time made it hard to sustain the heightened 'sense of community' over the following years. Community, we like to think, is there when we need it most, yet at a time in human history when people move around more than ever and when all kinds of unexpected crises seem to emerge, it is getting harder to sustain a sense of community. More than at any time in history, now community has to be consciously created and recreated.

The author of this chapter was not directly involved in any work on the *Guide to the proposed Basin Plan* produced by the Murray–Darling Basin Authority (MDBA 2010). He was, however, disturbed at the way in which the Guide was communicated to local basin communities that will be affected by new restrictions on water diversion and was not surprised at the angry response to the Guide. While the MDBA (2010:37) claimed that it had drawn on the 'best available biophysical and social science and knowledge' in drafting the Guide, it certainly did not draw on the best available knowledge relating to community consultation and 'engagement', and hence it missed an important opportunity to sell the case for water reform. Indeed, the word 'community' is used loosely and ambiguously in the literature of the MDBA and this shows neglect for the important work of Melbourne-based academic Susan Kenny in documenting the emergence of community-development practice in Australia over the past four decades (see Kenny 2006). Furthermore, the Guide should have been guided by work done outside Australia on the role of community in relation to natural-resource conservation (for example, Agrawal 1999). More serious consideration of the sociological literature on community and community development might have helped the MDBA avoid the mistakes it has made in putting the case for water reform to affected communities, and an investment in community development would change the dynamics in regard to community 'engagement' with water reform. The author draws on his recent research on the sustainability of local communities in Australia and in post-tsunami Sri Lanka to suggest new ways of thinking about the nature and role of community in addressing past mistakes in regard to natural-resource management.

The Murray–Darling Basin (MDB) communities were still recovering from the longest drought on record when the MDBA Guide was released. This was an ideal time to have a serious conversation about the need for reform because the need had become manifest and a crisis can often remind us of our need for community. At the same time, people can sometimes respond to threats and crises by becoming more parochial—that is, by retreating to rather narrow and often outdated conceptions of what constitutes the local community. The irony is that farmers and rural communities might be ahead of everyone else in understanding the need to eliminate wasteful practices in regard to water use and yet they have reacted badly to any attempt to 'impose' water reform 'from above'. We have seen that affected people and communities can react to 'rational arguments' for water reform with 'irrational' emotional responses and a kind of 'siege' mentality. But this should not come as a surprise. US psychologist Paul Slovic (for example, 1987, 2010) has long pointed out that perceptions of risk will often trigger 'irrational' emotional responses, and people working with vulnerable people and communities should know this. The failure to respond appropriately to 'irrational' responses will only compound the problem and give rise to reactive and parochial notions of community. Sadly, parochialism has become even more dangerous in a world of increasing global integration—faced with growing challenges of sustainability—and there is certainly an important role for agencies such as the MDBA to play in combating parochialism. In regard to water reform in the MDB, the MDBA and State and Federal governments clearly have to ensure that the burden of 'readjustment' is spread equitably across the basin. Yet people who have responsibility for engaging local communities in water reform need to have a much better understanding of how people and local communities will perceive and experience risk.

It does not help that the widespread flooding in early 2011 has created new confusion about the need for water reform in the Murray–Darling Basin. For example, in January 2011, the new Victorian Premier, Ted Baillieu, publicly contradicted Victorian Governor, David de Krester, in saying that the flood crisis did not confirm the dire warnings of climate scientists. Indeed, Baillieu claimed that the flooding flew in the face of earlier predictions about the increasing prevalence of drought, and he suggested that resilient Victorian communities already know how to cope at times of crisis. The problem is that climate scientists have predicted an increase in the frequency and intensity of 'extreme weather events', rather than a 'linear' increase in aridity, and the past can be only an approximate guide for such a daunting future. Having been thoroughly briefed on all this, Baillieu has little excuse for joining the ranks of the climate change 'sceptics'. Local communities, however, can be excused for feeling very confused about what the future might hold, and contradictory statements by public figures do not help.

The 2010–11 flooding did ease immediate pressure on water flows in the Murray–Darling Basin, yet this can be seen as providing an opportunity to proceed a little more deliberatively with the task of convincing local communities of the ongoing need for substantial reforms to water allocations. Flooding has provided an opportunity to talk of the need to be prepared for different kinds of extremes—rather than drought alone—because floodwaters might provide temporary relief for stressed ecosystems but they do little to put farming on a more sustainable basis. The recent, record drought might have blinded many to the ongoing threat of flooding, yet the real challenge is to be ready for anything that an increasingly fickle climate might throw at us. Now we can draw on recent, lived experiences of drought, devastating bushfires and widespread flooding to make the argument that we need to have all kinds of contingency plans in place for what could lie ahead. In this sense, droughts and floods can be linked—rather than counterposed to each other—and the good news is that this kind of link lies deep within the psyche of rural Australian communities. The Brisbane-based writer John Birmingham (2011) made this point rather well in his newspaper commentary on the flooding in his home state. It is no accident, he reminded his readers, that Australians have long taken to heart the sentiments expressed in Dorothea Mackellar's famous poem about life in the 'sunburnt country'. We might well turn to poetry to speak from the heart about what it means to live in this land, Birmingham wrote, because

> ...we love its extremes, as Dorothea Mackellar knew. Its droughts and flooding rains, its pitiless skies, its beauty and its terror. We have made a compact with the land, that we will remain forever wary of her harshest and most dangerous moods, as long as she will abide by our presence to at least some extent.

Maybe this expresses better than a rational discourse on the need for water reform the ways in which we need to rethink our relationships with nature in the 'wide, brown land'. It suggests that our very identity is bound up with the need to respect the tough conditions that can make life on the land unsustainable. We become a community, rather than a host of discouraged individuals, to the extent that we can make a compact with the land.

Seeking Community

On a global scale, several prominent sociologists have suggested that 'community' has become a search for a more secure sense of belonging in an increasingly insecure world (see, for example, Bauman 2001; Delanty 2003; Rose 1999). In a world of systemic flux and uncertainty, it has indeed become more difficult to pin down what the word community actually means, and many sociologists have noted that it can mean very different things to different people (for example, Sennett 1986; Walmsley 2006). As Cohen (1985) and Delanty (2003) have argued persuasively, however, the ambiguities embedded within the word community do not mean that we should turn our backs on the strong, even increasing, desire for a sense of belonging to community. The desire for community should, however, be seen as an aspiration rather than a description of particular social structures or relationships.

We tend to use the word community even more loosely in Australia than in countries such as Britain or the United States in that we can switch easily from talking about local communities to talking about things such as 'the Victorian community' or 'the Australian community'. Clearly, it is useful to talk about the diversity of local communities that live in the extensive Murray–Darling Basin, for example, and yet this use of the word community can also mask the diversity of lived experiences in settings that might range from the suburbs of Canberra to a remote community in western New South Wales. In one sense, the word community provides an entry point into some of the more intangible aspects of local life. On the other hand, the fact that the identity of any one community can always be contested means that the desire for a settled conception of community will always remain beyond reach—as an aspiration to strive for rather than a reality to fall back on. The dangers of parochialism can be countered by working towards the constant creation of more inclusive and more resilient communities. This conception of community formation—as distinct from community per se—helps to orient discussions of community towards the future rather than a romanticised notion of the past. It can also help people to collectively face up to the kinds of future challenges that might otherwise open up division and conflict. Narrow claims about the identity of any particular community will always create 'insiders' and 'outsiders', and that can be dangerous when the community is facing significant challenges.

We need to promote future-oriented conceptions of community rather than backward-looking, or parochial, ones. The notion of 'inter-generational equity'—which was central to the conception of environmental sustainability introduced by the famous Brundtland Commission report of 1987—helps to focus the mind on the future. We need to think more concretely, however, about how to make this abstract concept work within the lived experiences of the local communities that will be affected by water-reform policies. At the same time, we need to understand that parochialism is often borne of fear about the unknown. If the *desire* for community does reflect a search for a more secure sense of belonging in an increasingly insecure world, it is hardly surprising that claims about the identity of a community under some kind of threat will often be tinged with anxiety and emotion. Discussions about community are rarely 'rational' and that is just the way it needs to be. Indeed any attempt to turn the future of any particular community into an entirely rational discourse is likely to backfire.

The suggested point of departure, then, is that community is not a social structure but a set of contestable claims about identity and belonging in a rather insecure world. A sense of belonging to community is no longer the 'given' that it might have been 50 years ago. If we think of community as an ongoing process of formation and reformation—or creation and recreation—then parochialism will be kept in check provided we aim to construct an inclusive projection of a community identity. What makes the argument for inclusion even stronger is that diverse local communities are likely to be more resilient and adaptable because they will contain a greater store of knowledge and experience to draw from. You only have to think about the contribution that many immigrants have made to Australian society to understand that social and cultural diversity opens up new possibilities for local communities. Yet fearful communities will often turn against 'strangers' and 'outsiders'. Here again, the arguments for inclusion need to be grounded in the real-life experiences of many local communities rather than on the basis of an abstract argument.

As mentioned, a sense of community often emerges most strongly at a time of crisis and we see this time and again in the wake of natural disasters such as bushfires and floods. It is, however, hard to maintain that 'heightened' sense of community when the harsh realities of recovering from trauma set in or when things eventually return to 'normal'. We can all remember the times when a community has rallied to help those in need, and big community events—such as commemorations or festivals of various kinds—can help to sustain a sense of community. At the same time, it might be better to anticipate future crises and challenges and build an inclusive sense of community before it is put to the test. There is a clear role for local government in this regard, and the author was involved in a major study exploring the importance of community art for effective local government in Australia (Mulligan and Smith 2010).

An Inter-Generational Conception of Community

As the Brundtland Commission report put it, one way to focus the mind on the future is to think about the opportunities that our own children or grandchildren might have in comparison with the opportunities we had ourselves. Of course, a changing world is constantly creating new opportunities for younger people, but the notion of environmental sustainability—first introduced by the Brundtland report—raises big questions about the challenges that might arise from the depletion of the world's natural resources. When we bring this kind of thinking about the future down to the level of a local community it also raises questions about the forms of employment that might be available to people living in that particular community and this, in turn, raises questions about the resilience and adaptability of the local economy.

In earlier research on community wellbeing for the Victorian health promotion agency VicHealth, the author conducted fieldwork in four different local communities in Victoria, ranging from inner-urban St Kilda to the rural and regional community centred on Hamilton in the Western District of the State (see Mulligan et al. 2006). Perhaps the most positive case study looked at the community centred on Daylesford in the central highlands of Victoria. Not long before we began our research for VicHealth in this community, it had faced a major challenge when the State Government had announced its intention to phase out logging in the surrounding Wombat State Forest. Initially, this sparked division between people living in several small towns surrounding Daylesford, which had been historically dependent on the timber industry, and those living in Daylesford and Hepburn Springs who had become dependent on the burgeoning tourism industry. In part, this threatened to drive a wedge between the 'old community' of residents who had been born and bred in the area and the 'new community' of people who had moved into the area after the town started to experience a resurgence in the late 1970s. As the tensions were rising, however, the government invited the local community to come up with a 'community' plan for how to manage the Wombat forest as a community resource in the future, and this proved to be a circuit-breaker. It became a circuit-breaker because of the role played by two individuals from the 'rival' communities.

Peter O'Mara was a youth worker for the Hepburn Regional Health Service who had moved to the area from Melbourne because he was attracted by both the natural environment and the history of the area. He had a strong interest in the arts and had helped to popularise poetry readings at the local Cosy Cottage Café in Hepburn Springs. As a youth worker, Peter knew that many young people were not benefiting from the boom in local tourism and he was in touch with many families from the 'old community' who were feeling very anxious about the employment opportunities for their children. Peter put himself forward to chair the community consultations about the future of the Wombat forest and his credibility with families from the 'old community' encouraged some of them to come to the meetings and listen to what others had to say.

The other person who played a key role was a second-generation timber-mill owner, Jim Dwyer, who had already reached the conclusion that the timber industry was not the best future option for his own children. By changing the ways in which he sourced logs and processed the timber in his mill, Dwyer had already made adjustment that enabled his mill to continue when others had closed and he was able to show that there could be some ongoing employment for experienced timber workers. At the same time, he argued that the growth of tourism had given young people in the area more opportunities for employment and he agreed that conservation of 'old-growth' areas within the Wombat forest was important for maintaining and even expanding tourism opportunities. Dwyer was convinced by the argument that some loggers might even find a new career as knowledgeable forest guides. Between them, O'Mara and Dwyer were able to bring potentially warring communities together to work on a plan for future employment in the area that would aim to ensure employment opportunities for people who had previously worked in logging. Sadly, the State Government did not follow through on the particular proposals that came out of the community meetings—as it had promised—but the need to focus on future employment opportunities for the young had swayed the community in favour of forest conservation. A broad consensus was achieved that logging in the Wombat forest should be phased out provided alternative employment was found for timber workers who would become 'redundant'. O'Mara, in particular, then worked hard, and with considerable success, to make sure that the local shire council looked after the interests of former timber workers.

A second example of appealing to inter-generational sentiment relates to work carried out by researchers from the Globalism Research Centre in association with a community reference group in the Hamilton region of western Victoria. In this case, the local community had been rather divided about whether or not the threat of global climate change was real, and RMIT researchers were invited to run a scenarios-mapping exercise in order to grapple more fully with future uncertainties, including climate change predictions. The community reference group worked very hard to bring together a diverse array of community members in selecting about 40 people to participate in the two-day workshop that began with some of the modelling carried out by the CSIRO for the region and then worked through a number of plausible, yet challenging, future threats over a period of 30–50 years. A very experienced 'scenarist' was employed to run the workshop and, as expected, it resulted in the elaboration of a range of challenging scenarios for the region. The process took an interesting turn, however, when two experienced local writers—one of whom had participated in the workshop—took the outcomes of the workshop to create a single, complex scenario for the future into which they set four different stories focusing on how different characters might respond to the challenges (see Mulligan et al. 2009). Several of the characters who were integrated into these stories had been sketched out by participants in the scenarios workshop and the writers went back to some of the participants to make the characters and the situations in which they would find themselves more plausible, yet no less challenging to contemplate.

In part, this exercise forced the workshop participants to think of what the future might be like for younger people currently living in the region. In developing the four stories, the writers were interested in finding out what might make some people more resilient than others when facing unexpected challenges and they called their collection of stories 'Unexpected Sources of Hope'. The stories were able to deal with, for example, inter-generational issues within a farming family, and one story focused on a young man with artistic talent, raised by a struggling single mother, who was able to use his own 'survival skills' to become a very effective youth worker. In another story, the district was asked to take in an allocation of climate 'refugees' from an inundated area in southern Vietnam, and their settlement process was quite painful. Yet one young member of that community decided to stand for mayor 20 years on and she used her campaign speech to pay tribute to those people within the community who had gone out of their way to help the new settlers. We do not know if she succeeded in the election.

Finding Ways to Link the Local with the Global

Parochialism reflects, of course, a desire to hold at bay a number of disturbing 'outside' influences. This option is, however, becoming less and less tenable. Major changes in transport and communication technologies and the social and economic processes that are commonly referred to as 'globalisation' mean that local communities are more heavily integrated into global realities than at any other time in human history. Rural communities in the Murray–Darling Basin certainly understand that fluctuations in global markets for the things they produce can affect them very directly, and this point was driven home by the recent global financial crisis. No-one can hide from the consequences of global climate change, and future global shortages of oil will affect rural communities more than others. At the same time, globalisation also means that people have ready access to a vast store of information and ideas and enterprising people and communities can find new 'markets' for their ideas or products.

For reasons of history and geography, Australian society tends to be even more 'insular' than others. It is easy to imagine that we are far from the 'troubles' afflicting other parts of the world, and even within our own large land it is easy for local communities to believe that problems can be solved 'somewhere else'. Global realities might be making this kind of parochialism less tenable but more work needs to be done to overcome Australian insularity and convince local communities that there are great advantages in engaging consciously and directly with global challenges and influences.

The UK-based banker Nicholas Stern made this point forcibly in his now-famous 2007 'report' for the British Government on the economics of climate change. Indeed, Stern argued that those regions and nations that adapt early and proactively to the realities of climate change will find themselves at the cutting edge of new global industries. On the other hand, the economic cost of adapting will grow with the passage of time. Of course, not all economists agree with Stern's analysis, but there are already some examples of local Australian communities that are benefiting from being early adaptors, even if the benefits are rather uneven and sometimes unexpected.

Perhaps the most prominent example focuses on the Victorian town of Castlemaine and the surrounding Mt Alexander Shire. In this area a group of local community leaders got support from the CSIRO to introduce energy-saving strategies at the level of households and local businesses. While this can be seen as an energy-saving strategy that has been only partially successful, what was even more important was the way the initiative galvanised the local and regional communities and created new support for participating local businesses. New energy-efficiency businesses began to emerge and interested people decided they wanted to come to live in the area. Before the initiative, Castlemaine was already building a reputation for being a rather dynamic community, with a thriving arts sector. The Castlemaine 500 initiative—so named because it aimed to convince 500 households to adopt energy-saving technologies—raised the town's profile even further and created a new 'buzz' within the community. Of course, the downside of this has been a 'gentrification' of the area, resulting in a surge in real estate prices. As was the case with Daylesford earlier, this has created tensions between the 'old community' of residents and the newer settlers. Nevertheless, there has been a clear benefit to the local economy as a whole, even if action needs to be taken to ensure that the benefits are widely shared.

Researchers from the Globalism Research Centre (GRC) are involved with an initiative in another Victorian town—Coleraine—that has also managed to 'pull itself up by its bootstraps' in recent times. The Coleraine community responded to the closure of all commercial banks in the town by setting up its own successful community bank—under the auspices of the Bendigo Bank—and it is now working with the local water authority, Wannon Water, on an ambitious plan to use waste water from the sewage-treatment plant to supplement food production and improve environmental flows in the local waterways. Through the GRC, the Coleraine community invited an Indian botanist with great knowledge of the propagation and use of mushrooms for a residency in the town and, as a result, the waste-water project is now also looking at ways to propagate mushrooms as a food source and for use in 'bioremediation'. A Coleraine Enterprise Co-operative has been formed to build stronger community support for an ambitious waste-recycling plan in the town. While this project is still in its development phase, it has won support from the Southern Grampians Shire Council, which is using it as a model for community planning across the shire.

While it has become more possible for local communities to take advantage of global linkages it is never easy to 'go it alone'. The international Transition Towns network is a good example of a global initiative that can help local communities to become proactive about the future challenges of sustainability. It is interesting to note that the Transition Towns initiative, which began in the United Kingdom, was inspired by the principles of 'permaculture' that were first developed in Australia (Hopkins 2008). A range of local communities in Australia has adopted the aim to attain and maintain their status as a 'transition town' and that number is likely to grow.

Moving Beyond Localism to Achieve Equity in Responding to Sustainability Challenges

While parochialism is one danger, it is also unreasonable to assume that particular local communities should pay for the sins of the past in relation to natural-resource management. Obviously, many local communities in the Murray–Darling Basin felt that they were being asked to bear the brunt of the economic costs of the historic water-reform proposal when the MDBA Guide was released in late 2010. Much more needed to be done to develop alternative economic development and employment plans at the level of the communities that will be most affected. The MDBA Guide talked broadly of opportunities for alternative employment in a national economy that is still suffering from skills shortages and it outlined some of the economic benefits to be gained by making water use more sustainable. It soon became obvious, however, that such arguments did little or nothing to reassure local communities that their economic and employment interests had been seriously considered. The transition planning for the economic adjustments that would enable serious water reform took place at the wrong scale for affected communities and it was too easy for them to conclude that their specific interests had been 'sacrificed' for the sake of the environment.

The community-level transition planning that took place in the Daylesford area after the Victorian Government announced the phase-out of logging in the Wombat State Forest points to a very different way of engaging local communities in thinking about their alternatives because the conversation happened at a level that seemed real to the community. Of course, there are hundreds of local and regional basin communities facing an economic 'adjustment' in the wake of water reform and it would be a long and difficult process to engage them all in community-level economic planning. Furthermore, there is a clear need for a body such as the MDBA to ensure an equitable investment in all the affected basin communities. The long-term benefits in terms of 'bedding down' reforms that will have widespread community, and therefore political, support must, however, be considered. Priority could, and should, be given to local and regional communities facing the biggest economic 'adjustments', but the key argument here is that the process for engaging local communities in transition planning needs to proceed at the right scale and at the right pace. Such transition planning needs to be deliberate and inclusive rather than hasty and 'imposed'. Furthermore, the local communities need to be convinced that the national government is determined to make sure that they will not pay an undue 'cost' for rectifying historical mistakes in regard to the use of water resources. The flooding of early 2011 has sparked new calls for a national 'disaster fund'. Perhaps we need a national 'disaster and readjustment fund' to make sure that as a nation we can take responsibility for rebuilding, relocation and new forms of economic development within communities that are particularly vulnerable to natural disasters and to the kinds of reforms needed to ensure a much more sustainable use of the nation's natural resources.

There is a need for community-level transition planning and there is also a need for household-level 'readjustment'. The author was involved in a major study of post-tsunami recovery in Sri Lanka and India, which, in part, concluded that it is important to have a focus on household incomes rather than individual incomes (see Shaw et al. 2010) because this opens up the possibilities for households to diversify their sources of income. Less formal arrangements can be made to deliver better food security at a household level (for example, home gardening, community gardens and forms of bartering). More could be done to increase food security at a local and regional level in Australia and more could be done in general to look at ways in which the 'informal' economy can supplement the formal economy in delivering greater household food and income security.

As difficult as it appears to be at present, we need to build a stronger political consensus on the need to adequately address the long-term challenges of sustainability in Australia. There must be a role for 'civil society' in building that consensus and in trying to ensure that politicians do not continue to use the politics of fear for short-term political gain. In part, the battle needs to be won at the level of local communities, which might eventually pay a high cost for political inaction. Indeed, it is conceivable that rural communities, which currently sit on the margins of sustainability, could play a leading role in putting pressure on governments to act with greater foresight and courage. They would need to be assured, however, that the nation as a whole will foot the bill for necessary reform and readjustment. Armed with such a national pledge, community-development workers might have a bigger role to play in building the consensus for reform than the economists who tend to dominate such debates at present.

Australia has many talented and experienced community-development workers in local communities right across the country. As mentioned at the beginning of this chapter, Susan Kenny (2006) has effectively documented the emergence, since the 1970s, of skilful community-development practice. Furthermore, there are many more people doing this kind of work who might never use the term 'community development'—perhaps seeing their volunteer work as a form of 'service' to the community that enables them to feel personally 'connected'. The aforementioned study of post-tsunami community development in Sri Lanka and India (Mulligan and Nadarajah 2010) suggested that aptitude is even more important than experience for community-development work. This field of work is, however, not valued as highly as it should be in Australia. Good practice in community development is all about creating more inclusive local communities in order to make them more resilient and adaptable to future challenges. Hopefully, the mistakes made in the way the MDBA developed and released its Guide will serve as a timely reminder that community should never be taken for granted and that people with relevant experience and/or aptitude should be given adequate resources and time to ensure genuine community 'engagement' in meeting the challenges of long-term sustainability. In the end, it is important to remember what Gerard Delanty (2003:195) wrote: 'Community offers people what neither the state nor society can offer, namely a sense of belonging in an insecure world.'

Bibliography

Agrawal, A. 1999, 'Enchantment and disenchantment: the role of community in natural resource conservation', *World Development*, vol. 27, pp. 629–49.

Bauman, Z. 1999, *Culture as Praxis*, Sage Publications, Thousand Oaks, Calif.

Bauman, Z. 2001, *Community: Seeking safety in an insecure world*, Polity Press, Cambridge, UK.

Birmingham, J. 2011, 'Tough love for a tougher place', *Sydney Morning Herald*, 15–16 January 2011, p. 7.

Cohen, A. 1985, *The Symbolic Construction of Community*, Tavistock, London.

Delanty, G. 2003, *Community*, Routledge, London and New York.

Hopkins, R. 2008, *The Transition Handbook: From oil dependency to local resilience*, Green Books, Devon, UK.

Kenny, S. 2006, *Developing Communities for the Future*, [Third edition], Thomson, Melbourne.

Mulligan, M. and Nadarajah, Y. 2010, *Rebuilding Community in the Wake of the 2004 Tsunami: Lessons from Sri Lanka and India*, Globalism Research Centre, RMIT University, Melbourne, <www.rmit.edu.au/globalism/publications/reports>

Mulligan, M. and Smith, P. 2010, *Art, Governance and the Turn to Community: Putting art at the heart of local government*, Globalism Research Centre, RMIT University, Melbourne, <www.rmit.edu.au/globalism/publications/reports>

Mulligan, M., Humphery, K., James, P., Scanlon, C., Smith, P. and Welch, N. 2006, *Creating Community: Celebrations, arts and wellbeing within and across local communities*, Globalism Research Centre, RMIT University, Melbourne.

Mulligan, M., Nadarajah, Y., Smith, J.-A. and Zalchendler, Y. 2009, 'Community, scenarios and narratives of action: reflections on a case study in the Hamilton region of Victoria', in J. Martin, M. Rogers and C. Winter (eds), *Climate Change in Regional Australia: Social learning and adaptation*, Victorian Universities Regional Research Network Press, Ballarat, Vic.

Murray-Darling Basin Authority (MDBA) 2010, *Guide to the proposed Basin Plan*, Murray-Darling Basin Authority, Canberra.

Rose, N. 1999, *Powers of Freedom: Reframing political thought*, Cambridge University Press, UK.

Sennett, R. 1986, *The Fall of Public Man*, Faber & Faber, London.

Shaw, J., Mulligan, M., Nadarajah, Y., Mercer, D. and Ahmed, I. 2010, *Lessons from the Tsunami Recovery in Sri Lanka and India: Community, livelihoods, tourism and housing*, Globalism Research Centre, RMIT University, Melbourne, <www.rmit.edu.au/globalism/publications/reports>

Slovic, P. 1987, 'Perception of risk', *Science*, [New series] vol. 236, no. 4799, pp. 280–5.

Slovic, P. 2010, *The Feeling of Risk*, Earthscan, London.

Walmsley, J. 2006, 'Putting community in place', *Dialogue*, vol. 25, no. 1, pp. 5–12, Academy of Social Science in Australia, Canberra.

8. Why justice is important

Catherine Gross

Introduction

Rural communities in the Murray–Darling Basin reacted with outrage, anger and protests after the release of the Murray–Darling Basin Authority (MDBA) *Guide to the proposed Basin Plan* on 8 October 2010. Regional and national newspapers gave the controversy full coverage. They reported standing room only at a series of regional community information sessions organised by the MDBA. These meetings were characterised by large gatherings of people from rural communities at which symbolic gestures—such as burning copies of the Guide in the streets—were clear demonstrations of frustration and anger (Jopson and Arup 2010). The mood was succinctly expressed in regional newspaper headlines—for example, the black background and red-lettered headline 'Fight Back' on the front page of the Victorian-based rural newspaper *The Weekly Times*. This headline was surrounded by a collage of photographs depicting protest signs (such as 'This Means War'), conveying widespread disagreement with the proposals outlined by the MDBA (Hunt 2010:1).

The Basin Plan has been long anticipated. Water reform in Australia has been high on the national agenda for the past two decades (Connell 2007). Since 1994, water reforms have taken progressive steps, culminating in the *Water Act 2007*, which brought water reform in the Murray–Darling Basin under an independent authority—the MDBA—for the first time, and required the MDBA to publish a Basin Plan under which sustainable levels of water extraction would be attained. The MDBA acknowledges the importance of engaging with stakeholders in the development of the Basin Plan, with key stakeholder groups being involved (MDBA 2010). So, it could be asked, with this length of time for anticipation and preparation, what went wrong? Why was the publication of the Guide greeted with such vehement opposition and why did rural communities receive MDBA representatives so angrily? Were these protests, therefore, merely expressions of 'self-interest', or what is known as 'rent seeking' in the Australian vernacular? Are irrigation communities simply seeking to preserve their perceived property rights in water? Or, are the protests manifestations of a broader set of rural and regional concerns and issues, which includes the wellbeing of rural communities and the way agricultural communities are valued by Australian society? The debate about water in Australia is highly complex and is characterised by firmly held notions and positions, including perspectives implied by these questions (see, for example, Crase 2008; Hussey and Dovers 2007).

What is needed in situations in which there appears to be a 'stand-off' between various groups is a clearer understanding and appreciation of the perspectives held by stakeholders and an approach in which these can be understood and a useful dialogue can be developed and maintained. This chapter aims to provide some insights into the water-allocation debate through an exploration of perceptions of fairness and justice in two earlier social conflicts over water allocation in the Murray–Darling Basin (Gross 2008, 2010). The first centres on a 2006 NSW Government decision to cut the carry-over water allocation. This resulted in a community protest in the town of Deniliquin. The second social conflict is the community protest and campaign against the Victorian Government's North South Pipeline and Food Bowl Modernisation Project initiated in 2007. These two social conflicts were case studies in a larger research project exploring equitable resource allocation in environmental decision making (Gross 2010).

This chapter shows how perceived injustice in decision-making processes can be seen as a tangible harm to those involved and this in turn arouses a sense of injustice that impedes acceptance—not only of the decision, but also of the decision-making process itself. The chapter outlines several implications for decision makers concerning the need to recognise different types of justice in decision-making processes and the impact of perceived injustice on individuals and communities. The chapter makes the point that decision makers must recognise that not only outcomes need to be perceived as fair and equitable, but also decision-making processes and the way people are treated within decision-making processes.

Fairness and Justice

There is a sizeable body of research on fairness in natural-resource management in Australia and elsewhere (for example, Hunt and Haider 2001; Smith and McDonough 2001; Syme et al. 1999). There have, however, been many implementation challenges, as acknowledged by long-time Australian justice researchers Syme and Nancarrow (2008:242), who report that 'fairness is largely ignored despite the development in Australia of techniques to include it in a transparent fashion'. A key theme of fairness research in natural-resource management is the notion that decision-making processes that are perceived as fair are more likely to result in outcomes that are acceptable to those involved (Syme et al. 1999). In a study of water management in Europe, Hophmayer-Tokich and Krozer (2008) investigated aspects of public participation and found that better decisions could be achieved if a range of perspectives and viewpoints in defining the problems could be incorporated. They commented that 'people are willing to co-operate if they sense that they are being treated fairly' (Hophmayer-Tokich and Krozer 2008:255). Thus, considerations of fairness are important in participation processes and decision making.

There has been a significant amount of research on fairness and justice in the discipline of social psychology. The idea of procedural justice came about in the mid 1970s when researchers realised that it was not only the outcome, or distributive justice, that was important to people, but that processes were also important (Colquitt et al. 2001). This was in contrast with the dominant idea of the time that outcomes were more important. Procedural justice includes several independent criteria that are required in decision-making processes. These include being able to participate in the process and to have one's voice heard; being able to access the right amount of information (summary or detail as required); enough time to be able to conduct research, debate the issue, and form an opinion; a mechanism in which issues can be raised, discussed and responded to; and finally, but not least, being treated with respect (Tyler 2000). This last criterion is often referred to as 'interactional justice'. Interactional justice is seen as being important enough to separate out from procedural justice because of the value and importance that people place on how they are treated—whether or not this is within a decision-making process (Bies 2001). Interactional justice is a broad approach to interpersonal treatment—for example, while it includes respectful treatment as in courteous interactions, it also includes whether people are involved, or given standing, in discussions about matters in which they have an interest. In summary, there are three main constructs of justice: interactional justice, concerned with respectful treatment; procedural justice, concerned with elements of the decision-making process; and distributive justice, concerned with the fairness of outcomes.

There is a further distinction that needs to be made: between justice and injustice. While much thought and research have been devoted to the notion of justice—for example, in philosophy, jurisprudence and political philosophy—far less has been expended on the notion of injustice (Simon 1995; Shklar 1990). While the notion of justice is frequently seen as difficult to define, with different meanings in different contexts, the notion of injustice is seen as more tangible and a cause of actual harm. This is in contrast with justice, which is seen as more of an ideal state to aspire to. Thus, injustice comes *before* justice and justice becomes the 'corrective to injustice' (Wolgast 1987:146). Simon (1995:16) puts the case for investigating a theory of injustice and argues that injustice 'empirically, temporally, psychologically and morally' comes before justice and these two notions should not be considered as the two sides of a coin. They are separate from each other and should be considered that way. This perspective was clearly articulated by Cahn (1949:13), reflecting on his years in the legal system, who wrote that justice should be seen as an '*active process* of remedying or preventing what would arouse the sense of injustice' (emphasis in original). The experience of injustice, such as disrespectful treatment, can create harm and arouse a sense of injustice. Miller (2001:533) suggests that people perceive disrespectful treatment as an injustice because it 'deprives people of something to which they are entitled' and also 'subjects people to something they do not deserve'. Such explanations for why people react to perceived disrespectful treatment are important in the current context of water allocation in the Murray–Darling Basin.

A further area of justice is relevant in this context. This is the question of why people become engaged in matters of justice, or their motivation for justice. There are many theories about why people care about justice, and these can present a confusing picture. There are three broad groupings: that self-interest is the central motivation; that people care about justice because it indicates their status or value within a community or group; and that people care about justice because of its own intrinsic value, or as an end in itself (Montada 2003). The theory of self-interest has perhaps had the most influence as an explanation for why people care about justice: for material gain, to increase wellbeing or to protect one's livelihood. This theory has, however, been described as a myth because it has been seen to be overly dominant and because there are other valid explanations for why people seek justice (Tyler and Blader 2000). The second set of theories holds that people care about justice because it shows how they are valued within a group and involves their own self-worth within that group. The third area involves morality in which people can be motivated to achieve justice as an end in itself, perhaps as a means of ensuring stability in their social structure and environment (Lerner 1998). While these three groups have been viewed as competing models or theories, others see these as contextual motivations in which all can play a part depending on what types of disagreements or conflicts are taking place. Thus, it is possible that all of these motivations, to some degree, appear in individuals and groups within a community at the same time, depending on what type of injustice is perceived and how people perceive their current situation (Skitka 2009).

How people *experience* injustice has also been a topic of research. People have different experiences depending on their direct or indirect involvement and how they are affected by the perceived injustice. For example, there are the perspectives of the perceived victims, the perspectives of those responsible and the perspectives of bystanders—some of whom might be closely involved but are not directly affected and some of whom might be on the margins of the issue. In a social conflict, perceptions of injustice are by their nature subjective; the differing viewpoints of what is considered just or unjust are what lie at the heart of conflict and disagreement (Mikula 2005).

The Research

The previous section has outlined some different areas of justice, including considerations of justice in interpersonal treatment, decision-making processes and outcomes, the distinction between justice and injustice, and why people become engaged in matters of justice. From this brief review it is evident that there is a complex interplay of interactions associated with these justice areas between decision makers, communities and individuals in a particular social context. Perceptions of injustice and motivations for justice are likely to be complex and experienced at different levels by different individuals within and around the communities affected.

As mentioned above, the research described in this chapter is part of a larger body of research into fairness and justice in natural-resource management. The first part of the research explored a community's reaction to a proposed wind farm in terms of the perceived fairness of the consultation process (Gross 2007). That research found that the principles of procedural justice were important in determining people's acceptance of the proposed wind farm. Many perceived that the proposed wind farm had damaged the community's overall wellbeing even before a final decision was made, because it had divided the community, and winners and losers had emerged. Thus, for some, the decision-making process itself had created a perceived injustice even before the final outcome was known. From that research, a trans-disciplinary investigative framework was developed (Gross 2008), using justice constructs and exploring people's engagement with fairness and justice on three different levels: material (concerned with family and livelihood), social (concerned with social wellbeing) and personal (concerned with values and beliefs) (Skitka 2003).

The two case studies, introduced above, were chosen because there were clear issues of fairness and justice involved in both disputes, as seen by the protests in both states and formation of an opposition group, Plug the Pipe, in Victoria. Both conflicts were reported in metropolitan newspapers and both were involved with water allocation in the Murray–Darling Basin. The 2006 NSW carry-over water issue arose because the NSW Government cut a water allocation for irrigators in the Murray Irrigation District following an extended period of drought. This was unexpected, and communities were taken by surprise. The cuts included traded water and carry-over water that was on irrigators' water accounts at the time of the announcement. Irrigators were surprised because they believed that the water allocation was secure, being primarily carried over from the previous year's water allocation as part of an agreed risk-management strategy. A second cut took place in November, following consultation with irrigation representatives. The community protested against the cuts by organising a rally in Deniliquin. The NSW Government later responded with an Extraordinary Assistance Package for farmers who were suffering hardship as a result of the drought.

The second case study involves the combined Victorian project of the North South Pipeline and the Food Bowl Modernisation Project. The Victorian Government had agreed to fund upgrades to the ageing irrigation infrastructure in northern Victoria in return for a trade-off arrangement in which water would be diverted to Melbourne from the Goulburn River via the North South Pipeline. The dispute involved conflicting perspectives about the transfer of water from a rural area in drought to a coastal city that was perceived to have other options for water supply. Also disputed was the way the trade-off deal was perceived to have been constructed without adequate consultation between the government and differing sections of the communities affected by the deal.

Fieldwork for the research was carried out in New South Wales in early 2007 and in Victoria in late 2008. Semi-structured interviews were chosen as the most appropriate method in which a full range of perspectives of each social conflict could be explored. The interviews were based on a set of questions in which interviewees were asked to describe their involvement with the conflict, how it affected them materially, socially and personally, and their views on the fairness of the conflict from a process and outcome perspective. A networking approach to the interviews was used to select interviewees (by asking each interviewee to suggest other people to interview) and a variety of perspectives was achieved by selecting people from different stakeholder groupings. These included irrigators, dryland farmers, conservationists, business owners and employees, teachers, government agency employees and retirees. Confidentiality of the research was explained to all interviewees and was a key feature of the research. The interviews were recorded and the data transcribed and analysed.

Finding Injustice and Seeking Justice

Respect, Process and Outcomes

A wealth of information about fairness and justice was gathered from these interviews. The analysis presented here provides some insights into these social conflicts: a detailed analysis is provided in Gross (2010). The analysis uses a justice lens to explore people's perceptions of the decision-making processes, and their views of the impact of the proposed or actual outcomes on themselves and their community.

The first key finding in this research was that people talked about *un* fairness and *in* justice rather than fairness and justice. They talked about what was wrong within the social conflict and how it had already affected them. Many described their fears for the future following the events that had taken place. These perceptions of injustice varied according to people's circumstances, and whether they were directly or indirectly affected on a material, social or personal level. For example, those who had experienced the NSW irrigation cutbacks could be materially affected by the loss of water and impact on their business. In contrast, others in the community were affected only indirectly but were concerned about the plight of their neighbours, family or friends within the community who might be harder hit by the cuts. It was not, however, only the material impacts that were perceived to be unjust, but also the way the communities felt they had been treated.

Many members of the community in the Yea district in Victoria were concerned with the lack of communication and consultation by the Victorian Government in discussing the details and impact of the proposed North South Pipeline. Some were concerned with the impact of the pipeline on their agricultural business, whereas others were more concerned with the diversion of water from the Goulburn River and the impact on the environment and others relying on the Goulburn River for their business. These perceived injustices regarding process, or the lack of consultation and opportunities for discussion, also spread into personal concerns about the willingness of government agencies to provide information and how they conducted their business in these types of infrastructure developments. Major concerns in this area were the lack of available information about the proposed savings of water from the irrigation infrastructure upgrade and the lack of information on other options or any detailed justification for the diversion of water to Melbourne.

Thus, it can be seen that these perceived injustices were related not only to the proposed outcome, but also to decision-making processes and the way people were treated during those processes. These perceived injustices and others described by interviewees are summarised in Tables 8.1 and 8.2 and are listed in order of the type of justice construct perceived to be violated.

Table 8.1 Perceptions of injustice in the NSW study

Perception of injustice	Type of justice violated
Lack of respect in the way people were treated	Interactional justice
Lack of notification, information and involvement	Procedural justice
Perceived entitlement/property removed	Distributive justice
Burden of cutback inequitably distributed amongst irrigator groups	Distributive justice
Lack of recognition of actual harm done	Distributive justice
Equitable compensation not offered	Distributive justice
Inequity of Extraordinary Assistance Package	Distributive justice

Table 8.2 Perceptions of injustice in the Victorian study

Perception of injustice	Type of justice violated
Disdainful treatment of affected communities	Interactional justice
To break an election promise and lie in meetings	Interactional justice
Lack of consultation regarding initiation of pipeline and irrigation modernisation project	Procedural justice
Lack of information on pipeline and water savings	Procedural justice
To continue the process with inadequate consultation	Procedural justice
Lack of an environmental assessment to consider the combined impact of the pipeline and the irrigation upgrade	Procedural justice
Impact on the environment: removal of water from river system and loss of water flow	Distributive justice
Proposal based on unsatisfactory justification of 'need' when other options for increasing Melbourne's water supply had not been fully explored	Distributive justice
To remove water from irrigators who have paid for it through their annual fees and capital purchase	Distributive justice
To base the project on future water savings that are not validated and contested	Distributive justice

These tables show that perceived injustices arose in all three areas in both studies. In the Victorian study there was an emphasis on perceptions of injustice in the decision-making processes because interviews were conducted prior to significant infrastructure works taking place. In the NSW study the outcome, or decision, had already been put into action and perceived inequities associated with the government's Extraordinary Assistance Package had become apparent to some in the community.

Injustice as Harm to Community Wellbeing

> I put confidence under a personal banner. We are struggling to keep the confidence level up with all these changes…we need some sort of belief in the future.

> [I]t has hurt the community because it has destabilised the community… what underpins communities is partly the ability of people to be optimistic, to be positive about the future, therefore they are prepared to invest…yet that action has destabilised those sorts of things.

These two quotations from community leaders interviewed in the NSW study summarise their view of the lasting negative impact and harm that perceived unfair treatment can have on communities. These quotations reflect the importance of a strong belief in the future for the overall wellbeing of the community. Regional communities that rely on irrigation for a substantial part of their material livelihood also rely on confidence in their relationships with government. A significant adverse impact of the NSW carry-over water cutbacks study was that this was perceived as yet another blow to community confidence on top of a series of what were perceived to be unsatisfactory government–community interactions. A community leader who had spent a significant amount of time on committees dealing with government agencies lamented the fact that social impacts were routinely overlooked and that 'it is an awfully long haul to try and get governments to listen to what the real issues are for communities'.

Another recurring theme was the perceived injustice in the way irrigators are regarded in general by society. For example, a vegetable grower pointed out the irony of a news headline that had reported on the recent $10 billion plan put forward by the then Prime Minister, John Howard. The headline stated: 'Mr Howard is giving money to farmers to stop them wasting water.' In the vegetable grower's opinion, irrigators were held at the 'lowest level' when it came to water use, and this headline underscored the lack of understanding of the use of irrigation water in food production. Many community members in both case studies commented on the perceived rural–urban divide and what they believed was a lack of knowledge in society in general about how irrigation water is used and the value of different types of annual crops, such as rice, in making use of irrigation water when it is available.

Seeking Justice

As outlined in the 'Fairness and Justice' section above, there are many reasons why people are motivated to seek justice. In both studies individuals and communities were strongly motivated to voice their disagreement and to seek some sort of measure to restore the perceived injustice. It is possible to categorise some emerging themes with respect to the type of motive for justice and proposed measure to restore justice for each perception of injustice. A summary of some emerging themes from the NSW case study is outlined in Table 8.3.

Table 8.3 Community injustice framework: some emerging themes from the NSW study

Community perceptions of the issues and associated injustice	Proposed measure to restore justice	Possible underlying motive for justice
• Lack of recognition of impact • Carry-over water cutbacks as a violation of a property right • Lack of consultation • Lack of respect for local knowledge • Lack of recognition for contribution to society as an agricultural community	• Recognition of social impact • Equitable compensation for the water removed • Consultation that respects local knowledge and local perspectives • Government agencies to recognise value of agricultural production	• Protection of interests of community: mutual advantage • Protection of livelihood • Justice as an end in itself—to do the right thing • To maintain structural stability of the community and confidence in the future • Community to be valued for contribution to society

This summary shows a complex interplay within the community of perceived injustices and underlying motives to restore justice, depending on the perspective and involvement of the individual or group. This type of categorisation is useful in developing an understanding of the different perspectives held by differing individuals and groups within a social conflict.

Through the lens of injustice, this section has shown some perspectives and consequences of a variety of perceived injustices on individuals and communities. Calls for justice from such communities include that communities should be treated with respect, information should be provided, harm should be recognised, impacts should be understood and dealt with on a fair basis and wrongs should be righted.

Implications for Decision Makers

There are several implications for decision makers. The first is to recognise that there are different types of perceived injustice in decision making. Perceived injustices can arise in three main areas: in the way people are treated (interactional injustice); as a result of inadequacies in components of the decision-making process (procedural injustice); and as a consequence of the outcome or decision (distributive injustice). A consultation process that fails to deliver what people expect—such as real engagement in the process, being treated with respect, being given information that supports proposed decisions, time to understand and discuss information and implications—will be perceived as an injustice to those involved. The key point here is that consultation processes either can deliver fairness and justice if they include these elements, or can deliver perceived injustice if these elements are omitted.

A second implication is that a perceived injustice constitutes an actual harm to an individual or community, whether it is part of a process or an outcome. The harm can be manifested on a material, social or personal level. The perceived injustice can take the form of an outcome as part of a failure in the consultation process, such as a failure to be consulted or to consider the impact of a proposed decision. A harm resulting from a perceived injustice can have a significant long-term impact on individual and community confidence and belief in future security. Another aspect for decision makers to consider is that a harm resulting from a perceived injustice arouses a sense of injustice and incites opposition to the decision-making entity and proposed outcomes. Thus—and importantly for decision makers—*acceptance* of decisions is impeded by perceived injustice.

A third implication is that perceived injustices require justice in order to resolve them. If perceived injustices are not resolved then there is a build-up of perceived layers of injustice that not only results in communities feeling undervalued or badly treated, but also undermines their ability to meaningfully engage in government processes due to resentment and 'burn-out'.

Finally, a fourth implication that can be drawn from this analysis is that decision makers must understand their role in bringing out the different perspectives and viewpoints held by individuals and communities in decision-making processes. This analysis has shown that 'self-interest' is but one of many different perspectives in the water-allocation debate. Better outcomes could be achieved if these different perspectives could be acknowledged, respected and debated by society at large. This is why the term 'justice must be done and must be seen to be done' is important.

In conclusion, it is vital that decision makers understand the importance of the different types of justice when engaging with people in consultation processes and decision-making processes. Decision makers must understand the long-term consequences on communities and why acceptance of outcomes is harder to obtain if they fail to devise and deliver decision-making processes that those involved can perceive as just and fair.

Acknowledgments

Thanks to interviewees in New South Wales and Victoria for sharing their perceptions about fairness and justice. The research was funded by a Land & Water Australia scholarship.

References

Bies, R. J. 2001, 'Interactional (in)justice: the sacred and the profane', in J. Greenberg and R. Cropanzano (eds), *Advances in Organisational Justice*, Stanford University Press, Calif., pp. 89–118.

Cahn, E. N. 1949, *The Sense of Injustice: An anthropocentric view of law*, New York University Press, Manhattan.

Colquitt, J. K., Conlon, D. E., Wesson, M. J., Porter, C. O. and Ng, K. Y. 2001, 'Justice at the millennium: a meta-analytic review of 25 years of organisational justice research', *Journal of Applied Psychology*, vol. 86, no. 3, pp. 425–45.

Connell, D. 2007, *Water Politics in the Murray–Darling Basin*, The Federation Press, Sydney.

Crase, L. (ed.) 2008, *Water Policy in Australia: The impact of change and uncertainty*, Resources for the Future, Washington, DC.

Gross, C. 2007, 'Community perspectives of wind energy in Australia: the application of a justice and community fairness framework to increase social acceptance', *Energy Policy*, vol. 35, no. 5, pp. 2727–36.

Gross, C. 2008, 'A measure of fairness: an investigative framework to explore perceptions of fairness and justice in a real-life social conflict', *Human Ecology Review*, vol. 15, no. 2, pp. 130–40.

Gross, C. 2010, Water under the bridge: fairness and justice in environmental decision-making, PhD Thesis, The Australian National University, Canberra, <http://dspace-prod1.anu.edu.au/handle/1885/49419>

Hophmayer-Tokich, S. and Krozer, Y. 2008, 'Public participation in rural area water management: experiences from the North Sea countries in Europe', *Water International*, vol. 33, no. 2, pp. 243–57.

Hunt, P. 2010, 'Fight back', *The Weekly Times*, Wednesday, 20 October 2010.

Hunt, L. and Haider, W. 2001, 'Fair and effective decision-making in forest management planning', *Society and Natural Resources*, vol. 14, 873–87.

Hussey, K. and Dovers, S. (eds) 2007, *Managing Water for Australia: The social and institutional challenges*, CSIRO Publishing, Collingwood, Vic.

Jopson, D. and Arup, T. 2010, 'Irrigators vent fury at proposed water cuts', *Sydney Morning Herald*, 13 October 2010, viewed 24 October 2010, <http://www.smh.com.au/environment/water-issues/irrigators-vent-fury-at-proposed-water-cuts-20101013-16k41.html>

Lerner, M. J. 1998, 'The two forms of belief in a just world: some thoughts on why and how people care about justice', in L. Montada and M. J. Lerner (eds), *Responses to Victimisations and Belief in a Just World*, Plenum Press, New York, pp. 247–69.

Mikula, G. 2005, 'Some observations and critical thoughts about the present state of justice theory and research', in S. W. Gilliland (ed.), *What Motivates Fairness in Organisations?*, Information Age Publishing, Greenwich, Conn., pp. 197–210.

Miller, D. T. 2001, 'Disrespect and the experience of injustice', *Annual Review of Psychology*, vol. 52, p. 527.

Montada, L. 2003, 'Justice, equity, and fairness in human relations', in I. B. Weiner (ed.), *Handbook of Psychology. Volume 5*, Wiley, New York, pp. 537–68.

Murray–Darling Basin Authority (MDBA) 2010, *Guide to the proposed Basin Plan*, Murray–Darling Basin Authority, Canberra, viewed 24 October 2010, <http://www.mdba.gov.au/basin_plan>

Simon, T. W. 1995, *Democracy and Social Injustice: Law, politics, and philosophy*, Rowman & Littlefield Publishers, Lanham, Md.

Shklar, J. 1990, *The Faces of Injustice*, Yale University Press, New Haven, Conn.

Skitka, L. J. 2003, 'Of different minds: an accessible identity model of justice reasoning', *Personality and Social Psychology Review*, vol. 7, no. 4, pp. 286–97.

Skitka, L. J. 2009, 'Exploring the "lost and found" of justice theory and research', *Social Justice Research*, vol. 22, no. 1, pp. 98–116.

Smith, P. D. and McDonough, M. H. 2001, 'Beyond public participation: fairness in natural resource decision making', *Society & Natural Resources*, vol. 14, pp. 239–49.

Syme, G. and Nancarrow, B. 2008, 'The social and cultural aspects of sustainable water use', in L. Crase (ed.), *Water Policy in Australia: The impact of change and uncertainty*, Resources for the Future, Washington, DC, pp. 230–42.

Syme, G., Nancarrow, B. and McCreddin, J. A. 1999, 'Defining the components of fairness in the allocation of water to environmental and human uses', *Journal of Environmental Management*, vol. 57, no. 1, p. 51.

Tyler, R. T. 2000, 'Social justice: outcome and procedure', *International Journal of Psychology*, vol. 35, no. 2, pp. 117–25.

Tyler, R. T. and Blader, S. L. 2000, *Cooperation in Groups: Procedural justice, social identity, and behavioral engagement*, Psychology Press, Philadelphia.

Wolgast, E. H. 1987, *The Grammar of Justice*, Cornell University Press, Ithaca, NY.

9. Indigenous Water Management: Priorities for the next five years

Sue Jackson

Introduction

Indigenous systems of resource management coexist alongside and interact with the relatively recently introduced and rapidly transforming institutional systems of State land and water management. These latter systems encompass a mix of regulatory and market-based allocation mechanisms, incorporate scientific methods of resource assessment and management, and increasingly aspire to achieve transparency in water planning procedures, including opportunities for public participation in water management decisions. Much of the impetus for reform comes from the 2004 National Water Initiative (NWI), which for the first time in Australian water policy history explicitly recognised Indigenous rights and interests in water. Parties to the NWI have agreed that water-access entitlements and planning frameworks should recognise Indigenous needs (Jackson and Altman 2009).

A further impetus for reform of water management comes from the *Water Act 2007* (Cwlth), which requires that the Murray–Darling Basin Authority (MDBA) prepare a Basin Plan to set enforceable limits on the quantity of water that can be extracted from the Basin's ground and surface water resources. In fulfilling its obligations to assess the social and economic impacts of the Basin Plan, the MDBA will have regard for the social, cultural, Indigenous and other public benefit issues. According to Section 22(1) of the Act, the Basin Plan must contain a description of the water resources and the uses to which those resources are put, including by Indigenous people. The significance of Indigenous interests is also acknowledged in the legislative requirement for Indigenous representation on the Basin Community Committee and the establishment of an Indigenous Water Sub-Committee. Consideration of Indigenous interests is one of many plan objectives alongside requirements to implement international agreements, conserve Ramsar Convention sites and meet the water needs of ecological assets. One of the most influential international agreements is the Biodiversity Convention—one that obliges signatories to involve Indigenous people in biodiversity conservation and recognise their distinct values and knowledge systems. There are also international human and Indigenous rights instruments to which Australia is a signatory, including the UN Declaration on the Rights of Indigenous Peoples (United Nations 2008).

In January 2010, the CSIRO was engaged by the MDBA to undertake a scoping study of the impacts of changes in water availability on Indigenous communities of the Murray–Darling Basin (MDB). The MDBA requested a synthesis of current knowledge of Indigenous cultural, social and economic values of water and a descriptive characterisation of the potential impact of changes in water availability on Indigenous people. The report was to complement other socioeconomic assessments undertaken in early 2010 to optimise the outcomes of the Basin Plan. The study involved consultation with Indigenous groups, a review of the social science, legal and policy literature, and three case studies selected from across the MDB to illustrate the range of water-management issues facing Indigenous people and to describe water-management entitlements and activities undertaken by Indigenous organisations (Jackson et al. 2010).[1]

The scoping study sought to identify the most significant potential impacts arising from the introduction of the Basin Plan—those that would make the greatest difference to water availability decisions or require mitigation. The Plan could have a major impact on regional communities, particularly those groups reliant on irrigation water, through a reduction in consumptive water and by an increase in environmental water allocations. Communities, including Indigenous groups, might also be affected by the way in which the Plan is developed, the consultative approach taken and the degree to which governments respond to their concerns during plan implementation and continuing evaluation.

Time constraints limited the extent to which Indigenous participation could be built into the project design; nonetheless, the research team drew on Indigenous input and expertise at key points in the study. Indigenous representatives expressed a general desire for involvement in the preliminary study, viewing it as an opportunity to contribute to the Basin Plan, although it was realised that the scope and time frames set by the Plan schedule were unrealistic for a comprehensive study. Anxiety over future reductions in access to water and the difficulty of integrating Indigenous knowledge into water-planning processes were seen as critically important and cause for concern.[2]

The CSIRO study noted that there are many areas in which insufficient knowledge and paucity of data hamper efforts to measure specific socio-economic impacts of changes in water availability and to mitigate negative impacts or enhance positive impacts arising from the Basin Plan. Hence, a strong focus is given in the study's recommendations to improving the knowledge base, acting in a

1 The Nari Nari (Hay) and Ngemba (Brewarrina) cases describe two Aboriginal groups' efforts to access water under New South Wales' Water Sharing Plans and using the Indigenous Protected Areas model. The third case study, with the Yorta Yorta in Barmah–Millewa Forest, revealed the complexity of interrelated environmental planning frameworks and co-management agreements that affect the Yorta Yorta's engagement in water management in a cross-border Living Murray Icon Site. The full account of all three case studies can be found in the main report (Jackson et al 2010; <www.csiro.au/science/mdb/science>).

2 All concerns expressed to the project team during the study are reported in Jackson et al. (2010: section 3).

precautionary way, and advancing the scientific determination and evaluation of Indigenous water requirements. This effort should be given maximum priority during the next two to three years in order to prepare the MDBA and Indigenous communities to make more substantial advances for the second Basin Plan than might be achieved by the first, given the lack of evidence to inform policy and water allocation decisions. More intensive empirical research should integrate with other economic, social and hydrological modelling studies to provide a more rigorous assessment of impacts from changes to water availability across numerous social, cultural and economic dimensions and in all basin State jurisdictions.

This chapter first—very briefly—outlines the context and key features relevant to Indigenous water rights and interests and summarises the key water-management concerns articulated by Indigenous people. The next section outlines the potential socio-economic impacts that could occur with a reduction in water availability brought about by the introduction of sustainable diversion limits (SDLs). The CSIRO report made a number of recommendations to the MDBA. These are reframed in the final section as four potential priority actions for the next five years of basin management.

Indigenous People within the Murray–Darling Basin

In 2006 there were approximately 70 000 Indigenous people representing 3.5 per cent of the total basin population and about 15 per cent of the national Indigenous population (Taylor and Biddle 2004). There are clear differences in population distribution between the Indigenous and non-Indigenous population: Indigenous people are far less likely to reside in large centres and instead to reside in smaller settlements across the Basin. There are 35 discrete Indigenous communities. The largest share of the Basin's Indigenous population is resident within New South Wales (40 per cent), followed by Victoria (29 per cent) (Taylor and Biddle 2004). The Basin's Indigenous population has grown rapidly in recent years and this trend looks set to continue, with implications for the region's settlement pattern.

Indigenous people in the MDB have distinctly different demographic and socio-economic characteristics from the rest of the population. Compared with their non-Indigenous counterparts, Indigenous people have a much younger age profile and tend to have higher levels of disadvantage, with lower employment rates and income status (Taylor and Biddle 2004).[3] While Indigenous people are commonly employed in the government, health and service sectors, they are

3 For example, only 1.6 per cent of gross personal income accruing to adult residents in the Basin in 2001 went to Indigenous people despite the fact they represented 2.9 per cent of the adult population up to the age of sixty-five (Taylor and Biddle 2004).

less often employed in the two major industries in the Basin (agriculture and retailing) and seldom found in professional or management positions. Lack of data constrains the picture of their involvement in customary practices such as hunting, fishing and looking after sacred sites.

Across the MDB, Indigenous people use land and water resources in a variety of interrelated ways, including for subsistence use of wild resources, recreation and cultural practices. A range of social and cultural benefits is derived from these uses and interactions. Indigenous people also use water for economic purposes to support customary lifestyles and participate in water-based commercial activity. Participation in water management links many of these water uses; however, it also gives expression to other forms of Indigenous attachment to country, not least being the responsibility under Indigenous customary law to undertake management activities. It is this responsibility that forms the basis of the contemporary assertion of rights to be engaged and involved at all levels of river management and governance (Behrendt and Thompson 2004; Morgan et al. 2004).

According to a number of sources, changes to the region's river systems have eroded its capacity to meet the needs of Indigenous people (Morgan et al. 2004; Ward 2009). The literature contains many accounts of detrimental socioeconomic impacts arising from overdevelopment of the region's water resources and associated environmental impacts (for example, river regulation, seasonal changes to flows, salinity problems and land-use change) (Forward NRM and Arilla-Aboriginal Training and Development 2003; Weir 2009). Indigenous people raise the negative psycho-social effects of loss of control and inability to holistically manage their customary estates, to exercise custodial authority and to prevent further ecological degradation. The Murray Lower Darling River Indigenous Nations (MLDRIN) has also criticised the research and management community for failing to identify Indigenous values in studies of basin ecosystems and their water requirements (Morgan 2002). According to published accounts, this neglect has contributed to the marginalisation of Indigenous perspectives from allocation frameworks and the exclusion of Indigenous knowledge from formal management activities (see also Jackson 2009, 2011).

The Australian literature gives little attention to the economic implications of water scarcity for Indigenous people and little is known about Indigenous behavioural responses to changes in property-rights frameworks. The existence of Indigenous licence-holders (Altman and Arthur 2009)—particularly in New South Wales, where the demand for water for irrigation is high—suggests that the economic dimension is nonetheless an important one, perhaps more so in some basin subregions than others.

Despite the diversity in water uses and values, the literature describes commonalities in perspective and attitude and consistency in the vision for the Basin articulated by Indigenous people (Jackson et al. 2010). The vision for the River Murray is 'one of a healthy, living river system with natural flows and cycles' (Morgan et al. 2004:68). During consultations in the years following the establishment of the Murray–Darling Basin Commission (MDBC), Indigenous respondents stressed the critical importance of the Basin's river systems to social, economic and cultural life and the need for balance in meeting the needs of other stakeholders. Morgan et al. (2004) argue that Indigenous people have a shared interest with the environmental community to restore ecological functions to riverine systems.

There are numerous Indigenous groups with rights and interests in the land, waters and other resources of the Basin, including the Barkindji, Nari Nari, Kamilaroi, Ngarrindjeri and Yorta Yorta.[4] Across the region, Indigenous people have ownership rights to less than 0.2 per cent of the land area. A complex set of legal factors has restricted the number of Indigenous groups that have water rights recognised at law, the nature and extent of those legal rights, how much effective control any legal rights gives Indigenous rights-holders and the quantum of benefit derived from water-based enterprises on Indigenous land (Jackson 2011).

The history of water-resource development—particularly the priority of chronological possession of land and water rights—has made it difficult for Indigenous people to retain customary connection and attain legal rights to water bodies in recent native-title processes (Behrendt and Thompson 2004; McFarlane 2004). Subsequent legislative amendments have further narrowed the scope of native-title rights to water (Jackson and Altman 2009). Native-title statute and case law have established that where native title exists, there is a limited, non-exclusive and non-commercial right to use water without the need for a licence (Tan 1997). Jurisdictions appear to be waiting for native-title cases to be proven in the courts or resolved by negotiation before addressing water requirements for native title in plans, despite an NWI requirement that water plans take account of the possibility of native title (Jackson et al. 2009).[5]

Not surprisingly, therefore, Indigenous groups participating in the CSIRO study reported significant barriers to accessing water through allocations and thereby satisfying their water and related natural-resource management objectives. These barriers are of a legal, administrative, economic, institutional and

4 Morgan et al. (2004) estimate the presence of more than 30 Indigenous 'nations' within the Basin, each occupying core areas of land on either or both sides of each major watercourse and across catchments.

5 The CSIRO report (Jackson et al. 2010) describes the interaction of native-title law and water-resource law as well as the factors affecting Indigenous participation in basin State water planning, as do Behrendt and Thompson (2004); Jackson (2009); Rural Solutions (2008); and Tan (2009).

epistemological nature. Notwithstanding the many impediments, Indigenous organisations are undertaking water-management activities, some are using water as a key management tool in their conservation activities, and many others want to see environmental water directed to places and features of value or significance to their local communities.

The following section outlines three socioeconomic impacts that might arise from the Basin Plan, assuming it will bring about a substantial reallocation of water to the environment.

Potential Impacts from Changes to Water Availability

A number of potential impacts resulting from changes to water availability have been identified. A full social assessment of potential changes in water availability, however, has not been undertaken and a number of caveats need to be stated: 1) the assessment carried out was constrained by a tight project time line (approximately four months); 2) there are significant gaps in knowledge of Indigenous water use, participation in the water economy and relationships to water; and 3) there are uncertainties in the nature of the change under consideration; 4) assessments of need, social vulnerability and adaptive capacity to changes in environmental policy and water availability require demographic and socioeconomic information that was not readily available; and 5) the extent of the change (SDLs) and the likely spatial distribution of impacts were not known to the researchers, limiting their ability to predict geographic areas of greatest impact with certainty. With those limitations in mind, the following potential impacts were identified.

Enhanced Environmental Flows are Highly Likely to Generate Positive Impacts

Indigenous groups have articulated a vision for restoration of a healthy Murray–Darling system, indicating that improvements to the environmental condition of the Basin will be viewed positively by many Indigenous people. The benefits accruing to Indigenous people, however, could be enhanced if reforms are made to basin State water-planning processes and environmental water governance. In the absence of changes to the way that environmental water requirements are assessed and environmental water is managed, there is a risk that the full potential for increased environmental flows to substantially benefit Indigenous people will not be realised. Ways of mitigating that risk are discussed below.

Sustainable Diversion Limits (SDLs) Might Reduce Options for Economic Development

Indigenous people in the MDB use water to achieve multiple objectives, reflecting interdependencies between commercial, cultural and environmental uses, aspirations and values. Indigenous people play a small but active role in agriculture in the Basin: they own land, some have formal entitlements to water and some are employed in agriculture and related industries. The potential effect of water reductions on Indigenous settlement patterns is currently unknown (for example, intensifying the trend towards urbanisation). The disadvantaged status of Indigenous populations means that if their water entitlements are reduced sufficiently to have an impact on agricultural activity, they are likely to be particularly affected. The severity of this impact will, however, depend on the degree of Indigenous access to improved irrigation efficiency and the extent of Indigenous-owned agriculturally viable land in the areas most vulnerable to reduced water allocation.

The interdependencies between the water-based livelihood strategies employed by the Indigenous communities profiled in the CSIRO report should be taken into account. Water is both a resource that can be used as an economic asset and a feature of the cultural landscape that defines Aboriginal people's relationships to their customary lands, including their responsibilities to manage country. For example, one of the report's case studies shows that the Nari Nari people of the Murrumbidgee region use water as a management tool in the environmental and cultural restoration of their properties purchased by the Indigenous Land Corporation. Using five water licences—including a special-purpose Aboriginal Cultural Access Licence—the Nari Nari have designed a watering regime to achieve multiple non-market benefits. This group temporarily trades a high-security entitlement to underwrite the annual purchase of the water obtained under licence to meet their cultural and environmental objectives. This case raises many very interesting issues of equity and governance that are discussed fully in the report (Jackson et al. 2010).

Potential negative effects from changes in SDLs will need to be balanced against possible positive effects. For instance, reductions in water availability might increase the value of any high-security water entitlements held by Indigenous people or enterprises. In certain areas, the asset could also increase in value.

Risks to Indigenous Engagement from Basin Planning Processes, Including Setting SDLs

The strength of criticisms of previous water-planning and management processes reported in numerous sources suggests that the Basin planning *process* might have a negative effect on Indigenous people. Previously, Indigenous people have been critical of short planning time frames, the technical complexity of management measures such as SDLs, and the narrow ecological focus given to the definition of environmental assets or environmental water objectives. In all three cases reported in the CSIRO study, Indigenous groups have encountered considerable difficulty having their values recognised in environmental water management. Environmental water is not being directed to the features of greatest significance or value to Indigenous people interviewed for this study and for others (Weir 2009). Indigenous people interviewed during the case studies, as well as other consultations, report that they have had no opportunity to contribute to the identification of the Key Ecological Assets that will serve as a basis for the basin-wide Environmental Water Plan.

Priority Actions for Indigenous Water Management

This section outlines a set of priority actions that would improve Indigenous water management in the Basin over the next five years.

Define Indigenous Water Uses and Develop Methods to Specify Indigenous Water Requirements

One of the continuing challenges to water planning across Australia is to incorporate Indigenous issues more effectively, requiring—amongst a number of changes—that allocations are quantitatively defined in water plans (Jackson et al. 2009; National Water Commission 2008). Despite the existence of NWI guidelines for water plans to include consideration of Indigenous water use, it is rare to see a plan that specifically addresses Indigenous water requirements (Jackson 2009). The MDBA faced this challenge as it undertook its first step in developing the Basin Plan: to quantify the Basin's water resources and describe the uses to which they are put, including by Indigenous people.

Yet little is known about the current pattern of Indigenous water use in the Basin either for consumptive or for non-consumptive purposes. There are no systematic studies of water use within or across Indigenous groups, nor is there a comprehensive aggregated picture of the water requirements of the many distinct Indigenous communities where identification of critical human needs is reported to be an issue (Liz McNiven, Personal communication). There is scant detail on the Indigenous agricultural sector and its demand for water.[6]

6 A preliminary report on Indigenous access to commercial licences provides insight into consumptive-use rates at a coarse scale (Altman and Arthur 2009).

Non-consumptive water requirements are also not well understood because environmental-flow assessment methods have not adequately incorporated social assessments within their methodologies (Jackson 2006, 2008). For this reason, Indigenous groups are critical of the ecological research community and State water agencies for failing to address their values in environmental water management.

As described more fully in the CSIRO study, Indigenous people place great importance on the in-stream values that sustain customary life ways, and it is this interest that has motivated some Indigenous organisations to develop their own strategies for adapting concepts such as environmental flows to meet their water requirements (Weir 2009). A major issue articulated by some Indigenous groups in the MDB is the environmental impact of over-allocation of water and the perception that the ecological criteria upon which environmental flows or in-stream values are determined are too narrow (Jackson et al. 2010; Weir 2009). The intangible values that Indigenous people regard as critical to their sense of identity, cultural practices, spiritual beliefs, customary management practices and livelihoods are consistently raised as a challenge to the quantitative and competitive methods of resource allocation currently favoured by market-based reform programs (Jackson 2006).

Some traditional owners in the MDB advocate that a specific allocation distinct from environmental flows—a cultural flow—be made on the grounds of cultural differences. The Murray Lower Darling Rivers Indigenous Nations (MLDRIN) and the Northern Basin Aboriginal Nations (NBAN) define cultural flows as water entitlements that would be owned and controlled by Indigenous groups to improve their environmental, social, cultural and economic conditions. Winning widespread support for the 'cultural flow' concept will undoubtedly be contentious given the history of Indigenous resource rights in Australia (Altman 2004), and water scarcity might raise significant implementation obstacles. A further challenge lies in developing methods to accommodate the multiple values underpinning its definition and to governing its use, allocation and the distribution of any benefits. Although some Indigenous groups have expressed an interest in conducting feasibility studies of 'cultural flows' and measuring the benefits that might ensue, the concept is highly novel and if realised would represent a 'high-water mark' in Indigenous water management. Precedence might be found in efforts to meet Maori flow preferences in New Zealand (Tipa, Personal communication). Nonetheless, the literature does not provide a clear direction on appropriate methods and approaches to advance our understanding of the implications of separate allocations to meet Indigenous customary requirements. The MDBA could take the lead in fostering such efforts with a view to providing Indigenous water-allocation mechanisms in the second iteration of the Basin Plan.

Consideration should be given to setting benchmarks for accreditation of basin State water-resource plans that reflect Indigenous water needs. Systematic assessments of Indigenous water-use requirements might be achieved through the generation of Indigenous Water Management Plans for each catchment to achieve the objectives of the NWI. Such plans could include

- five-year targets for improved access to water
- assessments of the impact of SDLs on native-title interests
- identification of features, places and species of major significance for Indigenous people and their water requirements
- mechanisms to assist Indigenous representatives to receive, evaluate and prioritise applications for water or undertake water-related management
- monitoring and evaluation of outcomes of watering
- linkages between Indigenous representative structures, environmental groups, environmental water advisory groups, catchment management authorities (CMAs), and so on
- targets to improve Indigenous representation in water-management processes.

An Indigenous Water Fund might be established to resource these and other initiatives recommended in the full report.

Increase Benefits to Indigenous People through Improved Environmental Water Management

Traditional-owner groups express a strong desire to exercise authority, responsibility and control in the determination of allocations to meet their requirements. The benefits Indigenous people derive from increased environmental allocations could be enhanced if reforms are made to water-planning processes and environmental water governance so that they are more inclusive of Indigenous uses, values and priorities (for example, the MDBA's Environmental Water Plan). More inclusive environmental water management would allow differences to emerge in the priority given to the selection of environmental outcomes (for example, aquatic species of totemic significance, wetlands of value for customary use). Environmental water programs (for example, buybacks) need to be more accessible to Indigenous people and the opportunities for co-management with Indigenous natural-resource management (NRM) groups investigated, particularly in relation to Indigenous conservation activities (for example, joint management, Indigenous Protected Areas). To do so might require amendments to the *Water Act* given the narrowness of the definition of environmental outcomes to be achieved through the delivery of environmental water (that is, ecosystem function, biodiversity, water quality).

In exploring the feasibility of Indigenous-specific flows, the governance issues associated with a range of entitlement-holding and management models will need further investigation. Assistance will be needed to develop governance arrangements to support Indigenous management of water allocations. Investing in Indigenous capacity to contribute knowledge and manage environmental water offers one means of enhancing the potential benefits from greater access to water under the Basin Plan. This capacity could also spill over into other areas of natural-resource management (for example, national parks and Indigenous Protected Areas), and bring broader social and economic benefits (Hunt et al. 2009). Capacity should be built at, at least, two scales: the catchment level, where Indigenous groups need assistance to articulate their priorities and bring their knowledge to water assessment, administration and management processes; and at the regional scale, where northern and southern Indigenous alliances[7] need assistance to analyse data, reflect on trends in water access and participation in management, contribute to research on ways of quantifying Indigenous water requirements and to offer policy advice on overcoming barriers to Indigenous access to water, as well as Basin Plan implementation and adaptation. Close links between the two scales will improve the relevance of Indigenous policy contributions to local community needs and provide legitimacy for region-wide Indigenous contributions to Commonwealth water-management instruments such as the Environmental Water Plan.

Improve Indigenous Livelihood Outcomes from SDLs

Although Indigenous people are statistically under-represented in the allocation of water for commercial purposes, our understanding of Indigenous rates of participation in the water economy and degrees of water dependence is poor. Altman and Arthur (2009:9) conclude that in the absence of 'good knowledge of present or future water allocation and use, it is difficult to see how Indigenous users can be properly incorporated into planning or allocation processes'. Indigenous commercial water users need to be identified in order to determine whether SDLs will adversely affect the commercial viability of their enterprises. In such instances, Indigenous people should be made aware of any structural-adjustment programs developed by or in response to the Basin Plan. There is also considerable potential for structural change to present new opportunities for Indigenous people in emerging cultural and natural resource-based industries, such as payment for environmental services, stewardship arrangements, small-scale bush-foods businesses, and tourism.

7 The MLDRIN and the NBAN.

Improve the Knowledge Base for the Next Plan

A strong evidence base is needed to provide for a continual process of adaptive learning; information is needed to evaluate potential impacts prior to Basin Plan implementation and to develop adaptation responses, to monitor Plan implementation, and, in turn, to integrate revised needs into the next plan. The lack of quantitative data on Indigenous water use and sensitivity to change constrains water planning and, in particular, the ability to measure the impacts of SDLs. Effective mitigation strategies require monitoring to evaluate the impacts of the Basin Plan on Indigenous enterprises and communities during the life of the Plan.

Baseline socioeconomic and demographic data should be collected and a monitoring program designed to track changes in SDLs and impacts on Indigenous access to water and economic participation. To effectively and equitably manage water resources in the MDB there needs to be further research across a range of areas, including

1. baseline information—for example, on Indigenous commercial water use, socioeconomic regional profiles, and so on, which will also facilitate monitoring
2. understanding of the barriers to Indigenous participation in the water market
3. understanding of the 'cultural flows' concept articulated by Indigenous groups and how it aligns with environmental flows including under differing management models; a series of in-depth case studies that identifies Indigenous water requirements in all basin jurisdictions would advance this knowledge gap
4. techniques to quantify Indigenous environmental and cultural water use and evaluate benefits
5. water policy instruments to better accommodate Indigenous people's cultural, environmental and economic needs.

Conclusion

The Basin Plan and government responses to implementation and mitigation present a significant opportunity to address the longstanding neglect of Indigenous interests in water management and planning, as well as to markedly improve the extent to which Indigenous people benefit from water reforms, particularly from environmental water management. As water policy makers and planners seek to address the requirements of the *Water Act* and the NWI, they confront substantial gaps in our knowledge of Indigenous requirements for water and governance models to improve Indigenous water management. Research and dialogue are needed to further the conceptual and empirical understanding of Indigenous water requirements, values and governance, and, in doing so, to fully involve Indigenous people in subsequent policy development and decision making. In the short term, action is required from basin States to improve Indigenous access to water under water-resource plans and to adapt environmental water-management systems to better address Indigenous world views and relationships to country if the imperatives of the NWI are to be met. The effect of changes in water availability for Indigenous commercial uses—and any knock-on effects on non-consumptive uses—still requires better understanding and should be carefully monitored prior to and during Basin Plan implementation.

References

Altman, J. 2004, 'Economic development and Indigenous Australia: contestation over property, institutions and ideology', *Australian Journal of Agricultural and Resource Economics*, vol. 48, pp. 513–34.

Altman, J. and Arthur, B. 2009, *Water Licences and Allocations to Indigenous People for Commercial Purposes: An Australia-wide scoping exercise*, ANU Centre for Aboriginal Economic Policy Research, Canberra.

Behrendt, J. and Thompson, P. 2004, 'The recognition and protection of Aboriginal interests in New South Wales rivers', *Journal of Indigenous Policy*, vol. 3, pp. 37–140.

Forward NRM and Arrilla—Aboriginal Training and Development 2003, *Scoping Study for Indigenous Involvement in Natural Resource Management Decision-Making and the Integration of Indigenous Cultural Heritage Considerations into Relevant Murray–Darling Basin Commission Programs*, Murray–Darling Basin Commission, Canberra.

Hunt, J., Altman, J. and May, K. 2009, *Social benefits of Aboriginal engagement in natural resource management*, CAEPR Working Paper No. 60, ANU Centre for Aboriginal Economic Policy Research, Canberra.

Jackson, S. 2006, 'Compartmentalising culture: the articulation and consideration of Indigenous values in water resource management', *Australian Geographer*, vol. 37, no. 1, pp. 19–31.

Jackson, S. 2008, 'Recognition of Indigenous interests in Australian water resource management, with particular reference to environmental flow assessment', *Geography Compass (Environment & Society)*, vol. 2, no. 3, pp. 874–98.

Jackson, S. 2009, National Indigenous water planning forum: background paper on Indigenous participation in water planning and access to water, Unpublished report prepared for the National Water Commission, CSIRO, Darwin.

Jackson, S., P. Tan, and Altman J. 2009. *Indigenous Fresh Water Planning Forum: Proceedings, Outcomes and Recommendations.* A Report to the National Water Commission, Canberra, http://www.nwc.gov.au/resources/documents/Outcomes_of_NWC_Indigenous_Fresh_Water_Planning_Forum_outcomes2.pdf>

Jackson, S. 2011, 'Aboriginal access to water in Australia: opportunities and constraints', in Q. Grafton and K. Hussey (eds), *Water Resources, Planning and Management*, Cambridge University Press, UK, pp. 601–28.

Jackson, S. and Altman, J. 2009, 'Indigenous rights and water policy: perspectives from tropical north Australia', *Australian Indigenous Law Review*, vol. 13, no. 1, pp. 27–48.

Jackson, S., Moggridge, B. and Robinson, C. 2010, *The Effects of Change in Water Availability on Indigenous People of the Murray Darling Basin: A scoping study*, Murray–Darling Basin Authority, Canberra, <www.mdba.gov.au>

McFarlane, B. 2004, The National Water Initiative and acknowledging Indigenous interests in planning, Paper presented to the National Water Conference, Sydney, 29 November 2004.

Morgan, M. 2002, *MLDRIN Response to the Draft Water Sharing Plan for the NSW Murray Lower Darling Regulated River Water Source*, Prepared by the Murray Lower Darling Community Reference Committee.

Morgan, M., Strelein, L. and Weir, J. 2004, *Indigenous Rights to Water in the Murray Darling Basin*, Australian Institute of Aboriginal and Torres Strait Islander Studies, Canberra.

National Water Commission 2008, *Update of Progress in Water Reform: Input into the water sub group stocktake report*, National Water Commission, Canberra.

Rural Solutions 2008, *Aboriginal Access to Water Across Australia,* Department of Water, Land and Biodiversity Conservation, Government of South Australia, Adelaide.

Tan, P. 1997, 'Native title and freshwater resources', in B. Horrigan and S. Young (eds), *Commercial Implications of Native Title*, The Federation Press, Sydney, pp. 157–90.

Tan, P. 2009, *National Indigenous Water Planning Forum: A review of the legal basis for Indigenous access to water*, An unpublished report prepared for the National Water Commission, Canberra.

Taylor, J. and Biddle, N. 2004, *Indigenous people in the Murray–Darling Basin: a statistical profile*, Discussion Paper 264, ANU Centre for Aboriginal Economic Policy Research, Canberra.

United Nations 2008, *Declaration on the Rights of Indigenous Peoples*, United Nations, New York.

Ward, N. 2009, 'Good methodology travels: Australian case study', in T. Tobias (ed.), *Living Proof: The essential data-collection guide for Indigenous use-and-occupancy map surveys*, Ecotrust Canada and Union of British Columbia Indian Chiefs, Vancouver, pp. 339–64.

Weir, J. 2009, *Murray River Country: An ecological dialogue with traditional owners*, Aboriginal Studies Press, Canberra.

10. Water Planning and Dispossession

Jessica K Weir[1]

The policy failure of successive governments who over-allocated river water for agricultural consumption in the Murray–Darling Basin is now being addressed through the purchasing of 'environmental water' to boost river health. Already, 920 gigalitres of water has been purchased from farmers selling their consumptive-water entitlements to the Federal Government's Commonwealth Environmental Water Holder (CEWH). The Murray–Darling Basin Plan extends this work, proposing that an additional 2080 GL be purchased, to reach the target of 3000 GL. This planning is in line with the *Water Act 2007* (Cwlth), which directs that the Murray–Darling Basin Authority (MDBA) develop a water-management regime that returns water extraction to environmentally sustainable levels.

The intention of both the *Water Act* and the Basin Plan is to respond to river degradation by bringing 'balance' to economic water use and the health of the river country, although 3000 GL will restore river health only from poor to moderate (see Wahlquist, this volume). Such a deliberate response recognises that the prioritisation of water for agricultural production has neglected and impoverished our other river relationships. In this chapter, I explore the experiences of the traditional owners along the Murray River, particularly the elders who have witnessed dramatic changes to the river over their lifetimes. Continuing public and informal access to the inland rivers has provided important opportunities for traditional owners to enjoy those connections with their country that have persisted during the experience of colonisation. With the ecological degradation of the rivers, and the animals and plants they sustained, these enduring connections are being ruptured. Arguably, the traditional owners are experiencing this as a contemporary dispossession from their country.

---

1 This chapter is a revised and updated version of Weir, J. K. 2007, 'The traditional owner experience along the Murray River', in E. Potter, S. Mackenzie, A. Mackinnon and J. Mackay (eds), *Fresh Water: New perspectives on water in Australia*, Melbourne University Press, Carlton, Vic., pp. 44–58. The research for this chapter is supported by a research agreement arranged between myself and the Murray Lower Darling Rivers Indigenous Nations (MLDRIN). I particularly wish to acknowledge elders Richard Hunter and Agnes Rigney who are quoted in this article, and have since passed away. I thank MLDRIN for their support; however, the views expressed in this chapter are my own and do not represent the views of MLDRIN. My research was conducted as part of a PhD, funded by a Land & Water Australia scholarship, and undertaken at the Centre for Resource and Environmental Studies (now the Fenner School of Environment and Society) at The Australian National University. I thank Deborah Rose for comments made on an earlier draft of this chapter.

Changing Water Flows

River ecosystems in the Murray–Darling Basin are adapted to highly variable cycles of floods and droughts, and this variability has created a diversity of habitats in the river country (Young and Hillman 2001:92; Young et al. 2001:26–9). In the late nineteenth century, this river water was re-conceptualised as a resource for irrigated agricultural production. Large engineering projects began in the area during the 1920s. Since then, the rivers of the Murray–Darling Basin have been transformed by an extensive network of dams, weirs, locks, canals and pipes built to provide water to rural communities (Powell 1989). The dams now regulate the previously variable water flows by holding floodwaters back in large storages. Today, large water storages in the Basin have the capacity to store 35 000 GL (1 GL equals the amount of water in 1000 Olympic-sized swimming pools). Annually, more than 10 000 GL are diverted from the rivers, and 96 per cent of this diverted water is used by farmers for irrigation (Crabb 1997:26–9). The large storages also provide water for rural towns and communities to survive in extended periods of drought.

With the reduced flow of water in the rivers, the networks of floodplains, wetlands, swamps and soaks have dried out. Or, where water is permanently stored, forests and country have been drowned. Combined with land-use changes around agriculture, there has been a dramatic decline in the quality of water. Sediment loads in the river are now 41 times their natural rate, outbreaks of blue-green algae poison the water, and increased salinity of the soil and water have all reduced the capacity of the rivers to be a productive source of life (Gehrke et al. 2003:3; Proust 2003). The consequences of all this for the birds, fish and other animals have been profound: their numbers have declined dramatically and certain species have experienced local extinction. Birds that used to nest when the floodwater came are not breeding successfully (Leslie 2001). Native fish populations are only 10 per cent of what they used to be (MDBMC 2003:6).

The change from the variable inland-river rhythm of floods and droughts is not just an overall decline in the size, length and frequency of floods; the timing of the flow of water has also changed. Along the Murray in the riverine plains, the river used to flood in autumn and winter. Now, water is released into the Murray in the spring and summer to flow downstream to irrigators for their agricultural crops—effectively reversing the seasonal pattern of the river's flow (Young and Hillman 2001:108). This has led to the unseasonal flooding of the Barmah–Millewa forest—a large wetland on the Murray River—destroying grasslands and trees (Chong and Ladson 2003:162–3, 165).

By storing and diverting the river water, the dam builders have changed how the water flows across time and place, redistributing water over vast areas, entire communities and ecosystems (see also Mitchell 2002:21). Water management is not a discrete extraction of a resource from one location; rather, because of the capacity of water to move through landscapes and in seasonal rhythms, water management has consequences across basin ecosystems and through time. Because water is critical for all living things, there is now national concern about the scale of these consequences on peoples' lives, agricultural production, river health and species diversity. This concern includes losses in agricultural productivity from the increasing salt content of river water and agricultural land, in part also a result of land clearance for agriculture (Connell 2007, pp. 17-18). The movement and flux of water have always been part of inland-river life, but modern engineering projects have brought accelerated change in the form of large water extractions, challenging the capacity of ecosystems to adapt and survive. The degradation of the river country, and the policy failure of over-allocating river water, is now challenging the rural communities and livelihoods established next to the rivers.

The most dramatic changes to the rivers have occurred in the past 50 years, when the largest water storages were completed (Crabb 1997:29). This is within the lifetimes of the present generation of Aboriginal elders whose traditional country is in the Murray–Darling Basin. These changes are noted in the elders' dismay at the scale and speed of the ecological decline; as Mutti Mutti elder Mary Pappin (2004) has said: 'Such a short space of time! I can't take my grandchildren down to my favourite fishing spots and do what I used to do.'

Indigenous Peoples' Experiences in the Murray–Darling Basin

The fertility of the inland rivers has always drawn people to the region, in the past and today. The summer river flow of the Darling and the less seasonally variable flow of the Murray make locations adjacent to their banks attractive places to live. As hunting and gathering societies, traditional owners took advantage of seasonal fluctuations in resources (for example, the bogong moth feasts in the Snowy Mountains) (Flood 1976). In the arid and semi-arid lands, when the floods and rains came, the traditional owners were able to move camp from the rivers and out onto the plains (Allen 1980). In contrast, the Murray River Valley has also long been home to traditional owners living in river communities known for their high density and sedentary lifestyle (Webb 1984:168–70).

In 1788 the British colony formally arrived in Sydney, and, in the early nineteenth century, the Darling, Murray, Lachlan, Macquarie and Murrumbidgee rivers provided the new settlers with the first routes of exploration inland to what is now known as the Murray–Darling Basin. For the colonisers, tributary junctions, fords and ferry crossings became settlement nodes, and the flat, fertile floodplains with easy access to the water provided the best conditions for agriculture (Smith 2001:213). Marcia Langton (2002:46) has described how colonisation and conflict in Australia focused on water, as the new settlers spread 'from water source to water source'. Aboriginal people demonstrated their objection to the land and water seizures by waging wars of resistance (for example, Gammage 1986; Read 1988).

Prior to colonisation, the Murray–Darling Basin was Aboriginal land, with rights and responsibilities for land and natural resources held between individuals and communal groups (Berndt et al. 1993:135–49). Since colonisation the Basin continues to be identified by the traditional owners as their country; however, they now have to argue for their rights and responsibilities to be recognised within the introduced European systems of law and governance. Of the small parcels of land that were allocated as reserves or missions for the traditional owners in south-eastern Australia, the majority was revoked in the twentieth century (for example, Goodall 1996). As their country was and is highly prized for agricultural production by the new settlers, the traditional owners have experienced the pressure of an early and concentrated settlement. This long and intensive experience of colonisation has also meant that much knowledge has been lost between the generations of Indigenous people during this experience of deprivation, stress, adaptation and survival (Tonkinson 1993:xxii).

The settlers brought new technologies and knowledge with them, and, as the settlers and their institutions became more established, Aboriginal people adapted and found meaning in the life and work of what is now rural Australia. The new rural industries that were instrumental in the appropriation of traditional country were also opportunities for independence for Aboriginal people within the colonial system. The desire to become farmers was a stated aspiration in Aboriginal peoples' petitions to government for land in the nineteenth and twentieth centuries, including the separate applications made by the Ngarrindjeri, Yorta Yorta, and Taungerung and Woiwurrung peoples (Barwick 1972:21, 49; Bell 1998:107). Wamba Wamba men became famous for being 'big-gun' shearers (Hercus 1992:15). With the spread of irrigated fields, seasonal work for large groups provided the opportunity for Aboriginal people to re-establish kinship networks (Taylor 1988:227–8).

Today, the Murray–Darling Basin is home to more than 70 000 Indigenous people, which is 3.4 per cent of the total Basin population of two million people (Taylor and Biddle 2004). The disruption to traditional society since colonisation has created new identities among Aboriginal people, including 'historical people' and 'stolen generations'. Many Indigenous people continue, however, to primarily identify themselves and one another as traditional owners (Brennan et al. 2005:11). The traditional-owner identity is part of a long-held cultural, political and intellectual framework centred on the relationships people have with country, including the plants, animals and ecosystems of that country. There are at least 30 traditional-owner groups who have country within the Basin (MDBC n.d.:3).

These traditional identities have been transformed by the disruption and influence of colonialism, and today an important part of being a contemporary traditional owner involves building on and reviving cultural practices from earlier generations, such as teaching the local traditional languages and the performance of welcome ceremonies (Sutton 1995:47). Significantly, the traditional owners also continue to collect bush foods and medicines as a means for independence and as part of connecting with and affirming their identity (Behrendt and Thompson 2003:8; Clarke 2003). More generally, the traditional owners actively seek to increase their involvement in natural-resource and land-management policies and programs, such as pursuing joint-management arrangements over national park lands and establishing Indigenous advisory committees that engage with catchment management authorities.

Ecological Loss and the Extinction of Experience

Ethno-ecologist M. Kat Anderson has described the adverse impacts of the loss of ecosystems on the wellbeing of indigenous peoples in North America, revealing how interconnected ecological health is with the relationships indigenous people hold with their country (Anderson 1997). Indigenous peoples hold long-term associations with particular places and the species that live there, and, as a result, their identity, their ability to exercise their religion, their economic resources and their health are all intimately connected with the health of ecosystems. Anderson further argues that the perpetuation of indigenous knowledge about ecology requires access to that ecology. Due to these connections, the loss of ecosystems is also the loss of these dimensions of wellbeing for indigenous people. This argument about the connection between ecological health and indigenous peoples' wellbeing does not translate into a simple match between indigenous issues and environmental issues. The concerns indigenous people bring to environmental issues are much broader, and do not fit within the compartmentalism of environmental issues in Western knowledge frameworks (see further Braun 2002; Ingold 2000; Latour 2001; Strathern 1980; Weir 2009).

This distinction has important implications for the degree of correspondence between environmental water allocations and water flows deemed necessary by indigenous people to maintain country (Weir 2009).

In the Murray–Darling Basin, the recent construction of large water-distribution projects is having an impact on the way of life of the traditional owners of the river country. My experience is from fieldwork with an alliance of traditional owners from along the Murray River in the southern part of the Murray–Darling Basin, which is also where irrigation activity is focused.[2] Here, traditional owners are experiencing the degradation of river health within an intensively colonised part of Australia. The meandering inland rivers curve, loop and twist through landscapes that are now 'ordered' by agricultural production. Traditional country is fragmented, marked out by freehold and pastoral leaseholds, squares of irrigated fields, long lines of the roads and railways, and the life of the rural towns and communities. Here, the culture and lifestyle of traditional owners are part of modern intercultural Australian society (Merlan 2005).

As contemporary Indigenous peoples, the traditional owners I have worked with spoke to me about how they continue to collect bush foods and go fishing in the same places as their ancestors did. Such activities are not simply about recreation; they are about reaffirming relationships with country, connecting to intergenerational responsibilities, and practising traditional knowledge and skills. As Jason Behrendt and Peter Thompson have described, fishing is an opportunity for Aboriginal people to meditate on their relationship with country, without the need for colonial skills such as literacy, and in an activity in which Aboriginal knowledge is valued. Behrendt and Thompson (2003:8) write that going fishing is an important rest from colonisation: 'For a colonised people for whom political and geographical decolonisation are not realistic lifetime aspirations, opportunities to decolonise the mind are vital and life affirming.'

Traditional owners have spoken to me about how the enduring flow of the rivers connects them to their time scales and creation stories. Mary Pappin can imagine the Dreamtime when she is sitting by the river. The water, river, gum trees and wildlife make her feel that she is 'sitting right in it'. For Mary, this is a feeling of belonging: 'When I'm down by the river fishing or enjoying the bush, especially when I am resourcing it for bush foods, it just takes me straight back to the Dreamtime to continue on what my ancestors did, or nomads did, before the Europeans came' (Interview with author, 22 July 2004).

---

2 Here, the traditional owners have created an umbrella organisation called the Murray Lower Darling Rivers Indigenous Nations, which is formed by the Wiradjuri, Yorta Yorta, Taungurung, Wamba Wamba, Barapa Barapa, Mutti Mutti, Wadi Wadi, Latji Latji, Weragaia and Ngarrindjeri peoples (Weir 2009; Weir and Ross 2007). This alliance is a coordinated response to the devastation of the river country, so the traditional owners can better direct their engagements with government water management, law and policy in the Murray–Darling Basin.

In terms of ecological loss, I focus on the impact of water regulation on fishing and the gathering of bush foods, medicines and materials, because the traditional owners repeatedly expressed this loss to me during my fieldwork. This focus reflects where the impact of water regulation is being particularly felt by the traditional owners, and how the activity and expertise of going fishing, or collecting certain bush tucker, have survived the experience of colonisation. Elsewhere, I have written at greater length about how the impacts of water regulation are being felt more comprehensively by the traditional owners, including issues of identity, economy, health and spiritual beliefs (Weir 2009).

In their expressions of ecological loss, the elders first reflected on times past when their lives were connected to the rivers through the essential act of drinking the water and eating the plants and animals that also lived by the river. In South Australia, Ngarrindjeri elders Matt Rigney, Richard Hunter and Agnes Rigney spoke to me about drinking water straight from the Murray River and its lakes when they were young. In those days, said Richard and Agnes, the water was so clear you could see the bottom of the river (Interviews with author, 23 July 2004 and 21 July 2004, respectively). Agnes Rigney described to me how she grew up in a 'semi-traditional' lifestyle next to the Murray River at the Swan Reach mission. Here, the Murray supplemented mission food with fresh water, fish, yabbies and waterbirds. Upstream in the Barmah Forest, Yorta Yorta elder Henry Atkinson told me how his father and mother's father were able to live off and make a living from fishing for native fish, mussels, Murray crayfish and turtles (Interview with author, 7 August 2004).

In comparison with such experiences, Richard Hunter spoke about all the fish that have 'disappeared in my short time' or are now found only in small numbers, including the iconic Murray cod. Matt Rigney spoke about the loss of native grasses, rushes and reeds that used to filter the waterways and provide clean water. Henry Atkinson spoke about the loss of the floods, and the drying out of the red-gum forests. In central New South Wales, Wiradjuri elder Tony Peachy spoke to me about the scarcity of native fish in the Macquarie River, and how he had not caught a Murray cod in five years (Interview with author, 29 October 2004). Further south along the Murrumbidgee, Mary Pappin worried about the loss of native fish in relation to the proliferation of the European carp, and the loss of medicine and bush-tucker plants that used to grow on the former floodplains, or in the little creeks that previously flooded intermittently.

Agnes Rigney linked the loss of native animal and plant life in the river country today to the loss of what she called 'cultural living'. My understanding, from listening to Agnes, is that cultural living reaffirms continuities with country through the practising and passing on of cultural knowledge and experience. Agnes spoke of this in relation to how different it was when she was a child:

> I remember as a kid growing up in Loxton how clear the river was, the water was, and my father actually making us spears from bamboo and we used to walk down to the river and we used to spear the fish. And it is just sad what's happened to it now. That was a part of cultural living, connected to the river, that we can't really practise anymore. (Interview with author, 21 July 2004)

Today, muddy waters and the absence of native fish preclude this practice. The loss and decline of river ecosystems has reduced the range of current and future possible activities for the traditional owners, including the practising of skills and the revival of lapsed traditions. For example, a part of cultural living celebrated among traditional owners of the inland rivers is the skill of basket weaving from the grasses and reeds that grow along the rivers or in the wetlands. Down in the Coorong behind the Murray mouth, the successful revival of basket weaving by Ngarrindjeri women is now threatened by the salt levels in the water, which means that the freshwater reeds are harder to find locally (Corowa 2006). Upstream in New South Wales near Deniliquin, Mutti Mutti elder Jeanette Crew talked to me about how she and local Wamba Wamba women—her friends and relatives—want to revive the art of making woven grass baskets. But the swamps where the best grasses for the baskets used to grow in the Werai Forest have been left dry by the over-consumption of river water, making it very hard to find grasses for this activity.

The impact of all these ecological losses on cultural living—which is crucial for a culture centred on intergenerational relationships with country—is alienating the next generation from experiencing their country as healthy and alive, and the opportunity of learning traditional knowledge and skills. Yorta Yorta woman Monica Morgan was a child in the 1960s. Her elders taught her the ecological cues that would tell her when swan eggs would be available up in the Barmah–Millewa lakes. As Monica related: 'There was life...You'd sit there and they'd say, "Oh well the duckweed is coming down, that means the swan eggs are ready to go and be collected up in the lakes." So there were seasons happening' (Interview with author, 1 July 2004).

This is, however, a memory of the past. Today, the Murray floods out of season, and both the swans and the duckweed are rare. Monica is worried about the implications of this for Yorta Yorta children; she cannot teach them as she was taught, and the children are growing up without this knowledge and experience: 'So if I am seeing in just a short time—20, 30 years—the disappearance of things that I took for granted, and were a real reflection of nature, and also governed my life cycles, then what is going to be left for our children?' (Interview with author, 1 July 2004).

Robert Michael Pyle (1992) argues that the local loss of species is 'the extinction of experience'. When a species becomes extinct, all people lose direct, personal contact with that species. As time passes, the next generation does not have the opportunity to know the species, to interact with and identify with it. For Indigenous people, the local extinction of a species ruptures the transmission of intergenerational knowledge and skills because their knowledge is learnt and taught directly through experiences in country. Moreover, the ecological decline along the Murray is more than the local extinction of particular species. The over-extraction of water is affecting vast inland river ecosystems. By altering the distribution of water across time and place, the dam builders have altered the relationships between numerous traditional-owner groups and their country over thousands of kilometres. The twentieth-century changes to the water regime have broken connections passed down for a thousand generations, and instead offer the future diminishing and dying life.

The severe degradation of river ecosystems is a concern for all people living along the rivers. This chapter emphasises how the ecological loss and decline has particular consequences for the traditional owners of the river country. When the elders worry about not being able to go fishing or collecting bush tucker, they are concerned about the opportunity to continue being a traditional owner in a densely settled part of south-eastern Australia. Here, where the traditional owners do not have a large land base exclusively of their own, public access to the rivers, or mission lands, or informal agreements with private landholders have allowed for the continuance of cultural activities through the colonial experience. With the decline of the river ecosystems, these breaks in colonisation have been encapsulated and closed down. The alienation of the next generation from experiencing the life of the river country makes the elders feel that their cultural practice is threatened, and, I argue, compounds the experience of dispossession in south-eastern Australia.

The traditional owners are now part of intercultural rural society, and have benefited in part from the lifestyles and work opportunities provided by river regulation. Decisions about the scale and extent of water extraction have, however, been made without their consultation, or taking account of their interests, needs or rights—or the wider community more generally. Indeed, the foundation of water management, law and policy is premised on the exclusion of the pre-existing rights and interests of traditional owners. The language I have chosen to use—of dispossession—situates the ecological degradation of the river country within the colonial experience for the traditional owners. As Mary Pappin concludes: 'Somebody somewhere has made a decision on my behalf in taking that traditional resource away from me. From my way of looking at it, the Government is responsible for taking away the resource' (Pappin 2004). There is a growing body of research now addressing how to include the interests of Indigenous people in water planning, including the need to build Indigenous peoples and government capacity, and to address the failure to recognise Indigenous peoples' property rights in water (for example Jackson 2009, Jackson and Robinson 2009, Altman 2004).

Government decisions about water management—both the over-allocation of consumptive-water entitlements and the new policy of buying environmental water—have profound impacts on people because water is connected to so many important values. The impact of river degradation on Indigenous peoples' cultural life critically reveals this unique heritage that we are all losing when we lose our river ecologies.

The Murray–Darling Basin Plan's intention to return water to the river country, and restore river health, has the potential to address the traditional-owner experience of dispossession today and into the future. Traditional owners need to be part of that decision-making process, as the return of water to country is limited by vested interests in water consumption, and thus where and when small water amounts will be returned is determined by the priorities set by governments (see Jackson, this volume). Whilst the Basin Plan is a step forward to better recognise the primacy of river health, its modest goals to restore river health from poor to moderate reflect the continuing dominance of human consumptive needs in our water planning.

References

Allen, H. 1980, 'Aborigines of the western plains of New South Wales', in C. Haigh and W. Goldstein (eds), *The Aborigines of New South Wales*, New South Wales National Parks and Wildlife Service, Sydney, pp. 33–43.

Altman, J. 2004, 'Indigenous interests and water property rights', *Dialogue*, no. 23, pp. 29–33.

Anderson, M. K. 1997, 'California's endangered peoples and endangered ecosystems', *American Indian Culture and Research Journal*, vol. 21, p. 731.

Barwick, D. E. 1972, 'Coranderrk and Cumeroogunga: pioneers and policy', in T. S. Epstein and D. H. Penny (eds), *Opportunity and Response: Case studies in economic development*, Hurst and Co., London.

Behrendt, J. and Thompson, P. 2003, *The recognition and protection of Aboriginal interests in NSW rivers*, Occasional Paper 1008, Healthy Rivers Commission of New South Wales, Sydney.

Bell, D. 1998, *Ngarrindjeri Wurruwarrin: A world that is, was, and will be*, Spinifex Press, Melbourne.

Berndt, R. M., Berndt, D. H. and Stanton, J. E. 1993, *A World That Was: The Yaraldi of the Murray River and the lakes, South Australia*, Miegunyah Press, Melbourne.

Braun, B. 2002, *The Intemperate Rainforest: Nature, culture, and power on Canada's west coast*, University of Minnesota Press, Minneapolis and London.

Brennan, S., Behrendt, L., Strelein, L. and Williams, G. 2005, *Treaty*, The Federation Press, Leichhardt, NSW.

Chong, J. and Ladson, A. R. 2003, 'Analysis and management of unseasonal flooding in the Barmah–Millewa Forest, Australia', *River Research and Applications*, vol. 19, no. 2, pp. 161–80.

Clarke, P. A. 2003, 'Twentieth-century Aboriginal harvesting practices in the rural landscape of the Lower Murray, South Australia', *Records of the South Australian Museum*, vol. 36, pp. 83–107.

Connell, D. 2007, *Water politics in the Murray–Darling Basin*, The Federation Press, Sydney.

Corowa, M. 2006, 'Murray, life + death', *Message Stick*, ABC TV, Sydney.

Crabb, P. 1997, *Murray–Darling Basin Resources*, Murray–Darling Basin Commission, Canberra.

Flood, J. 1976, 'Man and ecology in the highlands of southeastern Australia: a case study', in N. Peterson (ed.), *Tribes and Boundaries in Australia*, Social Anthropology Series No. 10, Australian Institute of Aboriginal Studies, Canberra.

Gammage, W. 1986, *Narrandera Shire*, Narrandera Shire Council, NSW.

Gehrke, P., Gawne, B. and Cullen, P. 2003, *What is the Status of River Health in the Murray–Darling Basin?*, CSIRO Land and Water.

Goodall, H. 1996, *Invasion to Embassy: Land in Aboriginal politics in New South Wales, 1770–1972*, Allen & Unwin in association with Black Books, Sydney.

Hercus, L. A. 1992, *Wemba Wemba Dictionary*, Luise A. Hercus, Canberra.

Ingold, T. 2000, *The Perception of the Environment: Essays in livelihood, dwelling and skill*, Routledge, London.

Jackson, S. 2009, National Indigenous Water Planning Forum: Background paper on Indigenous participation in water planning and access to water, A report by CSIRO Sustainable Ecosystems, to the National Water Commission.

Jackson, S and Robinson, C. 2009, 'Indigenous participation in water planning and management', in Northern Australia Land and Water Science Review, CSIRO.

Langton, M. 2002, 'Freshwater', in *Background Briefing Papers: Indigenous rights to water*, Lingiari Foundation, Broome, WA.

Latour, B. 2001, *We Have Never Been Modern*, Harvard University Press, Cambridge, Mass.

Leslie, D. J. 2001, 'Effects of river management on colonially-nesting waterbirds in the Barmah–Millewa Forest, South-eastern Australia', *Regulated Rivers: Research & Management*, vol. 17, pp. 21–36.

Merlan, F. 2005, 'Explorations towards intercultural accounts of socio-cultural reproduction and change', *Oceania*, vol. 75, pp. 167–82.

Mitchell, T. 2002, *Rule of Experts: Egypt, techno-politics, modernity*, University of California Press, Berkeley and Los Angeles.

Murray–Darling Basin Commission (MDBC) n.d., *Indigenous Basin Wide Gathering: Conference papers, presentations and outcomes*, Murray–Darling Basin Commission, Canberra.

Murray–Darling Basin Ministerial Council (MDBMC) 2003, *Native Fish Strategy for the Murray–Darling Basin 2003–2013*, Murray–Darling Basin Commission, Canberra.

Pappin, M. 2004, Up the river forum, Message Sticks Festival, Sydney Opera House, May 2004.

Powell, J. M. 1989, *Watering the Garden State: Water, land and community in Victoria, 1834–1988*, Allen & Unwin, Sydney.

Proust, K. 2003, 'Ignoring the signals: irrigation salinity in New South Wales Australia', *Irrigation and Drainage*, vol. 52, pp. 39–49.

Pyle, R. M. 1992, 'Intimate relations and the extinction of experience', *Left Bank*, vol. 2, pp. 61–9.

Read, P. 1988, *A Hundred Years War: The Wiradjuri people and the state*, The Australian National University Press, Canberra.

Smith, D. I. 2001, *Water in Australia: Resources and management*, Oxford University Press, Melbourne.

Strathern, M. 1980, 'No nature, no culture: the Hagan case', in M. C. P. and M. Strathern (eds), *Nature, Culture and Gender*, Cambridge University Press, UK.

Sutton, P. 1995, *Country: Aboriginal boundaries and land ownership in Australia*, Aboriginal History, Canberra.

Taylor, P. (ed.) 1988, *After 200 Years: Photographic essays of Aboriginal and Islander Australia today*, Aboriginal Studies Press, Canberra.

Taylor, J. and Biddle, N. 2004, *Indigenous people in the Murray–Darling Basin: a statistical profile*, Discussion Paper No. 264, Centre for Aboriginal Economic Policy Research, Canberra.

Tonkinson, R. 1993, 'Introduction', in R. M. Berndt, D. H. Berndt and J. E. Stanton, *A World That Was: The Yaraldi of the Murray River and the lakes, South Australia*, Miegunyah Press, Melbourne.

Webb, S. 1984, 'Intensification, population and social change in southeastern Australia: the skeletal evidence', *Aboriginal History*, vol. 8, pp. 154–72.

Weir, J. 2009, *Murray River Country: An ecological dialogue with traditional owners*, Aboriginal Studies Press, Canberra.

Weir, J. and Ross, S. 2007, 'Murray Lower Darling rivers Indigenous nations', in F. Morphy and B. R. Smith (eds), *The Effects of Native Title*, ANU E Press, Canberra.

Young, W. J. and Hillman, T. J. 2001, 'A tale of two rivers', in W. J. Young (ed.), *Rivers as Ecological Systems: The Murray–Darling Basin*, Murray–Darling Basin Commission, Canberra.

Young, W. J., Schiller, C. B., Roberts, J. and Hillman, T. J. 2001, 'The rivers of the basin and how they work', in W. J. Young (ed.), *Rivers as Ecological Systems: The Murray–Darling Basin*, Murray–Darling Basin Commission, Canberra.

11. The Future of the Basin: Thriving or dying communities?

Chris Miller

The publication of the much-delayed *Guide to the proposed Basin Plan* by the Murray–Darling Basin Authority (MDBA 2010) should have seen it hailed as a landmark document. This was the opportunity to present for subsequent endorsement by the Commonwealth Government a robust basin-wide strategic plan, informed by the best available science. Such a plan would have secured the long-term environmental sustainability of this precious resource, ensured a sustainable future for irrigated agriculture—on which Australian food production depends—and offered the prospect of long-term sustainable livelihoods for basin communities, albeit a future with less water.

Instead, the publication of the Guide was met with howls of protest from irrigated-farming representatives and basin communities. The furore provoked an immediate loss of confidence by the MDBA in some critical aspects of its own work—notably, the socioeconomic data—and a declaration that further research was needed. This inexplicable retreat from the MDBA's own analysis led the Federal Government to question whether the Authority had met its legal obligations, seeking advice on the meaning of the *Water Act 2007* in relation to the balance between environmental entitlements and community social and economic impacts. The *Sydney Morning Herald* editorial (22 October 2010) was moved to declare that Tony Burke, the Minister for Environment, Sustainability, Water and Community, was 'running scared' from the first signs of concerted opposition.

The growing tension between the Minister for Water and the Chair of the MDBA led to the latter's resignation and replacement. Simon Crean, the Minister for the Regions, also announced a separate six-month Parliamentary Inquiry into the Socio-Economic Impacts of Water Reform, chaired by the Independent MP Tony Windsor. Meanwhile, the Water Minister, Tony Burke, attempted to create some daylight between the 'independent' Guide and any subsequent government decision, with the minister quoted as saying, 'It is not my Guide'. It is a sure sign of how bad the process is when a federal minister disowns recommendations not yet presented to him in the form of a draft Basin Plan and before the formal public consultation process had even begun. This was also a risky strategy for the Government as such actions could be interpreted as too hasty a rejection of recommendations based on the 'best available science' from an independent body established by the Government specifically to undertake this task. All this occurred within the first three months of publishing the Guide. Rarely can a document of this magnitude have been so quickly undermined—an achievement even more striking given the genesis of the Basin Plan with its strong bipartisan support for the principles of reform. Despite relatively rigorous hydrological and environmental science, the insufficiently robust social-science data and analysis have proven to be the Guide's Achilles heel and threaten to derail its progress.

A Reform of Complexity and Scale

The MDBA was very conscious of the enormity of the task it had been given, noting that in relation to scale and complexity it could find no other comparable exercise worldwide. Faced with this challenge, the MDBA acknowledged that its ability to marshal and apply the best scientific knowledge available would be critical to a successful outcome. Thus, the Guide notes:

> Planning of this scale and complexity has never been undertaken anywhere in the world. As a result it was important that the Authority brought together the best available data, modelling and scientific knowledge to support decision making...The independent reviews confirm that the approaches being taken...represent the best available (biophysical and social) science and knowledge...[but] there is much scope for further work and additional data capture into the future. (MDBA 2010:37)

The Guide makes clear the objectives of the Basin Plan in terms of ensuring the long-term management of the Basin's water resources, as 'redressing the degraded ecological health of the Basin while optimising the social, economic and environmental outcomes' (MDBA 2010:5) and providing 'a clear transition path for entitlement holders and communities through the period from plan adaptation to implementation at local level' (p. 7).

Unfortunately, the Authority has failed to live up to the requirements of a task that undoubtedly demanded some fortitude when confronted with predictably hostile interest groups. Instead, it was unable to provide a robust defence of its findings and proposals or to engage stakeholders in constructive debate. Indeed, the Chair began to give ground on the Authority's bottom line in sustainable diversion limits (SDLs) after a mere week of facing sometimes large, hostile and generally negative reaction from public consultation meetings that had attracted audiences of up to 5500 people. Anger and defiance from those most immediately affected by the proposals should have been anticipated, but questioning the validity of its own social-science data, especially the potential socioeconomic impacts of reform, only expanded the space in which outrage and criticism could flourish.

This chapter is concerned with one area of major weakness in the Guide—namely, the lack of attention given to what it describes as 'mitigation strategies' and its failure to go beyond mere mitigation measures to enable those communities most affected by water reform to make long-term adaptations to ensure a sustainable future. The use of the concept of 'mitigation' reveals the poverty of the MDBA's thinking in relation to potential socioeconomic impacts. It is clearly not concerned with how communities might go about transforming and diversifying their local economies faced with a contracting irrigated-agriculture industry. Rather, its perspective is restricted to managing the scale of such impacts so as to arrive at some 'acceptable' level. It seeks only to leave basin communities much as they were—albeit in a diminished or depleted form. Further, even within the limited bounds of mitigation, the MDBA assumed irrigated agriculture to be a healthy industry prior to the requirements of water reform. In fact, there were some major structural fault lines in the industry evident long before the so-called millennium drought. In failing to identify these fault lines, any future-oriented proposals were going to fall well short of what is required. This chapter highlights both why this approach is a major weakness and how a failure to include a robust investment plan for the future of regional communities puts at risk the water-reform agenda. Finally, it offers an approach to transition that could yet provide the means by which to keep the reform process on track.

The Place of Social and Economic Impacts in Water Reform

The Guide makes clear that the Basin Plan must take cognisance of social and economic impacts whilst ensuring a sustainable environment. Thus, it notes: 'The Water Act 2007 requires that in meeting the additional environmental water needs…the Authority must optimise social, economic and environmental outcomes, and at a minimum, the impacts need to be well understood' (MDBA 2010:39).

Despite the not inconsiderable amount of social-science research commissioned in preparing the Guide, the MDBA concluded that this remains insufficient to enable it to make what it describes as finely tuned decisions even after it had sought additional advice from a range of sources. This overly cautious viewpoint is stated at numerous points throughout the Guide and is summarised thus: 'the existing social and economic evidence base…was not considered adequate to undertake the required assessment at a fine degree of resolution' (MDBA 2010:39); and again: 'Of the evidence available to the Authority, the social and economic evidence is the weakest' (p. 197).

We are assured, however, that this assessment was reached only after extensive efforts. Thus, it notes:

> To overcome the weakness...the Authority sought advice from industry, community and government stakeholders...commissioned a wide range of studies...Work to develop regional profiles was also undertaken and a comprehensive data store of available social and economic information was compiled...A synthesis of current knowledge of the concepts of community resilience was also undertaken...with an emphasis on understanding the drivers of change...economic modelling was undertaken...to estimate the direct impacts on agricultural industries of various scenarios of reductions in water availability...work to understand the effects of changes in water availability on Aboriginal communities...was undertaken by CSIRO. (MDBA 2010:39)

But in the final analysis, the MDBA (2010:197) is forced to conclude that it 'has been unable to identify any consistent social and economic data that allows analysis of the flow-through impacts beyond the farm gate to the broader local economy and social fabric of the Basin'.

The MDBA (2010:xxviii) is clear in its disappointment in the exactness of the current social-science database and its desire to see further comprehensive assessments undertaken at community and industry levels. Nevertheless, it remains confident in its judgment about the scale of these impacts on community sustainability. The Guide acknowledges that some of the work undertaken might not provide sufficient insight into what is a complex set of dynamic interactions at a community and industry level. Yet when combined, the evidence is said to provide a solid foundation for consideration of the socioeconomic issues relevant to determining a SDL regime (MDBA 2010:41): 'the studies' findings provide an understanding of the social and economic capacities of communities and improve information on the upper and lower bounds of likely effects' (p. 84).

Indeed, the Authority suggests that the socioeconomic impacts 'may be overstated as they do not take account of the potential for farming innovation that might increase yields' (p. 88).

Ironically, whatever the perceived shortcomings of the existing data, research published more than 10 years ago by Neil Barr from Victoria's Department of Natural Resources and Environment, using data for the period 1986–96, highlighted structural fault lines within irrigated agriculture, and agriculture more broadly, that have little to do with water reform. These include the declining numbers of farm establishments, farm families and farmers; the loss of young people from agriculture and from basin communities; the ageing profile of farmers; insufficient productivity gains for a majority of farms to compensate for the compression in terms of trade; low incomes amongst many farmers and increasing dependency on off-farm income; the loss of 'entrepreneurial' farmers with mid-sized farms through increased investment-driven debt and the high costs and risks associated with entry into agriculture. Had the MDBA explored these data, it might have been persuaded to further investigate the health of the industry. It might also have recognised the need to disentangle the socioeconomic impacts of water reform from the more deeply rooted structural issues confronting irrigated agriculture. Such an examination might also have drawn the MDBA to conclude that what was required was a thoroughgoing investment strategy for regional communities that took account of, but went beyond, specific water reform-related impacts.

More recent research (Grafton et al. 2010) reinforced the sense that basin communities already struggle with major structural problems of their own. The research identified higher than average suicide rates amongst farmers, growing referrals to mental health services well beyond the capacity of service providers, increasing levels of reported domestic violence and crippling levels of household debt. In addition, there was some evidence of a growing level of antisocial behaviour amongst young people and low levels of educational attainment. The recruitment and retention of essential professionals such as medical and healthcare workers, social workers and teachers are longstanding problems. Some communities are already showing signs of further decline as indicated by lower housing market values and closures within the retail and service sectors. There is evidence too of an outward migration of those with highly valued skills and expertise and without whom there will be only a limited and restricted future for some basin communities. There can be little merit in a Basin Plan that fails to address such core structural issues. There will be no future if young people cannot see a future in the Basin but continue to leave for the metropolitan areas. Irrigated farming will continue to be an unattractive option until the mental health of farmers and farming families is addressed.

There is no doubting that socioeconomic data at a community level could be strengthened. Equally, that communities with sufficiently diverse economies will be 'relatively resilient' so as to navigate the transition (MDBA 2010:xxi). Yet, it can also be assumed that for some communities in some catchments, the impacts will be severe, and the long-term survival of some smaller communities heavily dependent upon water-based industries might be at risk. Further, all communities faced with large reductions in irrigated water will be required to make major adjustments. Some will be faced with the transformation of their local economy. Such transformations are concerned not simply with the minimisation of job losses—important though that might be—but in re-imagining quite different ways in which the community can generate wealth.

Thus, the overall volume of total potential job losses as a consequence of reduced SDLs might be relatively small when viewed in a national context and could theoretically be absorbed in what continues to be a buoyant labour market, facing if anything a skills shortage and generating a job surplus. Yet such new employment opportunities do not always match the skill mix of those losing their jobs, nor are they always in the right locations, and people are not always in a position to relocate. In many regional communities, the impacts of job losses are likely to reverberate across the whole local economy. These are, after all, small, interdependent economies retaining a large proportion of locally generated income, and therefore big employment losses in a major industry such as irrigated agriculture will have real consequences for jobs in other sectors, including those not obviously dependent upon irrigated farming. Thus, the Guide notes that

> [t]he agriculture industry in the MDB provides an annual average of $15b worth of produce to the national economy…Indirectly agricultural activity is also a key economic driver of local industries and regional activities that support small and medium enterprises and employment across the Basin. For example, around one-third of people employed in manufacturing are employed in food products industries, representing a further 30,000 employees. (MDBA 2010:21)

Further, it (MDBA 2010:96) points out that between 50 and 70 per cent of farm expenditure is in nearby towns, with a further 20–30 per cent in regional centres, estimating that more than 75 per cent of total farm expenditures are recycled back into the regional economies. Similarly, with some 65 000 farms across the Basin, the Guide acknowledges the dependency of small and medium-sized businesses on agricultural revenues and irrigator expenditures, as well as the loss to the rateable base of local government services. These likely follow-on effects will have an impact on the overall level of confidence in the future of such communities, such that the skilled and mobile workers might relocate to seek better, more secure employment elsewhere, local investment capital might do likewise and external capital will be disinclined to invest. It is this type of logic that informs the MDBA in judging that it cannot go beyond a 4000 GL/yr SDL. Such assessments are, however, at best predictions based on what might happen if there are no other counter measures put in place. Such interventions could enable communities to plan for reform and transition towards a more diversified sustainable economy.

A Flawed Analysis

It is difficult to see how the MDBA can justify paying so little attention to community transitional strategies. For those living with the prospect of reform, mitigation is the minimal response to an impact assessment. With sufficient advance warning and planned change, actions can be taken to remove or minimise the dependency on whatever is the source of the impact. The MDBA (2010:xiv) was indeed aware of the importance of such transitional arrangements, noting 'it is essential that effective transitional arrangements be put in place to help businesses and individual water entitlement holders adjust to change, and why action must continue to be taken to maintain strong and prosperous regional communities'.

Again, it highlights what it sees as a critical issue: 'Depending on the local communities' capacity to adapt...the short-term social and economic impacts on some communities and regions could be severe without structural adjustment... more and specific targeted assistance could be considered by government' (MDBA 2010:xxvii).

Yet despite such concern and acknowledgment that '[e]ffective transitional arrangements...will be essential', the MDBA (2010) offers only two very limited mechanisms—neither of which matches its own declared sense of urgency, plus some half-hearted reassurance:

> The Water for the Future program is likely to affect the eventual impacts...by providing additional water savings...and by providing regional economic stimulus. (pp. 120–1)
>
> For transitional or interim water resource plans that cease less than five years after the date of the Basin Plan being adopted...the Authority proposes that SDLs be phased in over a period of five years, allowing water users and communities more time to adjust to the new arrangements...
>
> The Authority recognises that there will also be benefits that will flow through to communities because of improved environmental conditions. (p. 124)

Even by its own admission, the MDBA accepts that revenue from selling water entitlements might not recirculate through local communities but could instead be used to discharge debt, to which one might add relocation, retirement or inheritance gifts. What we are left with is a series of assertions that are either insufficiently developed or unsubstantiated. First, we are told the data on the social and economic impacts could be improved. Next, we are informed that transitional arrangements are critical in determining actual impacts but little is offered by way of such interventions. Finally, despite reservations over the data, no consideration can be given to water reductions of more than 4000 GL/yr due to the apparently unreliable predicated impacts. Thus, it states: 'The Authority recognises that there is no formula for determining the optimal result and will do this by applying its judgement' (MDBA 2010:107).

Also, 'the Authority has made a number of critical judgements...reductions that exceed 4,000GL/y...would not represent an optimisation of the economic, social and environmental outcomes under the Water Act' (p. xxi).

There is indeed no formula but there are processes available that better determine the outcomes. The MDBA is, however, unable to argue this point with any real conviction, and indeed the whole chapter in the Guide on socioeconomic impacts is scattered with conditional statements with plenty of 'mights', 'coulds' and 'maybes'. The crucial role given to 'judgment' in making determinations for the social and economic impacts stands in sharp contrast with the emphasis given to peer-reviewed science in the determination of the SDLs. The MDBA (2010:81) is aware of the shortcomings in its analysis, noting: 'This chapter does not include analysis of the potential contribution to mitigation measures. The impacts presented in this chapter are likely to be larger than the final impacts.'

In adopting an approach that ignores the potential for regional development via transformational strategies, the MDBA exposed itself to avoidable criticism. It was immediately required to defend itself from farming and irrigation industry organisations in particular but also from many angry farmers, growers and community members about the veracity of its analysis.

The Response to Social and Economic Impacts

Despite its own lack of confidence in the data, especially its capacity to address community-specific impacts, the MDBA does offer some aggregate assessments. Thus, it notes that

> [r]egions with a relatively higher dependence on lower-value irrigated agriculture would experience a larger reduction in economic activity. The social fabric of some towns and communities may be significantly affected, particularly in the near term. The capacity of towns to adapt is likely to vary widely, and would be influenced by factors such as the size of the community, the diversity of its economic base, its demographic mix and its proximity to other large regional towns. Some communities may be permanently changed by the reduction in diversion limits. (MDBA 2010:81)

Part of the problem, however, is not in the availability of data but in the capacity of the MDBA to both analyse data and identify relevant types of available data. For example, in determining relative community success in adaptation to reduced water (MDBA 2010:xxxiii), the primary focus is on water-related indicators to the exclusion of other equally important signposts such as the minimisation of long-term job losses, the level and speed of job replacement, the maximum number of communities retaining current levels of or growth in population, producing an equivalent contribution to regional/national economies, community capacity to retain young people and retaining those 'pull' factors to attract new migrants, and so on. It is the failure to explore the range of available transitional arrangements that leads the MDBA to perhaps its most important decision to restrict consideration of water diversions to 4000 GL/yr. Thus, it argues:

> In light of the severity of this impact on specific sectors and communities the Authority has judged that…it can only consider Basin-wide reductions of between 3,000–4,000 GL/y for the Basin…reductions in current diversions above 4,000 GL/y have been judged to be beyond the range of acceptable reductions. (MDBA 2010:82)

The judgment is based on data that suggest a decline in gross value of irrigated agricultural production of about 13–17 per cent or $0.8–1.1 billion per year, or between 1.1 per cent and 1.5 per cent reductions in permanent long-term regional production (MDBA 2010:120). No account is taken of the potential for this lost value to be ameliorated by equivalent or greater gains in other parts of the Basin economy, either by developing existing industries or through the introduction of new products and services. In other words, this is a judgment made only on the basis of the impacts of reducing irrigated agriculture and an assumption that there will be no new economic development. The failure to plan for such change will only increase the likelihood of this scenario being realised. Thus, the role of 'judgment' becomes central to social and economic impacts, in contrast with the role of science in respect of environmental requirements.

One explanation for the MDBA's failure to consider transitional strategies lies perhaps in its equally flawed claim that change is driven by the actions of multiple individuals acting independently. Thus, the MDBA (2010:83) asserts that 'economic and community change is driven by the adjustment decisions of individuals. In response to a reduction in diversions, it is irrigators' decisions that will drive change in communities where irrigated agriculture is a large part of the economy.'

Whilst it is true that individuals will indeed make decisions and in doing so will take account of a range of factors, the context or framework in which those decisions are taken is determined by more powerful forces, such as government, national and global corporations, the financial services sector and global markets. Indeed, the Plan itself is an obvious example of governments generating change. Again, the MDBA (2010:xiv) appears to have chosen to ignore its own assessment that previous government interventions have been critical in shaping the conditions of today:

> The Basin and its communities are facing significant challenges and risks…the direct result of the actions of successive governments over the history of the Basin…failed to strike a balance between meeting the needs of the environment and those of a growing economy and population…
>
> Many towns in the Basin have grown throughout many years of government policy that encouraged water use and regional development. (p. 82)

In making this rather limited assertion that individuals are the ones in the driver's seat of change, the MDBA can claim that as it is unable to predict how so many diverse individuals will behave there is little point in dwelling on this aspect. Had it followed the logic of its assessment that there are other more influential forces at work in creating the conditions in which individuals act, and the importance of government in this process, it would have considered the range of transitional strategies adopted elsewhere, supported by governments and global agencies such as the United Nations, and evaluated the outcomes. Instead there is an extraordinary absence of any applied social science in the Guide despite an apparent awareness of the limitations of its own proposed transitional arrangements: 'These transitional arrangements in themselves may not be sufficient and action may be needed by all levels of government to maintain prosperous, resilient regional communities' (MDBA 2010:xx).

The Authority further compounds the problem by demonstrating its own inability to construct good social-science analysis. There is evidence of this in its justification not to explore impacts in more detail by attempting to categorise the diversity within the farming sector (MDBA 2010:88). We are offered six somewhat dubious categories or farming 'types'. Thus, it is suggested that one type of farmer is one who is 'concerned and uncertain about the future of irrigated and dryland agriculture in the Basin'. Now it might be the case that there are indeed still farmers who are unconcerned and certain about the future following the worst droughts on record, but one would hope that these are few in number and would probably be better off in some other line of work.

Next, we are offered those who are well prepared for lower water entitlements against those who are not; those who are optimistic and/or have strong connections with the community against those who are pessimistic and have weak connections; those who are close to retirement and those who are younger; those who are financially stable and those financially exposed due to the extended drought and surviving on exceptional circumstances payments and off-farm income. Such categories are, however, awash with flawed argument. It is, for example, erroneous to suggest that high off-farm income somehow makes for financial insecurity. Rather, the opposite could be the case. With secure off-farm income, households can make better decisions and better manage inherent farming turbulence. Equally, it is not at all clear why a personal characteristic such as optimism should be lumped together with strong community or farm connections. These are unrelated, as connectivity is relational and dynamic and speaks as much to something desired as to something already achieved. It might also reflect something that was once strong and important but has been weakened and is no longer important. Nor is it necessarily one or the other—strong or weak—but is in fact a relative state.

Further, although the meanings are not explained, to have strong community connections is probably something the MDBA thinks of in terms of intergenerational longevity, and indeed this might have been a better category to adopt. The main point is, however, that this does not reflect a 'type' of farmer (there might be some farmers who do not have and do not want to have strong community connections), but rather it describes both an existing relationship and something that can be altered. These are not types of farmers but rather in the main they describe different sets of circumstances in which farmers might sometimes find themselves.

One factor that cannot be altered is that of age. So a better starting point of analysis would be to consider those who are close to retirement and those who are relatively young farmers. In each of these categories, the extent to which someone is financially secure and efficient will make a difference to how they decide on a future direction. Other more useful categories might relate to the size of farm or the nature and diversity of production. In other words, the categories of farming types identified by the MDBA do indeed convey a sense in which it is difficult to envisage a rational policy response, given such muddled categories that do not lend themselves to analysis.

Community Transitional Strategies

One of the weaknesses in using only a set of social and economic assessment indicators is that, while useful as a predictive tool that provides stakeholders with an opportunity to consider the potential repercussions of any impending change, they are less robust in anticipating how people might act. Increasingly, it is recognised that identifying such indicators cannot rely simply on objective criteria determined outside the community but must engage the communities in shaping the data (Young et al. 2002), and that any criteria should be open to scrutiny and revision through community dialogue.

Other models are also available, including the 'Sustainable Livelihoods' approach (Scoones 1998) that is popular in the context of international development (Brocklesby and Fisher 2003). It has been adopted by various governments—including in the United Kingdom, and some Canadian Provinces, notably British Columbia and Quebec (Markey et al. 2009)—and is recognised by the UN International Fund for Agricultural Development. The UK Department for International Development (DFID 1999:1) suggests that the framework—which is to be applied in a participatory manner—is to 'help stakeholders with different perspectives to engage in structured and coherent debate about the many factors that affect livelihoods, their relative importance and the way in which they interact' in order to identify 'appropriate entry points for support of livelihoods'.

A third, albeit derivative, approach is that of an 'Adaptive Capacity Index' (Stenekes et al. 2010), which requires, according to Nelson et al. (2007:iv), 'the continual process of inventing, adapting and adopting…to anticipate and respond to change'. Rather than focusing on impacts and responses that might mitigate the worst of these, the Adaptive Capacity Index plays to the potential for agency and creative change. It gives emphasis to the opportunities within any situation, however undesirable, and our emergent capacity to respond to stress (Sietchiping 2006). As with all theories that prioritise agency, it is useful to also take account of potential barriers to adaptation, or what might explain people's failure to exploit these capacities. Finally, recent attention has focused on the concept of community resilience that again is focused on community resources and adaptive capacities (Maguire and Cartwright 2008).

Rather than explore the potential application of these well-respected approaches, the MDBA opted for a narrow, economically driven model of adaptation. Thus, it draws upon work undertaken by Frontier Economics (2010) to highlight five aspects of what it refers to as 'adjustment pressures'. These are: market, social, technological, government policy and environmental pressures. Such factors are, however, not felt simply as pressures to adjust, as they can both shape and channel the direction of adjustment, as well as blocking off certain avenues. This is a particularly important consideration when we examine what individuals might take into consideration when making a decision.

The Guide—again, drawing from the same source—is focused most extensively on business-related factors, such as expected profitability, outlook, financial position, business objectives, risk aversion, understanding and uncertainty, and strategic behaviour whilst also including 'perceptions, attitudes and ethics'.

This ignores other potentially influential factors, including commitment to community and place, investment in future generations, lifestyle, position in the life cycle and opportunities to move into new areas of expanding economic activity. The MDBA locates the influence of 'alternative economic opportunities' in a box designed to show how the severity of impacts will vary alongside the 'extent of aggregate/cumulative adjustments' made by individuals. In other words, the adopted model has individuals acting as purely rational economic planners who make decisions in the context of a number of adjustment pressures. The cumulative effect of such individual decisions is what is assumed will motivate industry-level structural change and this is then offset in terms of its severity by a number of other factors, including the potential for alternative opportunities and a community's capacity to respond to change.

In reality, life is more complex, messy and dynamic than this suggests. Individual decisions are not made solely on economic grounds, nor are they always 'rational'. Decisions about a specific industry are not made in isolation or without reference to wider economic and non-economic considerations both actual and potential. For example, much is made in explaining the behaviour of national and global markets of the role of 'market confidence', and there is no reason to suspect that this is less so at a local level. Yet confidence in current and future prospects is intangible and multidimensional, and includes emotional states of mind. At a local level, it is difficult for any community to sustain a sense of self-confidence and optimism when faced with what feels like overwhelming negative external pressure on a core industry—a sector that defines its identity as well as sustaining its economy. Optimism in the future is especially difficult to sustain in a context in which a community has only just emerged from what has been the worst drought in recorded history. In such circumstances, communities need the reassurance of government and might also need some external intervention to enable them to rebuild community resilience and adaptive capacities. Such circumstances highlight the importance of early intervention transitional strategies in anticipation of major social and economic impacts following policy reform. The MDBA is not unaware of such complexities, stating:

> [T]he Authority recognised the range of complex and inter-related factors that will exert influence on the ultimate outcome...
>
> The short-term economic effects...depend on the particular circumstances of the Basin's businesses and individuals and their capacities to adapt... they will respond in different ways to the transitional support that is provided to enable Basin communities to adjust. (MDBA 2010:94)

An evaluation of transitional strategies practised elsewhere—designed in response to predicted social and economic impacts and to enable communities to make adjustments to ensure future sustainable livelihoods—would have led to a far greater range of options than what is offered in the Guide. Such considerations are well within the remit of the MDBA, charged as it is with securing environmental sustainability whilst optimising social and economic outcomes. Instead, the Authority walks away from this responsibility, handing it to others—namely, State and Federal governments—and in doing so appears to reduce the problem to the mere provision of social facilities. Thus:

> The success with which communities transition will be shaped by the continued provision of community services…and the on-going activities of community clubs, sporting clubs and other community connections. Sustaining the social fabric of communities will be in part determined by the economic adjustments…and the strength of communities' social capital will in turn also shape communities' ongoing economic success… the Authority has put significant weight on the policy settings of Basin governments as a critical determinant of the long-term future of Basin communities. (MDBA 2010:159)

The Authority cannot have it both ways. Either it treats seriously its requirement to optimise the social and economic outcomes of basin communities or it declares that this is beyond its brief or capacity. If it is part of its brief then it is responsible for a thoroughgoing analysis of both social and economic data but also of those strategies that enable communities to move forward when faced with major change. Anything less than this is a failure to fulfil its responsibilities. More importantly, it produces only a partial representation of what is a complex public concern, leaving critical options unexplored and generating unsubstantiated and inadequate recommendations that do an injustice to those communities most affected by water reform.

Rather than assisting in the process of adaptation and adjustment, the adopted approach is likely only to further exacerbate the social and economic impacts as communities experience prolonged and gradual decline, removing any sense of urgency and leading to a gradual exit from the community of human, social and financial capital. In contrast, the introduction of a community-adjustment strategy, referred to here as the 'Thriving Communities model', could provide a framework to enable communities to make the transition not simply by coping with social and economic impacts but by identifying a sustainable future for the communities of the Basin.

Thriving Communities Model[1]

In outlining this approach to adaptation, it is first of all important to remember that it is not the first time that basin communities—in the context of nation building—have been called upon to plan for the future. For example, in 1945, the Australian Broadcasting Commission (now Corporation, ABC) produced a pamphlet in response to its radio program series entitled *Communities Can Do It—Make a plan*, calling on basin and regional communities to work together to identify local needs and build communities for the future.

Well before the publication of the *Guide to the proposed Basin Plan*, basin communities had been actively discussing how to live with less water. Local initiatives have sprung up to act on emergent ideas, although such actions have been small scale, localised and dispersed. The Guide makes no reference to such endeavours so any outcomes are not factored into the equation. Further, it is important to acknowledge that basin communities are diverse and complex so that there is not one but many stories to tell about what are the critical issues facing a particular community. Not only are there multiple perspectives and interests, some will be in conflict or tension with those of others. While some interest groups are well organised and resourced, confident and articulate, others will not be, and many will have remained silent, dispersed or excluded from the discussion. Even amongst the well-organised interest groups there is not a single voice or standpoint, although minority perspectives might have found it hard to be heard. This is especially so when people are represented by formal organisations. Again, based upon recent research (Grafton et al. 2010), there is substantial evidence that the resilience and adaptive capacity of basin communities have been greatly diminished by having to cope with 10 years of extreme dry conditions. The research also showed that there are currently in some parts of the Basin very low levels of trust in relation to government and to the MDBA in particular.

There is also an overwhelming sense in which the knowledge, skill experience, know-how and capacities of basin communities have not been recognised or respected and their views and perspectives have not been sought or heard. This failure to draw upon the wealth of knowledge in the Basin—to place it alongside expert scientific knowledge—is not simply a technical deficiency but will impede the implementation of any transitional strategy that ultimately depends on trust and collaboration between communities and government. Again, the research highlighted what might be described as a withdrawal from public affairs, with fewer people volunteering and low levels of participation in public meetings. In

1 An earlier version of the Thriving Communities Model was developed in conjunction with my Flinders University colleague Fiona Verity. It was first presented in Wentworth Group of Concerned Scientists et al. (2010).

other words, the current levels of human, social and cultural capital within the Basin—critical in any change process—are yet another element missing from the assessment of social and economic impacts.

Any adaptation model has first to accept water reform as a whole-of-community issue and must involve previously excluded groups, such as young people, new migrants, women and Indigenous people. Second, community adaptation cannot be left to the vagaries of multiple individual decisions, but requires extensive community engagement over a sustained period, which acknowledges community specificity and provides for the maximisation of community control over the determination of its future. Despite differences within communities, there is evidence of a shared goal of securing a sustainable future with less water—a goal that brings together the concepts of growth and development. The objectives of any such adaptation strategy could be expected to include

- retention of local capital
- attraction of new capital investment
- fostering of the development of new products and services for a more diversified economy
- extension of community ownership of local services and enterprises
- encouragement of new migration
- retention of young people
- building human, social and cultural capital
- collaboration with other basin communities
- partnerships with local, State and Federal governments
- integration with local natural-resource management.

Government can be expected to play a critical role in the success of such an approach in at least five critical aspects. Key to this would be the creation of an investment fund for the future of basin communities to be allocated initially to those catchments most affected by the new SDLs, with the amount per catchment to reflect the percentage of water required for the environment.

Next, government needs to make available expert technical advice, especially in relation to developing new business opportunities. The emerging regional development agencies could provide the vehicle for such advice—although better still, communities could be given the opportunity to identify what they thought would be the most appropriate institutional structure. The key point, however, would be to ensure that the advice was provided by those who had firsthand experience in whatever was the field of knowledge. This could be ensured through a program of 'community secondments' from the business and financial sectors.

Third—and again, critical to the success of any adaptation strategy—would be the appointment of a Basin Task Force to work within and alongside those basin communities most badly affected. Faced with a reform of this magnitude and given the specificity of basin communities and the diversity within, as well as between, them, highly skilled development practitioners are essential in bringing together and navigating this whole-of-community approach. A reform of national significance that seeks to redress previous government policies with major implications for regional Australia justifies the mobilisation of the most experienced and skilled workers able to work with such complexity and uncertainty.

Next, the Federal Government would be expected to work in partnership with basin communities as well as State and local governments, and new time-limited structures might be required specifically for this purpose. Finally, government would be responsible for mapping out the framework, the principles, structures and governance arrangements essential for the management of such a way forward.

The establishment of an investment fund is not a request for new or additional funds, but rather for a review and reallocation of monies already set aside for rural development and water reform. In particular there is a need to review the $5.8 billion currently available for irrigation infrastructure improvements under the Water for the Future program. While the Federal Government is already committed in principle to State governments to about $3.1 billion worth of projects, few of these are at an advanced stage of development and none has as yet been given final approval. There are at least three reasons why we need to revisit these earlier decisions in the light of the Guide.

First, many basin farmers and growers say themselves that they are already highly efficient, having invested heavily using their own resources during the drought. They would not have access to these funds that somewhat perversely would benefit those so-called industry laggards who might not deserve such a subsidy. Second, we now know that infrastructure projects of this nature are inefficient in securing additional water and that buybacks are a cheaper option. Indeed the Treasury in its 'red book' to the incoming government advised that such investment should be avoided except where it can be shown that the public benefit is greater than that secured through water buybacks. Third, time has passed since these decisions were taken and we are now more aware of the need to ensure the long-term sustainability of basin communities so we need a broader definition of what is understood by 'infrastructure investments'. In redefining the purposes of support, further improvements to irrigation infrastructure would not be excluded where that can be shown to generate a good return.

To what purposes might such an investment fund be used? Keeping in mind the objectives outlined above, there are many possible areas for investment, including the following: the design and testing of new products and services, including drawing up business plans and marketing strategies; the creation of rural laboratories for the development of innovative solutions; supporting individuals with scholarships for tertiary education or training; the development of arts, cultural and heritage projects; the extension of the community-service sector including education, health care, social care and libraries; support for local community initiatives; investment in connectivity infrastructure, including road, rail and air services but also telecommunications. Similarly, a number of mechanisms are available in allocating the funds including small development grants, low-interest loans, co-investment with local private capital in venture-capital projects, and direct investment in community-owned local resources including property, industry and facilities. Regardless of the nature of the investment or the chosen mechanism, all projects seeking support would be required to meet viability benchmarks against a triple bottom line of economic, social and environmental sustainability.

Basin communities must be at the heart of any adaptation strategy. Maximum community engagement, along with community control, with all the uncertainties and messiness implied, in determining as far as possible its future direction is the key to any successful adaptation approach (Pepperdine 2000). The role of communities is to determine the local structures that will work best for them rather than being expected to accommodate structures that have been imposed upon them. Such structures would, however, be expected to ensure maximum participation and deliberation across the community, including previously excluded groups. Structures would need to be not only accessible but also transparent in decision making, accountable both to the community and to government, and genuinely community owned. Communities would be expected to mobilise local leadership and facilitate the emergence of new leadership capacities, linking up across the Basin into what might be described as a basin-wide Community Leaders' Forum to ensure mutual learning and enhanced strategic decision making. Communities would be expected to bring together local knowledge, know-how and experience to be placed alongside expert scientific, environmental and social-science impact knowledge. The combination of such knowledge and the identification of market opportunities provide the basis on which to generate Local Adjustment Action Plans that would shape the allocation of the Investment Fund. As with any rigorous change model, Thriving Communities would be underpinned by ongoing action-oriented research to ensure that the learning can be iterative and that past practice is shared with other basin communities.

The model outlined here works from within communities. It adopts an integrated approach that works towards achieving sustainable livelihoods but begins with current and potential adaptive capacities and therefore with the history of those communities. Armed with the best scientific knowledge, it aims to build a platform for shared knowledge that includes local knowledge. The latter will provide critical accounts of community history, will cast light on the fine detail of local context, reveal current relationships, concerns and possibilities and identify resources for change. This is the work of the Task Force.

All this is vital if any community change and adaptation process is to succeed. The approach seeks to identify shared interests but is alive to the unequal distribution of impacts and unfair outcomes and acknowledges community diversity, differentiation and conflict. It affirms and works from existing community assets to further build local capacities and to strengthen community resilience. In pursuing these objectives, it seeks maximum participation in decision making through deliberative dialogues. A helpful metaphor for such deliberative dialogues is that of the kitchen table. Not only does this indicate the number of people that can comfortably engage in the process, it evokes that sense of an unfolding dialogue as people join the discussion for a period and then leave to do something else before returning. The time away from the table is an opportunity for reflection, a time that allows one to return having learnt something as a result and possibly modified one's standpoint. But the kitchen table also evokes that feeling of conviviality, care for those present and a desire to understand their perspective; it is the place where we prepare and share food together. Such cross-community kitchen-table dialogues need then to be joined together in larger community forums.

Would such a model work in practice? For there to be any hope of success, some preconditions would be necessary. Governments must first recognise and demonstrate respect for community knowledge, know-how, ideas, creativity and capacities. Governments need also to trust communities to deliver sound outcomes and must devote substantial resources to the task. In addition, all those involved need to be committed to finding more effective ways of talking with each other about complex issues in uncertain times. There are, however, no magic bullets, no short-term fixes.

What is required is long-term, sustained and incremental investment in rural futures. It needs to be built upon a vision for development that creates a framework of confidence in the future. It needs to deliver realistic and comprehensive plans for economic restructuring. Critically, such a model needs to be institutionally embedded.

Examples from elsewhere can provide confidence in such an approach. For a comprehensive regional policy, that of Quebec's National Rural Policy (<www.mamr.gouv.qc.ca>) offers a good starting point. For an example of more local practices in circumstances similar to the Basin, the Canadian Province of British Columbia (BC 2004) has for more than 30 years invested in regional trusts, as a key vehicle is regional social and economic development. For example, the Columbia Basin Trust, established in 1994, received a C$295 million endowment from the Province plus an additional C$2 million a year over 16 years, C$45 million of which is used as an investment fund for community benefit. The Community Futures Program (<www.communityfutures.ca>), with more than 90 initiatives across the Provinces of British Columbia, Alberta, Saskatchewan and Manitoba, has been running since the mid 1980s. British Columbia has some 34 Community Futures programs, which, during the period 1995–2010, issued a total of 11 442 loans, worth C$330.8 million, which leveraged in another C$618.9 million and created 44 680 jobs—an average of just less than 3000 per year. Whilst this does not tell us about the nature of such jobs or how many jobs were lost during that same period or what might have happened without the program in place, it nevertheless demonstrates that such local investments can be an important source of economic and social rejuvenation.

More than two million people live and work in the Murray–Darling Basin and it is home to many thriving communities that are critical to Australia's food production. The Basin is also facing an environmental crisis that demands fundamental changes to the way we manage and use natural resources. In making the required transition, the challenge is how to optimise social and economic outcomes by ensuring thriving and sustainable communities into the future. An approach such as the one outlined here provides an opportunity to secure this result.

References

British Columbia (BC) 2004, *The BC Heartlands Economic Strategy: A place to revitalize our entire economy*, Province of British Columbia, Victoria.

Brocklesby, M. and Fisher, E. 2003, 'Community development in sustainable livelihoods approaches: an introduction', *Community Development Journal*, vol. 38, no. 3 (July), pp. 181–98.

Department for International Development (DFID) 1999, *Sustainable Livelihoods Guidance Sheets 1*, Department for International Development, London.

Frontier Economics 2010, *Structural Adjustment Pressures Affecting Irrigated Agriculture in the Murray–Darling Basin*, Murray–Darling Basin Authority, Canberra.

Grafton, Q., Miller, C., Duvnjak, A., Jared, D., Jiang, Q., Nikolakis, W., Ryan, P., Verity, F., Ward, P. and Zutshi, M. 2010, *Potential Water Quality and Quantity Impacts in the Murray–Darling Basin from Communities and Industry Responding to Climate Change*, Murray–Darling Basin Authority, Canberra.

Maguire, B. and Cartwright, S. 2008, 'Assessing a community's capacity to manage change: A resilience approach to social assessment, Bureau of Rural Sciences, Canberra.

Markey, S., Halseth, G. and Mason, D. 2009, 'Contradictions in hinterland development: challenging the local development ideal in northern British Columbia', *Community Development Journal*, vol. 44, no. 2 (April), pp. 209–29.

Murray-Darling Basin Authority (MDBA) 2010, *Guide to the proposed Basin Plan. Volume 1: Overview*, Murray–Darling Basin Authority, Canberra.

Nelson, R., Brown, P., Darbas, T., Kokic, P. and Cody, K. 2007, *The potential to map the adaptive capacity of Australian land managers for NRM policy using ABS data*, CSE49, 2007, National Land and Water Resources Audit, Canberra.

Pepperdine, S. 2000, *Social Indicators of Rural Community Sustainability: An example from the Woady Yaloak catchment*, Department of Geography and Environmental Studies, The University of Melbourne, Vic.

Scoones, I. 1998, *Sustainable rural livelihoods: a framework for analysis*, IDS Working Paper 72, International Development Studies, University of Sussex, Brighton.

Sietchiping, R. 2006, 'Applying an index of adaptive capacity to climate change in north-western Victoria, Australia', *Applied GIS*, vol. 2, no. 3 (16.1–16.28).

Stenekes, N., Kancans, R., Stayner, R., Reeve, I. and Coleman, M. 2010, *Indicators of Community Vulnerability and Adaptive Capacity Across the Murray–Darling Basin—A focus on irrigation in agriculture*, Australian Bureau of Rural Sciences and Institute for Rural Futures, University of New England, Canberra.

Wentworth Group of Concerned Scientists in association with Grafton, Q., Kowalick, I., Miller, C., Stubbs, T., Verity, F. and Walker, K. 2010, *Sustainable Diversions in the Murray–Darling Basin: An analysis of the option for achieving a sustainable diversion limit in the Murray–Darling Basin*, Wentworth Group of Concerned Scientists, Sydney.

Young, M., Young, D., Hamiliton, A. and Bright, M. 2002, *A preliminary assessment of the economic and social implications of environmental flow scenarios for the Murray River system: a report prepared for the Murray Darling Basin Commission*, July, CSIRO Land and Water Policy and Economic Research Unit and PIRSA Rural Solutions, Canberra.

Part III. Legal issues

12. A Sustainable Murray–Darling Basin: The legal challenges

Douglas Fisher

Introduction

The range of legal instruments informing how the Murray–Darling Basin (MDB) is managed is extensive. Some provide guidance; a number indicate strategies and policies; some assume the form of protectable rights and enforceable duties. What has emerged is a complicated and sophisticated web of interacting normative arrangements. These include

- several international agreements including those concerning wetlands, biodiversity and climate change
- the Constitution of the Commonwealth
- the *Water Act 2007* of the Commonwealth
- the *Murray–Darling Basin Agreement* scheduled to the Act
- State water entitlements stated in the Agreement
- Commonwealth environmental water holdings under the Act
- the Murray–Darling Basin Plan
- water-resource plans under the Act or State or Territorial water legislation
- State and Territorial water legislation
- water entitlements and water rights under State or Territorial water legislation.

It is particularly significant that many of these instruments are to an increasing extent incorporating protectable rights and enforceable duties applicable to both the public and the private sectors. This is the inevitable consequence of what the National Water Initiative (NWI) set out to achieve—namely: a nationally compatible market, and a regulatory and planning-based system of managing surface and groundwater resources that optimises economic, social and environmental outcomes.

Constitutional Perspectives

The constitutional validity of each of the provisions of the *Water Act* depends upon one or more of a number of sources of legislative capacity. These include

- matters referred to the Commonwealth by a State
- trade and commerce among the States
- trading corporations
- external affairs.

The Act carefully states the source of legislative power for each set of provisions. The exercise of the external affairs power involves the enactment of a law that constitutes the implementation by Australia of an obligation imposed on it by an international agreement. Hence, the substance of the international obligation indicates the scope of the provision able to be enacted by the Commonwealth. The market-based rules and the trading rules are sourced in the interstate trade and commerce power and the trading corporations power. The matters referred by the relevant States to the Commonwealth include: first, the powers, functions and duties conferred on Commonwealth agencies that relate to MDB water resources and are conferred by or under the *MDB Agreement*; and second, the management of basin water resources to meet critical human water needs (*Water Act 2007* [Cwlth], s. 18B[9] 'referred subject matters').

These matters relate to the water resources but not the natural resources of the Basin. How the natural resources of the Basin are managed is likely to be a reflection of Australia's international obligations implemented through the external affairs power. While there is nothing to prevent a referring State from terminating its reference, the Act specifically enables a referring State by a provision of a law to declare a matter to be an excluded matter in relation to the Act. While there is no certainty in these matters, there is a potentially close relationship between international agreements to which Australia is a party, the Constitution, and specific provisions in the Act, the *MDB Agreement* and the Basin Plan.

The Enforceability of Instruments

The capacity of a provision in an instrument to be implemented and enforced depends upon its status within the legal system and the form in which the instrumental provision is expressed. A specific international obligation is enforceable in the relevant international forum. It supports the exercise of the external affairs power by the Commonwealth if the conditions indicated by the High Court are satisfied. A provision in the Act is enforceable in accordance with the procedures stated in the Act and provided the provision satisfies the conditions for a judicially protectable right or a judicially enforceable duty. The Act is a legislative instrument with that status.

But what is the status of the *MDB Agreement* and of the Basin Plan? The Agreement as a schedule to the *Water Act* is part of the Act and prima facie has the status of legislation. The Agreement is formally amended when the instrument that amends the schedule is registered as a legislative instrument.[1] This by implication confirms that the Agreement itself is a legislative instrument. The *Water Act* (s. 33[1][a]) itself states that the Basin Plan is a legislative instrument. While the Act, the Agreement and the Plan might well be legislation in this sense, in practice, it will be the nature and function of the relevant provision and the form in which it is expressed that will determine whether and to what extent the provision is able to be enforced. Does the provision state a purpose or object, a principle, a strategy or policy, a procedure, a power, a right or a duty? The Act, the Agreement and the Plan (in prospect) are replete with examples of all of these.

The Objects of the Act

One of the objects of the *Water Act* (s. 3[b], [c]) is to give effect to relevant international agreements and in so doing to promote the use and management of the Basin's water resources in a way that optimises economic, social and environmental outcomes. In addition, and without limiting the scope of this object, there are three particularly relevant objects (*Water Act 2007*, s. 3[d]). The first is to ensure the return to environmentally sustainable levels of extraction for water resources that are over-allocated or overused. The second is to protect, restore and provide for the ecological values and ecosystem services of the Basin. The third—subject to the first two—is to maximise the net economic returns to the Australian community from the use and management of the Basin's water resources. The use of the expression 'subject to' means that the maximisation of net economic returns is conditional upon the return to environmentally sustainable levels of extraction. Significantly, the optimisation of economic, social and environmental outcomes is linked to giving effect to relevant international agreements, whereas the return to environmentally sustainable levels of extraction is not so linked. In addition, it is noteworthy that optimisation involves a qualitative judgment while maximisation involves a quantitative judgment. In any event, the object of the Act that most closely resembles a duty is to ensure the return to environmentally sustainable levels of extraction.

1 *Murray–Darling Basin Agreement*, art. 5(2) as included in Schedule 1 to the *Water Act 2007* (Cwlth).

The Purpose and Development of the Plan

The purpose of the Basin Plan is linked to the objects of the *Water Act*. In addition, however, the Plan is expressly required to provide for

- the establishment and enforcement of environmentally sustainable limits on the quantities of surface water and groundwater that may be taken from basin water resources (*Water Act 2007*, s. 20[b])
- the use and management of the Basin's water resources in a way that optimises economic, social and environmental outcomes (s. 20[d]).

These are expressed in such a way that they are tantamount to obligations. While the provision in relation to environmentally sustainable limits is in effect a specific obligation, the provision in relation to the optimisation of economic, social and environmental outcomes is formulated as a goal or a result.

The development of the Plan is itself controlled by the Act. In developing the Plan there is an obligation to act on the basis of the best available scientific knowledge and socioeconomic analysis (*Water Act 2007*, s. 21[4][b]), and there is a duty to have regard, among others, to

- the consumptive and other economic uses of basin water resources (s. 21[4][c][ii])
- social, cultural, Indigenous and other public-benefit issues (s. 21[4][c][v]).

Consequently, economic, social, cultural and Indigenous issues are of mandatory relevance in developing the Plan.

The Content of the Plan

A wide range of matters must be included in the Basin Plan. These include the long-term average sustainable diversion limits (SDLs) and temporary diversion limits (*Water Act 2007*, s. 22[1], Items 6, 7). It is specifically provided that a long-term average SDL for the Basin's water resources, for the water resources of a particular water-resource plan area or for a particular part of those water resources must reflect an environmentally sustainable level of take (*Water Act 2007*, s. 23[1]). This itself appears to be a reflection of the stated object of the Act to return to environmentally sustainable levels of extraction (s. 3[d][i]). The cumulative effect of these provisions is for SDLs to ensure a return to environmentally sustainable levels of extraction. These specific provisions are directed at environmental sustainability rather than the trilogy of environmental, economic and social sustainability.

Temporary diversion limits are linked to what is described as a temporary diversion provision (s. 22[1], Item 7). The purpose of a temporary diversion provision is to provide for a transitional period to minimise social and economic impacts when the long-term average SDL is lower than the long-term average quantity of water that has in fact been taken (s. 24[1]). The Act accordingly contemplates that the establishment and enforcement of environmentally sustainable limits are likely to have social and economic impacts. These are therefore not only relevant in the development of the Plan, but also addressed through the mechanism of a temporary diversion provision.

The Relationship Between Objects, Purposes and Obligations

The *Water Act* does not give explicit priority to any of the three outcomes: economic, social or environmental. Each provision in the Act, however, must be given effect not only according to its own terms but also in the context of the provisions in the Act stating the objects of the Act and the purpose and content of the Basin Plan. For example, the specific provisions relating to SDLs have effect according to their own specific terms but in the context of the optimisation of economic, social and environmental outcomes.

Although these statements of objects and purpose are not in the form of traditionally enforceable rules, they perform a critical function in the overall system of governance. These statements inform and direct how decisions are to be made at all levels within the system, including those about SDLs and temporary diversion limits. But environmentally sustainable limits, long-term or temporary, are established only in accordance with the specific requirements set out in the Act—not an easy task.

The overall but formally unstated goal is ecologically sustainable development. It brings together the economic, the social and the environmental factors involved in decision making. It is a simple idea but difficult to implement in practice. The ecologically sustainable development of the water resources and natural resources of the Basin is the outcome expected—but not required—by this integrated set of normative arrangements. What is the relationship between these objects, these purposes and these requirements? Compliance with specific requirements stands at the threshold to the optimisation of economic, social and environmental outcomes. Such compliance is almost a condition to be satisfied before attention turns to optimisation of outcomes.

Sustainable Diversion Limits

Assuming that the ultimate goal of these arrangements is the ecologically sustainable development of the water resources as well as the natural resources of the Basin, how is this to be done? One of the specific objects of the Act (s. 3[d][i])—already discussed—is to ensure the return to environmentally sustainable levels of extraction for water resources that have been over-allocated or overused. The principal method for doing this is the inclusion in the Basin Plan of long-term average SDLs. The determination of such a limit—also already discussed—is required to reflect an environmentally sustainable level of extraction (*Water Act 2007*, s. 23[1]). The term 'reflect' is unusual in this context. It probably means to return to an environmentally sustainable level of extraction—the terms used in the relevant statement of objects of the Act. Without question, the reference is to environmental sustainability. In this respect, the Plan is concerned only with environmental sustainability.

The requirements for a water-resource plan for a water-resource plan area are different. A water-resource plan for a water-resource plan area must include requirements in relation to

- the long-term annual diversion limit for the water resources of the water-resource plan area (*Water Act 2007*, s. 22[3][b])
- the sustainable use and management of the water resources of the water-resource plan area within that diversion limit (s. 22[3][c]).

The water-resource plan thus must address the sustainable use and management of the water resources of the relevant area in general but in the context of the SDLs stated in the Plan. Those in the Plan are intended to bring about a return to an environmentally sustainable level of extraction. But the water-resource plan is directed at the sustainable use and management of the water resources in general without any specific reference to environmental sustainability. This presumably means the use and management of water resources in ways that are economically and socially sustainable, as well as environmentally sustainable. This raises the issue of how water resources are to be managed so as to optimise simultaneously economic, social and environmental consequences.

The *Guide to the proposed Basin Plan* (MDBA 2010) indicates how this is done. An environmentally sustainable level of extraction is one that will not compromise the environmental-water requirements of key environmental assets, key ecosystem functions, the productive base, and key environmental consequences for the water resource (MDBA 2010:103). In setting this level, the economic, social and environmental consequences are stated to be optimised and the net economic returns maximised (p. 103). The Guide (pp. 106, 107) indicates how the optimisation of economic, social and environmental results is to be achieved. There are three aspects.

- To meet key environmental outcomes at the Basin level.
- To satisfy the environmental requirements of each catchment at the catchment level.
- To minimise social and economic impacts on basin communities and industries (MDBA 2010:101).

This approach to optimisation picks up the concepts and terms underpinning the Act. The minimisation of social and economic impacts is undoubtedly relevant. For this purpose, impacts are defined as predicted consequences. The Act is directed at the optimisation not only of environmental but also of economic and social outcomes. For this purpose, outcomes are intended consequences. In this way, optimisation of these outcomes goes beyond the minimisation of social and economic impacts and in the direction of the achievement of economic, social and environmental outcomes. But, SDLs for the Basin's water resources are directed at environmentally sustainable levels of extraction—a specific and narrower focus. In any event, environmental sustainability in this sense is a critical part of these overall arrangements. It is within this framework that economic sustainability and social sustainability need to be accommodated within a framework of environmental sustainability. This is an extremely difficult task, which is replete with conflicting interests.

Critical Human Water Needs

The way in which critical human water needs (critical needs) are incorporated within this framework for the governance of the Basin's water resources demonstrates the interdependence of these several instruments. The *Water Act* makes what might be described as a statement of policy—although it is described as a fact—in relation to the preparation of the Basin Plan. In such terms, the Plan must be prepared having regard to the fact that the Commonwealth and the States have agreed

- that critical human water needs are the highest-priority water use for communities who are dependent on basin water resources
- in particular, that to give effect to this priority in the River Murray system, conveyance water will receive first priority from the water available in the system (*Water Act 2007*, s. 86A[1]).

It is accordingly the policy of the Commonwealth and the Basin States to give the highest priority to satisfying critical needs in managing the water in the River Murray system. This policy is implemented by the imposition of a duty that the Basin Plan must include a statement of the amount of water required in each referring basin State (but not Queensland) to meet these critical needs (s. 86B[1]).

The Plan must in addition address the possibility that there will be insufficient water to meet these critical needs. This includes a duty to specify the conditions for the commencement of Tier 2 or Tier 3 water-sharing arrangements in place of Tier 1 water-sharing arrangements among the Basin States (*Water Act 2007*, ss. 86D, 86E). It is neither the Act nor the Plan but the *MDB Agreement* that describes the States' water-sharing arrangements.[2] When Tier 2 and Tier 3 water-sharing arrangements are in place, the distribution of waters to the Basin States is governed by the schedule for water-sharing required to be prepared by the MDBA and approved by the Ministerial Council. This schedule must be prepared on the basis that the highest priority is to be given to the satisfaction of critical needs (MDB Agreement, art. 135[8][a], [b]). There is an additional statement of policy in the Agreement but not apparently reflected in the Act—namely, each state-contracting government will be responsible for meeting critical needs in its State and will decide how water from its entitlement is used (MDB Agreement art. 135[8][c]). Critical needs are thus met in accordance with the arrangements in the Act, the Basin Plan, the *MDB Agreement* and the water-sharing schedule that is part of the Agreement. These arrangements specify a range of relevant duties. It is, however, the responsibility of each State to meet critical needs and to decide how its water entitlement is to be used. This responsibility is circumscribed by the accompanying sets of rules in these instruments.

State Water Entitlements

What has emerged so far is a set of rules of various kinds that relates to the direction of these governance arrangements, the matters required to be addressed by the Basin Plan, and the rules according to which water must be shared among the Basin States to ensure critical needs are met. The State water-sharing arrangements included in the *MDB Agreement* represent potential restrictions upon the States' entitlements to water conferred by the Agreement itself. A State water entitlement is described as the entitlement of a State to water determined in accordance with the relevant part of the Agreement (MDB Agreement art. 2 'State water entitlement'). The relevant part includes the State water-sharing arrangements already discussed. The focus now returns to the entitlements of the Basin States rather than the restrictions on these entitlements.

The Agreement specifies the monthly quantities of River Murray water that South Australia is entitled to receive (MDB Agreement art. 88). In addition, South Australia may store any part of its monthly entitlement for the purpose of meeting the critical needs of the SA community in the upper River Murray

2 *Murray–Darling Basin Agreement*, arts 131–4 as included in Schedule 1 to the *Water Act 2007* (Cwlth).

storages provided this does not affect the availability for New South Wales or Victoria (art. 91). The Agreement confers upon New South Wales and Victoria an entitlement to use the specified amounts of water at the several locations in the river systems described in the Agreement (arts 94, 95). New South Wales and Victoria are, however, under a duty to provide in equal proportions South Australia's entitlement from the water available under these arrangements to New South Wales and Victoria (art. 96).

The Operational Management of the Basin's Water Resources

The responsibility for the operational management of the waters distributed in accordance to these entitlements lies with the MDBA and the agencies of the States. But the operational management of these entitlements is regulated by a complex set of rules in the *MDB Agreement*. These relate in particular to rules for water accounting, the periods of special accounting and for accounting for South Australia's storage right (MDB Agreement, arts 105–30). These entitlements are subject to the range of rules related to critical needs already discussed. What is significant about the nature of these entitlements is this: South Australia is entitled as the downstream State to receive water from the upstream States and to store water in storages located in the upstream States. New South Wales and Victoria as upstream States are entitled to use water from the specified locations in the upstream States. On the face of it, these entitlements are in the form of protectable rights, but their exercise is subject to the range of duties imposed upon the Basin States and the MDBA in relation to the operational management of these entitlements. What is important is that these comprise sets of entitlements that correlate with sets of duties.

The Agreement makes it clear that, whatever the actual entitlements of a State are on a day-to-day basis, the responsibility for managing these entitlements on a day-to-day or indeed a longer basis lies with the States. While the *Water Act*, the Basin Plan and the *MDB Agreement* might be seen as the overarching elements of this system of governance, they are intrinsically linked to the arrangements for managing water resources within the States. This raises the very specific and rather difficult question of the relationship between these overarching arrangements and the rules for managing water resources in the States.

The legal arrangements in the States and Territories vary substantially in terms of detail. This might in certain circumstances be critical; however, the broad pattern across the States and Territories is similar. The right to control water and water resources is vested in the State or the Territory. For the most part, access to and use of such water requires authorisation either directly by the legislation or by a grant under the legislation. Once a water right exists or has been conferred in these ways, its exercise is subject to the relevant rules in the legislation, in the plans and in support of the market arrangements introduced as part of this overall governance framework. For example, the right of an agriculturalist to receive water for irrigation from an irrigation infrastructure operator is subject in the first instance to the rules that comprise the law of the State or Territory but also to the rules set out in the Basin Plan, which is part of the overarching framework. The relationship between these various sets of rules needs to be examined from two perspectives. One arises in the context of the potential liabilities of the holders of water rights under the legislation of the States and Territories. The other arises in the context of environmental water rights held by the Commonwealth Environmental Water Holder (CEWH), and this will be considered first.

Commonwealth Environmental Water Holdings

One of the functions of the CEWH is to manage the Commonwealth's holdings (*Water Act 2007*, s. 105[1][a]). These are water-access rights, water-delivery rights, irrigation rights and interests in such rights held by the Commonwealth (s. 108[1]). The management of these holdings includes buying, selling and dealing in water and in water rights and in making water available from these holdings (s. 105[2][a], [b], [d]). There is a specific duty to manage these holdings in accordance with the environmental watering plan (part of the Basin Plan), any other relevant plan, any operating rules and any environmental watering schedules to which the CEWH is a party (s. 105[4]).

The broad purpose of the environmental watering plan is to protect and restore the environmental assets of the Basin (s. 28[1][d], [e]), and the functions of the CEWH are directed at the protection and restoration of the environmental assets of the Basin and other areas outside the Basin where the Commonwealth holds water (s. 105[3]). Specifically, the CEWH is under a duty not to dispose of water or of environmental-water holdings during a water-accounting period unless the water or the holding is not required to meet the objectives of the environmental watering plan, any other relevant plan or any applicable environmental watering schedules (s. 106[1]). Rights in relation to environmental water may be held by anyone—not only by the Commonwealth. The common restriction is for them to be used to achieve environmental objectives. The same rules apply to the operational management of environmental water, whether the rights are held by the Commonwealth or by someone else. The Commonwealth is, however, in a privileged position in one respect (s. 110). It is exempt from the laws of a Basin State that

- prevent a non-landowner from using water available under a water-access right
- require a non-landowner to hold a licence to use the water.

This exemption applies only to water used to protect declared Ramsar Convention wetlands or water-dependent ecosystems that support species or ecological communities protected by the environmental protection legislation of the Commonwealth. Subject to these exceptions, the same rules apply to all holders of rights in relation to environmental water.

Enforcement Criteria

The *Water Act*, the Basin Plan and its associated water-resource plans confer rights and impose duties upon a range of public-sector and private-sector decision makers and operators. The rules relating to the formulation and content of plans are directed mostly at the Commonwealth and its agencies. Some of the rules—particularly those in the plans—are directed at the private sector to the extent that the plans confer rights, provide for the conferment of rights, impose duties or provide for the imposition of duties. These rights and duties tend to be operational rather than managerial. How to comply with these duties depends upon how the duty is formulated and the standard or criterion it incorporates. The range of standards or criteria is considerable. To whom do they apply?

In principle, everyone is subject to the rules of law that apply in the particular circumstances in question. The *Water Act* (s. 12[1]) specifically states that it binds the Crown in each of its capacities. Neither the Crown—an agency of the Commonwealth—nor an agency of a State is, however, liable to be prosecuted or subjected to civil proceedings for a civil penalty (s. 12[2]). But agencies of the Commonwealth or of a State of certain kinds are not granted this immunity. The kinds of agency not exempt from this immunity—and liable to prosecution—are companies or body corporates in which the Commonwealth or a State has a controlling interest or an agency of a State appointed for a public purpose but operating primarily on a commercial basis (s. 12[4]). Broadly, therefore, agencies operating as corporations in a commercial context are exposed to the full range of enforcement mechanisms.

Then there are the compliance criteria applying to the rules in the Basin Plan, in a water-resource plan and to critical needs. First, there are the criteria for compliance with the rules in the Basin Plan and in a water-resource plan. The obligation imposed upon the MDBA and the other agencies of the Commonwealth is to perform their functions and exercise their powers consistently with and in a manner that gives effect to the Basin Plan (*Water Act 2007*, s. 34[1]). This does not apply to the preparation of the Basin Plan itself (s. 34[2]). The criteria are twofold. The first is consistency with the Basin Plan and the second is in a manner that gives effect to the Basin Plan rather than to give effect to the Plan itself.

Different compliance criteria apply to these institutions and persons

- the Basin Officials Committee
- an agency of a basin State
- an operating authority
- an infrastructure operator
- the holder of a water-access right (*Water Act 2007*, s. 35[1]).

Again, there are two compliance criteria. The first is not to do an act if the act is inconsistent with the Basin Plan (s. 35[1][a]). The second is not to fail to do an act if the failure to do that act is inconsistent with the Plan (s. 35[1][b]). Again, the test is consistency. But in this case, it is restricted to an act or failure to do an act. There is thus one set of criteria for the Commonwealth and its agencies and another set for the other group of persons.

The compliance criteria in relation to critical human water needs are the same but the subject of their application is different. The criteria in relation to the MDBA and other agencies of the Commonwealth are consistency with the matters included in the Basin Plan and in a manner that gives effect to the matters included in the Plan (*Water Act 2007*, s. 86G[1]). In relation to the other agencies and persons, it is only consistency. In this case, it is consistency between the act of the person in question and the matters included in the Plan (s. 86H[1]). The compliance criteria in relation to critical needs do not apply to the performance of a function or the doing of an act or a failure to do an act where that affects State water-sharing arrangements (ss. 86G[2], 86H[4]). The only circumstance in which these compliance criteria apply in relation to critical needs is where one of two conditions is satisfied

- where the Ministerial Council has agreed to the Basin Plan applying to the act or failure
- where the provisions of the Basin Plan required to deal with critical needs have been taken to be a schedule to the *MDB Agreement* in accordance with the Agreement (*Water Act 2007*, ss. 86G[2], 86H[4]).

The test of consistency with the Basin Plan and the test in a manner that gives effect to the Plan afford a degree of flexibility in their application. In other words, there is no clear and absolute duty to comply with the rule. While such tests are not unprecedented, it has never been easy to be certain about their application in practice. It is possible that the meaning of the test of consistency will become an issue. This is because it is the test that applies to operating authorities, infrastructure operators and holders of water-access rights. They might be in the private sector or the public sector. It is likely that the rights conferred upon them and the duties imposed upon them originated in instruments granted under the laws of the State or the Territory. These rights and duties might be inconsistent with the rules stated in the Basin Plan. Under the *Water Act*, the holder of a water-access right is under a duty not to do an act if the act is inconsistent with the Plan. To do so would activate some of the enforcement mechanisms available under the Act.

Enforcement Mechanisms

So much for duties and compliance criteria in respect of these duties. Who has the capacity to protect these rights and enforce these duties? The *Water Act* (s. 137) confers the responsibilities for doing so upon either the MDBA, the Australian Competition and Consumer Commission (ACCC) or the relevant minister. One set of enforcement mechanisms applies to a contravention of the Act, of the regulations under the Act, and of water-charge rules or water-market rules (s. 136). The enforcement mechanisms are

- injunctions granted by the court
- declarations made by the court
- civil penalty orders made by the court
- sanctions imposed by the court in criminal proceedings
- infringement notices
- enforceable undertakings (*Water Act 2007*, ss. 140–64).

These enforcement mechanisms are of general application and reflect the traditional approach to enforcement.

Enforcement notices are directed quite specifically at the management of the water resources of the Basin (*Water Act 2007*, s. 165). Specifically, this includes the Basin Plan and water-resource plans. The power to issue an enforcement notice is conferred upon the MDBA. It is available if there has been a contravention or there is likely to be a contravention of a provision of the Act relating to the management of the Basin's water resources in general. The power is also available in circumstances that reflect the compliance criteria already discussed in relation to the Plan and water-resource plans. That is where a person's conduct is inconsistent with the Basin Plan or a water-resource plan, is prejudicing its effectiveness or implementation, or is having an adverse effect on its effectiveness or implementation. In addition, it is available if the person is omitting to perform an act contrary to the compliance criteria of consistency or effectiveness (*Water Act 2007*, s. 165[1]).

It is non-compliance with the compliance criteria that lies at the foundation of the power to issue an enforcement notice. An enforcement notice enables the MDBA to direct a person to take the actions specified in the notice for all or any of a range of purposes (*Water Act 2007*, s. 165[2]). These include the remediation of any adverse consequences on the health or continued availability of basin water resources (s. 165[2][c]). More specifically, the MDBA is authorised to direct a person not to exercise some or all of the water rights that the person holds (s. 165[3]). The sanction for a breach of an enforcement notice is a civil penalty (s. 166).

While the Act enables members of the public to take part in the planning processes, there appears to be no provision for them to take part in enforcement proceedings. Accordingly, if a member of the public feels aggrieved by a decision made under the Act then the only option, it would appear, is to raise the issue in judicial review proceedings. Given the nature of judicial review proceedings, this could be difficult. The applicant needs to show standing and that there is an issue capable of adjudication by the court. On the face of it, this might be the only way in which a person issued with an enforcement notice may seek a determination of its validity. In effect, the Act provides for enforcement but not dispute resolution.

The Applicable Law

The final question is what law applies to any particular set of circumstances: the law of the Commonwealth or the law of the State or Territory? The principle embedded in the Constitution is clear. The law of a State is invalid to the extent that it is inconsistent with a law of the Commonwealth. This assumes that each law is otherwise valid. The principle behind the *Water Act* is equally clear. The Act of the Commonwealth is not intended to exclude or limit the concurrent operation of any law of a State (*Water Act 2007*, s. 250B[1]). It is a matter of some practical importance. A water right granted by a State may have attached to it a quantity of water determined in accordance with a State water plan. The quantity of water so determined might not be the same as the quantity of water determined in accordance with a water-resource plan in conjunction with the Basin Plan under the Act of the Commonwealth. How is this resolved? The Act deals with it in a number of different ways.

First, if a law of a referring State declares that a provision of a State law displaces a provision of a Commonwealth law, then the provision of the Commonwealth law does not prohibit the act or impose any liability if the provision of the State law specifically permits, authorises or requires the doing of the act (*Water Act 2007*, s. 250D[3], [4]). Second, the operation of a provision of the Commonwealth law is restricted to the extent necessary to ensure that no inconsistency arises between the provisions of the Commonwealth and State laws (s. 250D[5]). Third, the Act authorises the enactment of regulations that modify the operation of the Act itself so that a provision of the Commonwealth law does not apply to a matter dealt with by a law of a referring State or that no inconsistency arises between the two (s. 250E). Fourth, the provisions of the Commonwealth Act do not apply if a provision of a law of a referring State declares the matter to be an excluded matter for this purpose (s. 250C). Otherwise, the intention is for the law of the Commonwealth and the law of the State to have concurrent operation.

The *Water Act* itself provides an example. If the Basin Plan provides for a maximum quantity of water that may be taken from the water resources of a particular water-resource plan area, it is not intended to exclude or limit the concurrent operation of a State law that provides for the same or a lower maximum quantity of water that may be taken from those water resources (s. 40). In other words, the purpose of the provision in the Basin Plan is effectively achieved by the relevant provisions of a State law. There are, however, likely to be a number of situations when the solution might not be so simple. While the Act itself addresses the issue of the relevant law to be applied, the answer to any particular problem will be found in a very careful analysis of the detailed provisions of all of the relevant legal instruments operative under both Commonwealth and State or Territory law.

Conclusion

The achievement of the objects of the *Water Act 2007* will depend to some extent upon how this complicated and interrelated set of normative arrangements is realised, interpreted, implemented and enforced by those who formulate and implement the plans, exercise the rights conferred on them and discharge the duties imposed on them. It will be the public sector and the private sector working together that ensures success or not. The structure that has emerged is—as intended by the National Water Initiative—a combination of plans, regulation and market mechanisms. Regulation has always been a feature of water-resources governance. To some extent, plans have also been a feature of the system To this there has been added markets. These markets are themselves regulated. So there is now a complex amalgam of protectable rights and enforceable duties across the spectrum of these three functions. How one function is performed will have an impact upon the performance of the other functions and vice versa. There is now a rule-based system rather than an administration-based system. The rules apply to the private sector or to the public sector or to both. Most of these rules have been formulated in ways that are potentially enforceable through the legal system.

The optimisation of economic, social and environmental outcomes is one of the goals—intended consequences—of these arrangements. This is no easy task, even in legal terms. Another of the goals is—almost in the form of a duty—to return to environmentally sustainable levels of extraction for water resources that have been over-allocated or overused. The Basin Plan is critical in this respect. It must include the maximum long-term annual average quantities of water that can be taken on an environmentally sustainable basis from the Basin's water resources as a whole and from the water resources or particular parts of the water resources of each water-resource plan area. This is a threshold responsibility and once this has been determined then the optimal economic, social and environmental outcomes can be identified and then achieved. In performing all of the functions, consideration of economic, social and environmental impacts—predicted consequences—is mandatory.

The simultaneous achievement of these three outcomes in this way will be achieved or not by bringing together all the individual structural elements of the system in a rational and coherent way. Individual rights are affected. The way in which these rights are exercised is restricted. The commercial value of these rights might be reduced. Contraventions of the rules are sanctioned. There might be unexpected economic, social and environmental consequences. It is a structure that has been created by the legislature. The legislature has seen fit to move from an administration-based system to a rules-based system. Coming to grips with this system is the major challenge facing the sustainable governance of the Murray–Darling Basin.

Reference

Murray-Darling Basin Authority (MDBA) 2010, *Guide to the proposed Basin Plan. Volume 1: Overview*, Murray–Darling Basin Authority, Canberra.

13. Evidentiary Issues with the Implementation of the Sustainability Duty of Care in the Basin Plan

Jennifer McKay

Introduction

This chapter explores some of the possible legal ramifications of implementing the sustainability duty of care in the Murray–Darling Basin Plan. Australia—having pushed to the limits all available power bases in the Federal Constitution—was able to achieve a political settlement with the States in 2008. This enabled the Federal government to use power refered from the States under section 51(37) of the Constitution. These unique sets of legal arrangements were marshalled to enact the *Water Act 2007* with its requirements to draft the first Basin Plan for sustainable diversion of surface and groundwater in the Murray–Darling Basin (MDB) area of four States. The Basin Plan will be known as the Murray–Darling Basin Plan. The *Guide to the proposed Basin Plan* (released in October 2010) is the locus for community consultation on 'the quality of the data and evidence used and the analyses undertaken' (MDBA 2010:ix).

These political accommodations to create the *Water Act* and the Basin Plan occurred during the final year of a decade of drought and after the patchy implementation of two previous Federal Government initiatives: by the Council of Australian Governments (COAG) in 1994 and in 2004—both known as the National Water Initiative (NWI). The *Water Act 2007* places a sustainability duty of care and imbeds innovative standards of care in one law—in Section 21—applicable to all State governments in relation to sustainable diversion of surface and groundwater. The sustainability duty is expressed as the following: 'to achieve the national interest and the standard of care is to...implement relevant international agreements and to achieve ESD' as defined in Section 4(2). The duty is imposed on each State to draft regional water plans to achieve the sustainability duty, and then the Murray–Darling Basin Authority (MDBA) is empowered to accredit or adopt these—or not—according to the above standards.

This chapter will examine issues in the most potentially litigious areas of dispute in relation to the sustainability duty and standard of care. It will look at how a plaintiff may discharge the burden and the social conflicts that might be generated between different stakeholders. It will consider sustainable diversion limits (SDLs), and the tests around national interest and relevant international agreements, and will consider legal evidentiary issues in all three.

It concludes that a duty to cooperate needs to be inserted in the *Water Act 2007*, and that the power to do this would come from *inter alia* relevant international agreements such as the 1997 UN Watercourses Convention and the recent International Law Commission's Transboundary Aquifer articles. This duty would apply to the States in the drafting of their water plans. Such a duty should also be inserted in relevant State laws to apply to landholders.

The Sustainability Duty of Care in the *Water Act 2007* and the *Guide to the proposed Basin Plan*

The *Water Act 2007* created an independent body—the Murray–Darling Basin Authority (MDBA)—and charged it with developing a Basin Plan for the consideration of the relevant Commonwealth minister (MDBA 2010). The sustainability duty of care was expressed as: 'the Authority is to determine the volume of water required to maintain and restore environmental assets, using best available science and the principles of ecologically sustainable development. Subsequently, the Authority addressed the optimisation of environmental, social and economic outcomes' (MDBA 2010:iii).

This was built on earlier reforms, and in fact embeds the 2004 COAG reforms known as the NWI. These punctuated and hence changed the State governments' introspective water use, management cultures and law regimes (McKay 2010). The pre-existing introspective State law regimes have been weakly cooperative, and they promoted economic and social development.

These law regimes were punctuated with requirements for the achievement of ecologically sustainable development. This was a broad requirement. The various State law formulations included: the precautionary principle; inter and intra generational equity; integration of long and short-term environmental, economic and social considerations; water plans in regions; separation of water from land to create water markets; incorporation of the private sector in water supply; environmental allocations in water plans to maintain the health and viability of river systems and groundwater basins (COAG 1994:Attachment A Clause 4[d]; Commonwealth of Australia 2004); capacity sharing of consumptive pools of water (after environmental needs have been catered for); and full cost recovery. The 2004 reforms were specific and required improved environmental-management practices, completely returning all currently over-allocated or overused systems to environmentally sustainable levels of extraction and managing surface and groundwater as a single source. The NWI defined environmental outcomes as maintaining ecosystem function through periodic inundation of floodplains, wetlands biodiversity, water-quality and river-health targets (Commonwealth of Australia 2004:Schedule B).

The *Water Act 2007* arose out of the dissatisfaction by the political leader of the day (Prime Minister, John Howard) with the implementation of the above reforms, despite financial incentives being used to drive reforms—that is, payment to the States for achieving the above. It is fair to say that there have always been disputes between the parties about the meaning of the terms in the NWI and also the Water Act 2007.

The *Water Act 2007* was built, then, against a background where implementation issues were known to be a problem and the communication of science to the policy makers was fraught (Tomlinson and Davis 2010). Many of these problems arose from lack of knowledge but also from lack of coordination between the States. The *Water Act 2007* continued, however, to be innovative and created several new terms, such as long-term sustainable diversion limits (SDLs). These are long-term average volumes of water that can be diverted for consumptive use (irrigation, town-water supplies, industry, and so on) after the environment has received what it requires (MDBA 2010:103). The *Water Act 2007* (s. 23[1]) refers to this standard as the 'environmentally sustainable level of take', and it requires that this be established using the best available science, and also so that social, economic and environmental goals are met. The MDBA considers that the SDL is a judgment about how best to balance these requirements.

The Guide to the Proposed Basin Plan specified tat the reductions in take (interceptions and diversions) should be between 3,000 and 7.600 GL. It was recommended, however, that the reduction in take whould be no more than 4,000 GL because of socio-economic considerations. Even the 3,000GL proposed reduction resulted in considerable community angst and was also criticised for not taking into account social and economic effects (McKay 2010). The main dispute was over the assertions of projected job losses. The MDBA stated 800, but the NSW Farmers' Federation predicted *a massive depopulation of the region*. The MDBA took legal advice on the basis of the 3000 GL figure and suggested that this figure could be too high because of 'greater consideration of the social and economic needs of rural communities. This was welcomed by irrigators but left environmentalists fuming. The advice would force a review with less returned to the river for environmental purposes' (Ker 2010).

A key question is, how will these judgments be made? In a court case how will the words in the *Water Act 2007* be interpreted and the judgments of the

MDBA be evaluated? The evidentiary standard for SDLs is the best available science, which *prima facie* is a high objective standard and is one bound to cause controversy on the issues of 'best' and 'available'.

This term has been used in several US acts but not any other Australian acts. For example, in the United States, National Standard 2 of the *Magnuson–Stevens Fishery Conservation and Management Act* states that conservation and management measures shall be based on 'the best scientific information, available'. Further, the US Environmental Protection Agency (USEPA) has emphasised the role of best available science in implementing the *Clean Water Act 1997* (Sullivan et al. 2006). US discussions of best available science include these factors

- a clear statement of objectives
- a conceptual model, which is a framework for characterising systems, making predictions, and testing hypotheses
- a good experimental design and a standardised method for collecting data
- statistical rigour and sound logic for analysis and interpretation
- clear documentation of methods, results, and conclusions
- peer review.

All of these issues may be litigated. The availability of science is a further complication. The MDB Guide was based on specially commissioned data and studies that were subsequently made available. Other sources of scientific knowledge are, however, clearly available through the Internet, and include refereed journal articles, grey literature and the opinions of experts.

The legal issues in the use of scientific evidence in courts have been discussed by Edmond and Mercer (1997) in relation to the weight judges should apply to the various types of scientific knowledge (such as above). The context discussed by Mercer was the judicial use of scientific evidence regarding causation of birth defects in the United States. In fact, the cases have ranked available scientific studies according to how they were generated—independently peer reviewed, replicated or produced for the litigation.

All of these considerations would be applied to any litigation taken by a State in relation to the Basin Plan and have already arisen in the several State cases on the scientific bases for litigation when State water plans have reduced the amount of water able to be allocated on ecologically sustainable development (ESD) grounds. The US courts recognised the primacy of published epidemiological studies (Dietz and Stern 1998; Edmond and Mercer 1997:666).

The judges will require some intellectual dexterity to wrestle with the content of the water plan and the complex range of science, such as epidemiological studies, which could result in litigation. As demonstrated by the US litigation cases, the choices made are amenable to a political analysis, and the level of the court, the issue and other considerations will be relevant.

The ESD Implementation in State Water Plans and the Standard-of-Care Evidentiary Issues with Accreditation of State Water Plans

An ESD standard has been placed in State laws that have required regional water plans in agricultural regions for several years. Under State laws, large water users such as individual companies are also required to produce water-efficiency plans (Gunningham and Sinclair 2010).

In many cases, there was social conflict between users in the regional water-planning areas as plans had been drafted to implement reduced water allocations to achieve ESD. Hence, there has been considerable State litigation, which will assist the High Court and State Supreme Courts in dealing with these new issues. These decisions were, however, made under State laws with different formulations of the words used to describe the ESD standard. These decisions are not binding on the High Court, which will be hearing these matters under the *Water Act*.

The several decisions have tended to uphold the adoption of the science that suggests a river or aquifer is over-allocated. The courts have looked into the processes of notification and required natural justice to the water-holder but in the end have generally reduced the amount of water that an individual may hold—even if this would destroy their business. A case in point is Elandes in the Environment Resources and Development Court of South Australia (*Elandes Nominees Pty Ltd vs Minister for Water Resources* [2002] SAERDC 130), in which reduced allocations to achieve ESD in an aquifer meant that one grower had to leave the almond industry. Another case is Harvey in New South Wales, where the plan reduced water allocations by nearly 50 per cent in an aquifer. The dispute involved evidence on the quality of the science. The science was tested, and ultimately a scheme was devised to reduce water allocation not *pro rata* (as is common in decisions), but according to a scheme that recognised past use.

The *Water Act* incorporates a process whereby a State-based water-resources plan that provides for management of water is accredited under Section 63—or adopted under Section 69.

Figure 13.1 The Water Act 2007 and the oversight of some basin plans-why not the entire country?

Note: Concept 2010 - Jennifer McKay, drawn by D Blaiklock, copyright held (one of two)

The accreditation requires evaluation of the evidence used by the State to support the ultimate level of water that is deemed to be available. This will rehearse some of the above considerations, but in an administrative-law context in the first instance. There is potential for a contest on the scientific bases for the decisions and an examination of the consultation processes undertaken by the State when drafting the plan. Hence, there are multiple avenues for evidentiary assessments of the underlying documents.

With regard to adoption or non-adoption of a State plan, the reasons for the non-adoption decision would also need to be reasonable and this would incorporate administrative-law standards of fairness and natural justice at the suit of a State. All of this litigation would take place in the High Court and is likely to develop the rules of evidence.

Legal Duty of Care: The national interest

The national interest is not a well-litigated term in Australian law. It appears in the objects Clause (s. 3[a]). The objects of this Act are '(a) to enable the Commonwealth, in conjunction with the Basin States, to manage the Basin water resources in the national interest'.

If it is accepted that the SDL is in the national interest then the Commonwealth minister will need to give reasons for not accrediting or adopting a State plan if the reason is that the plan is not in the national interest. The minister could form a view that a plan in one region takes too much water and hence is not in the national interest.

If we accept that SDLs are in the national interest then evidence will need to be adduced by the minister to provide the reasonable grounds for the non-adoption or non-accreditation of the State plan. This will be a highly political question and hotly contested. The community in the Murray–Darling Basin is not static, and the individuals in it belong to a matrix of regional communities, occupational associations and socioeconomic segments. The reaction to the Basin Plan shows that social change does not come about just by government decree but depends on changes in human relationships among individuals (Saunders 1999). The socio-political question could be stated thus: will the plan transform the ESD dictate into a popular national belief? The early hostile reactions—including the public burning of the Guide in New South Wales—suggests that the process to transform ESD into a national belief has a long way to go.

One final point on the national aspect of the test: in law, the national aspect may apply only to a region, but politically this will prove very difficult. Growers in half of Victoria will be subject to the national interest test and the rest merely subject to State laws. This will mean two sets of rules and will increase community angst.

The Legal Standard of Care: Relevant international agreements

> Section 21 provides the general basis on which the Basin Plan is to be developed. The *Water Act* directly incorporates several relevant treaties and makes allowance for inclusion of others by the term 'relevant'. It is unlikely that relevance will provide much cause for legal dispute; however, this direct incorporation of the treaties brings Section 15AB(2)(d) of the *Acts Interpretation Act 1901*(Cwlth), which permits recourse to 'any treaty or other international instrument that is referred to in the Act' as extrinsic material that may be used to confirm the ordinary meaning of the text (s. 15AB[1][a]) or where there is ambiguity or the ordinary meaning would lead to an absurd result (s. 15AB[1][b]). Clearly, what is not considered ambiguous will be an issue in specific cases and this will cause evidentiary issues to arise.

Apart from the treaties listed, there are two texts of relevance: the UN Convention on the Non-Navigational Uses of International Watercourses 1997 and the 2010 UN International Law Commission Law of Transboundary Aquifers. McCaffrey (2010) has long argued that the 1997 Convention is customary international law and hence applies to all nations.

In the drafting of the *Water Act*, the 1997 Convention was not referred to at all, although many of the concepts in the 1997 Convention do accord with ESD. The 1997 Convention and the Transboundary Aquifer law include an innovative topic, and it is suggested that the *Water Act* include this. The concept covers the general duty to cooperate. The general obligation it states in the transboundary aquifer rules is as follows:

> Aquifer States shall co-operate on the basis of sovereign equality, territorial integrity, sustainable development, mutual benefit and good faith in order to attain equitable and reasonable utilization and appropriate protection of their transboundary aquifers or aquifer systems. States are enjoined to establish joint mechanism of cooperation. (Article 7)

It is recognised that treaties may have an indirect impact on domestic Australian law prior to their implementation (such as in the Teoh case), therefore the above documents could also have an impact. Irrespective of the debatable position in law, the concept of joint cooperation mechanisms is sound and could arise in a de facto sense through the creation of the Basin Plan.

Conclusions

This chapter has outlined some of the key issues where the social conflicts over the potential Murray–Darling Basin Plan are likely to reach a court for judicial settlement. This will bring with it a host of evidentiary issues related to the best available science and the concepts of national interest, and the ambiguity in these words might fail to be guided by international standards. Clearly, some sectors of the community think the plan has emphasised the environment too much, while others hold the opposite view.

When cases actually come to the High Court, some of these ambiguities will be settled, but the conflict-resolution processes will not end there, as it is always possible to change the Act in the event of too much social conflict. At the moment, it seems that the sustainability duty of care is being tested on the Australian community over this period (until 2012) (AAP 2010), and stakeholder views are to be considered in the new final Plan. One way to assist this would be to highlight the importance of cooperation over the management of shared basin resources and to outline and study examples of cooperative arrangements.

Bibliography

Australian Associated Press (AAP) 2010, 'Basin authority needs more time for plan', *The Age*, 2 November 2010.

Commonwealth of Australia 2004, *Intergovernmental Agreement on a National Water Initiative*, Commonwealth of Australia, Canberra.

Council of Australian Governments (COAG) 1994, *Water Resources Policy Communique and Report of the Working Group on Water Resources Policy*, Council of Australian Governments, Canberra.

Dietz, T. and Stern, P. C. 1998, 'Science, values and biodiversity', *BioScience*, vol. 48, pp. 441–4.

Edmond, G. 1998, 'Negotiating the meaning of a "scientific" experiment during a murder trial and some limits to legal deconstruction for the public understanding of law and science', *Sydney Law Review*, vol. 20, no. 3, p. 361.

Edmond, G. and Mercer, D. 1997, 'The secret life of (mass) torts: the "Bendectin Litigation" and the construction of law–science knowledges', *University of New South Wales Law Journal*, p. 666–706.

Gunningham, N. and Sinclair, D. 2010, 'Water efficiency plans', *Policy and Practice*, vol. 27 EPLJ, p. 331.

Hassan, F. A. 2010, *Water history for our times*, IHP Essays on Water History No. 2, International Hydrological Programme, UNESCO, Paris.

International Law Commission (ILC) 2010, *Law of Transboundary Aquifers*, International Law Commission, United Nations, New York.

Ker, P. 2010, 'Murray Darling back flip', *The Age*, 27 October 2010.

McCaffrey, S. 2010, *Sovereignty and Cooperative Management of Shared Water Resources in a Time of Shrinking Availability: The role of international law in ISARM 2010 transboundary aquifers—challenges and new directions*, 6 December 2010, UNESCO, Paris.

McKay, J. 2010, 'Some examples of the integration of environmental, economic and social considerations into decision making: the jurisprudence of facts and context', in D. Freestone and M. Nijhoff (eds), *Legal Aspects of Sustainable Development. Volume 7*, Leiden, Boston.

Murray–Darling Basin Authority (MBDA) 2010, *Guide to the proposed Basin Plan*, Murray–Darling Basin Authority, Canberra.

Saunders, H. H. 1999, *A Public Peace Process Sustaining Dialogue to Transform Racial and Ethnic Conflicts*, Palgrave, New York.

Sullivan, P. J., Acheson, J. M., Angermeier, P. L., Faast, T., Flemma, J., Jones, C. M., Knudsen, E. E., Minello, T. J., Secor, D. H., Wunderlich, R., and Zanetell, B. A. (2006). 'Defining and implementing best available science for fisheries and environmental science, policy and management.' *Fisheries* Vol 31(9), pp. 460-465.

The Age 2010, 'NSW farmers want Murray water plan pulled', *The Age*, 27 October 2010.

Tomlinson, M. and Davis, R. 2010, 'Integrating aquatic science and policy for improved water management in Australia', *Marine and Freshwater Research*, vol. 61, pp. 808–13.

United Nations 1997, *United Nations Convention on the Law of the Non-Navigational Uses of International Watercourses*, UN Doc. A/Res/51/229, United Nations, New York.

Cases

Elandes Nominess Pty Ltd vs Minister for Water Resources [2002] SAERDC 130

Harvey & Anor vs Minister Administering the Water Management Act 2000 [2008] NSWLR 165.

Teoh vs Minister for Immigration, Local Government and Ethnic Affairs (1994) 121 ALR 436 at 443

Part IV. Economics

14. Economic Costs and Benefits of the Proposed Basin Plan

R. Quentin Grafton

> If the focus does not swing back towards considering water required for the environment, then the nation risks irretrievably damaging the attributes of the Basin that enable it to be so productive.
>
> — Murray–Darling Basin Authority (MDBA 2010a:iv)

Introduction

This chapter examines the cost and benefits of increasing environmental flows in the Murray–Darling Basin (MDB) relative to the recent past, and the costs of business as usual of not increasing flows. The starting point for the analysis is the *Guide to the proposed Basin Plan. Volume 1: Overview* released on 8 October 2010 by the Murray–Darling Basin Authority (MDBA). The Guide is a consultation document that tries to give effect to the *Water Act 2007* (Section 3, paragraph [d])—namely

1. to ensure the return to environmentally sustainable levels of extraction for water resources that are over-allocated or overused
2. to protect, restore and provide for the ecological values and ecosystem services of the Murray–Darling Basin.[1]

To achieve the objects of the *Water Act 2007*, the MDBA estimates that there needs to be between 3000 GL/yr and 7600 GL/yr of extra volumes of water for the environment. These volumes are specified as long-term averages such that in periods of sustained low inflows the volumes allocated to the environment would be less than average, while in periods of sustained above-normal flows the volumes allocated to the environment would be greater.

The Guide specifies a range of basin-wide volumes for the environment that are paralleled by specific environmental volumes in the principal catchments of

1 The *Water Act 2007* was preceded by the National Water Initiative (NWI), which has been agreed to by all governments in the Murray–Darling Basin and the Australian Government. The NWI commits the signatories to 'complete the return of currently overallocated or overused systems to environmentally sustainable levels of extraction' (Paragraph 23, iv) and to a timetable whereby 'States and territories agree that substantial progress will be made by 2010 towards adjusting all overallocated and overused systems' (Paragraph 44).

the Basin. Given that the proposed environmental volumes in most catchments within the MDB exceed current environmental flows in terms of long-term averages, there will need to be a reduction in water interceptions and/or extractions from watercourses in most parts of the Basin.

The proposed permissible water extractions for the MDB are called sustainable diversion limits (SDLs) and are specified for 19 regions. As currently proposed in the Guide, SDLs 'must reflect an environmentally sustainable level of take' (*Water Act 2007*, Section 23, Paragraph [1]), and, in many catchments, are less than long-term average extractions. The level of SDLs—and what should be the associated increase in volumes of water for the environment—is at the core of the debate about the proposed Basin Plan. We examine these issues in terms of the marginal benefit and the marginal cost of increased volumes of water for the environment and the costs of business as usual.

Costs of Business as Usual

Norris (2010) provides an excellent overview of the costs of business as usual. Even before the drought, the *Snapshot of the Murray–Darling Basin River Condition* (Norris et al. 2001) documented that 40 per cent of the river length had biota that was significantly impaired and 95 per cent had degraded environmental condition. This finding is supported by the first Sustainable Rivers Audit, released in 2008, which rated the health of 23 river valleys in the MDB. The 2008 audit found that 20 of 23 river valleys were in very poor (13) or poor (seven) health (Davies et al. 2008).

The poor state of the rivers in the MDB is a direct result of overuse that has occurred from the over-allocation of water entitlements to irrigators in decades past. The CSIRO (2008) calculates that the relative level of surface-water use in the MDB is extremely high and is likely to deteriorate, based on median climate change projections to 2030 under current water-planning arrangements, as shown in Figure 14.1. The adverse environmental effects arise not only from reduced environmental flows, but also from a change in the timing of flows. Two examples of this 'timing' problem are presented in Figure 14.2 for the Goulburn–Broken and the Murrumbidgee rivers. The green line in Figure 14.2 represents pre-development flows while the blue line represents historical flows with irrigation development. In the case of the Goulburn–Broken rivers, peak flows now occur in January–February instead of August–September.

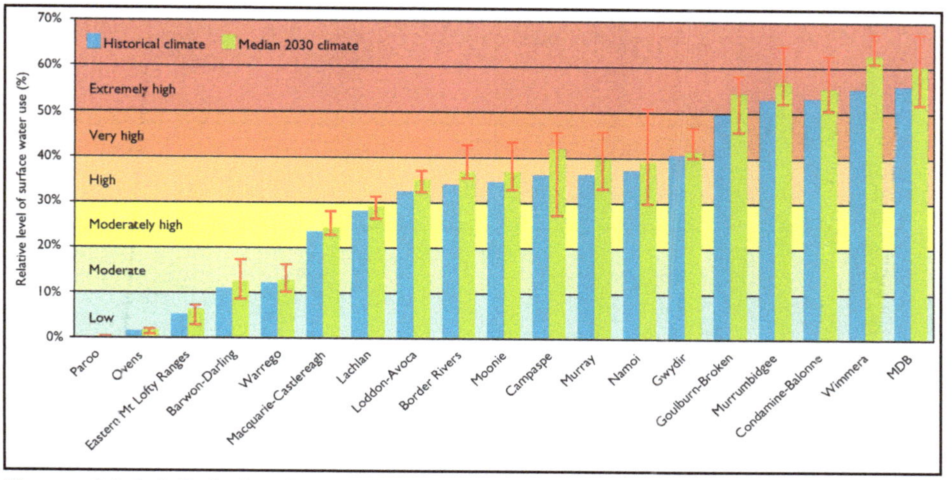

Figure 14.1 1 Relative level of water use in 18 regions and the entire Murray–Darling Basin

Source: CSIRO (2008:33).

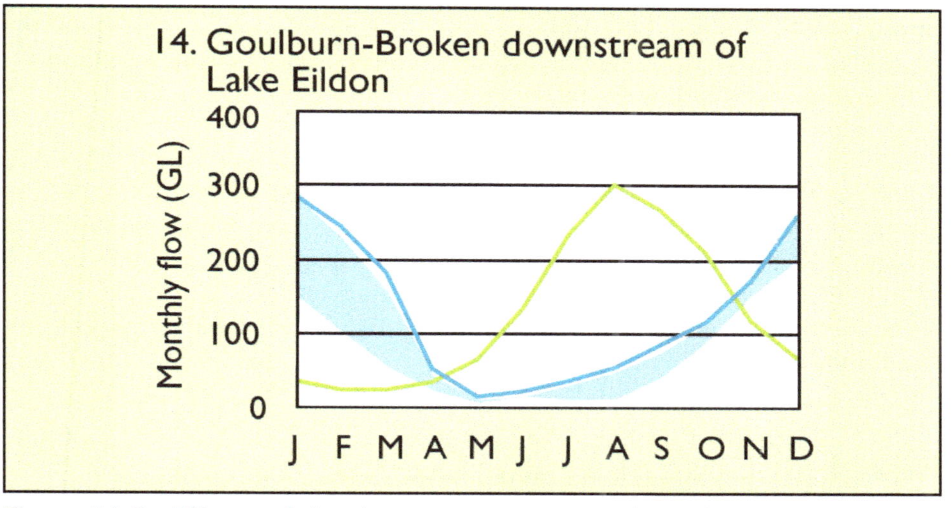

Figure 14.2a Effects of development on seasonal flows in the Goulburn–Broken and the Murrumbidgee rivers

Source: CSIRO (2008:54).

On the basis of recent climate (1997–2006) and with existing water-resource plans, or business as usual, average surface-water availability in the Murray–Darling Basin would be 27 per cent less than long-term average surface-water availability. Total surface-water diversions, however, would be just 13 per cent less while total end-of-system flows would fall by about half, or about four times as much as extractions (CSIRO 2008:59). This is because 'current surface water sharing arrangements in the MDB would generally protect consumptive water users…but offer little protection for riverine environments' (CSIRO 2008:8).

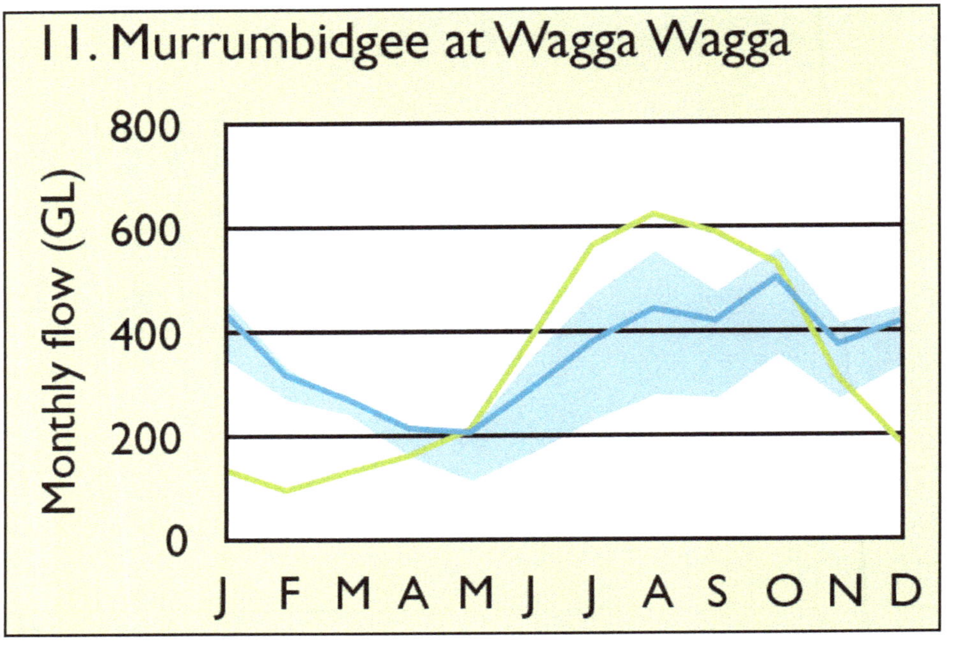

Figure 14.2b Effects of development on seasonal flows in the Goulburn–Broken and the Murrumbidgee rivers

Source: CSIRO (2008:54).

These current water-planning arrangements are largely responsible for the almost zero flows at the River Murray mouth over the past decade, as shown in Figure 14.3. Prior to the development of irrigation in the Basin, zero flows at the Murray mouth were estimated to have occurred just one year in 100, on average. Using historical climate and business-as-usual water-planning arrangements, this is expected to occur 40 years every century (CSIRO 2008:5).

Marginal Cost of Water Reform

The Guide provides a summary of the possible costs to irrigated agriculture and basin employment from reductions in current diversion limits from between 3000 and 4000 GL/yr (MDBA 2010a:147) based on modelling undertaken by the Australian Bureau of Agricultural and Resource Economics–Bureau of Rural Sciences (ABARE–BRS 2010a). For the Basin as a whole, these reductions result in: 1) a lowering of forgone profits in irrigated agriculture of between 6 per cent and 9 per cent; 2) a fall in the gross value of irrigated agriculture (GVIAP) of between 13 and 17 per cent; and 3) a decline in basin employment of between 0.09 and 0.12 per cent.

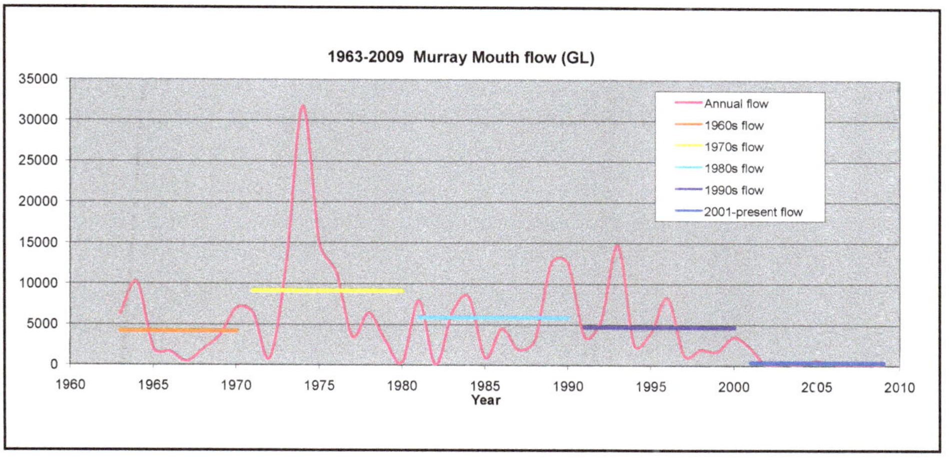

Figure 14.3 Flows at the River Murray mouth (GL/yr)

ABARE–BRS calculate the costs of reductions in current diversion limits of 3500 GL/yr under two scenarios: the implementation of the SDLs only, without offsetting government programs (Scenario 1); and with the additional funds provided to irrigators under the Water for the Future package, including the Australian Government commitment to bridge the remaining 'gap' by purchasing water entitlements from willing sellers should the desired diversion limit be less than current diversion limits (Scenario 2). Their results are obtained from a model of irrigated agriculture (Water Trade Model) that generates changes in GVIAP that links to a seven-region economic model of the MDB (AusRegion Model), as shown in Figure 14.4. Under Scenario 2, and with a 3500 GL/yr reduction in current diversion limits, the GVIAP in the Basin is predicted to fall by 10.1 per cent and profit in irrigated agriculture by 4.6 per cent. Employment in the Basin is expected to increase by 2018–19 as a result of the interventions evaluated under Scenario 2, but not with Scenario 1.

Other modelling of the effects of reductions in long-term average diversions of between 3000 and 4000 GL/yr using the Term-H20 general equilibrium model provides similar results. Namely, if irrigators are fully compensated for reduced water diversions then the impact basin wide on real gross domestic product (GDP) is small and negative, but aggregate household consumption is weakly positive (Wittwer 2010).

The ABARE–BRS and Term-H20 model results indicate only moderate impacts, basin wide, of large reductions in watercourse diversions. These findings are consistent with other models (ABARE–BRS 2010a:87–91) such as the findings of the Wentworth Group of Concerned Scientists, who predict much smaller proportional reductions in profits for a given decline in irrigated-water use. Mallawaarachchi et al. (2010:2) also find that if current cap diversion was reduced by 37 per cent then gross agricultural returns would fall by 16 per cent, and the loss in regional economic surplus—not accounting for the economic benefits from the sale of water entitlements—would also be 16 per cent. The small impacts basin wide, however, mask larger impacts in regions more dependent on irrigated broadacre agriculture, primarily located in the upper catchments of the Basin. For instance, ABARE–BRS (2010b:25) predict the largest absolute reductions in GVIAP will occur in the southern Basin, particularly in the Murrumbidgee, Murray (New South Wales), Murray (Victoria) and Goulburn–Broken rivers regions.

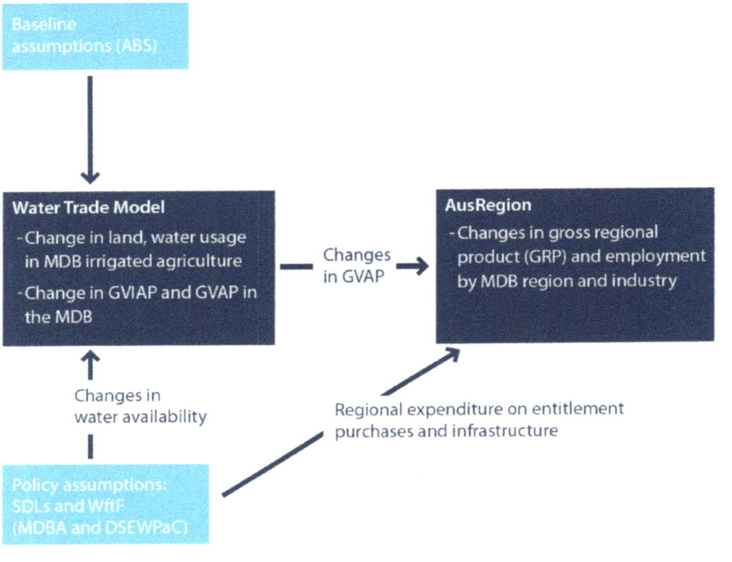

Figure 14.4 ABARE–BRS modelling of effects of reduced watercourse extractions in the Murray–Darling Basin

Source: ABARE–BRS (2010b:14).

The model results of ABARE–BRS and others (Grafton and Jiang Forthcoming; Mallawaarachchi et al. 2010) are supported by what actually happened in the recent drought. Over the period 2000–01 to 2007–08 irrigated surface-water diversions fell by about 70 per cent, yet the GVIAP in nominal terms declined by less than 1 per cent (Australian Bureau of Statistics 2010). This is because irrigators have been able to substitute other inputs (such as purchased fodder) and because water trading has reallocated water from lower to higher value irrigation uses (National Water Commission 2010:56). Indeed, some of the worst drought-affected communities (Griffith and Shepparton) actually grew in size from 2004 to 2008. While water trade has played an important role in helping to maintain GVIAP during the drought, shifts in water use due to water trade do not explain changes in population, unemployment or average incomes across regions in the Basin (National Water Commission 2010:66–7).

In the past decade-long drought, farmers faced the twin hardship of increased water demand for their crops and uncompensated reductions in water allocations. In contrast, the August 2010 commitment by the Gillard Government to bridge the gap between current diversions and SDLs by purchasing only from willing sellers provides full compensation for any reduced allocations to water entitlements. Thus, with the transition to SDLs proposed in the Guide, farmers can plan their operations with a great deal more certainty.

A comprehensive review of the economic impacts from $1.5 billion recovery of water entitlements by the Australian Government for the period 2008–09 to 2010–11 finds that water-entitlement sales would increase disposable incomes and consumption in regions where sales occur (Hone et al. 2010:32–3). This does not imply that there will not be negative impacts from reduced water extraction as currently proposed, but it does stress that these effects will be: 1) highly localised; 2) almost exclusively in irrigation-dependent communities primarily dependent on lower-value irrigation activities (such as irrigated broadacre agriculture); and 3) incurred by those who do *not* own water entitlements, but are, nevertheless, directly and indirectly dependent on irrigated agriculture for their livelihood.

Marginal Benefit of Water Reform

Some of the key benefits from increased environmental flows cannot be measured in the marketplace in terms of traded goods and services. To help overcome this information problem, economists have developed methods for non-market valuation (Champ et al. 2003). A frequently used technique is choice modelling, whereby respondents are provided with a set of choices in a questionnaire that includes environmental attributes (kilometres of healthy riverbank vegetation, frequency of bird-breeding events, and so on) and a financial cost associated with each choice. This allows the researcher to calculate a willingness to pay for marginal changes in the defined environmental attributes, which can then be aggregated to the population as a whole, based on the reported characteristics of the respondents.

Several choice-modelling exercises have been undertaken in the MDB and these are summarised by Bennett (2010). These values are site and time specific. For NSW rivers, Morrison and Bennett (2004) estimate that respondents, on average, are willing to pay between $1.25 and $2.61 to ensure a 1 per cent increase in the length of healthy native vegetation along their local rivers.

The most recent literature on the willingness to pay for marginal improvements in aquatic or riverine environments within the MDB is summarised by Morrison and Hatton-MacDonald (2010). They use estimates from existing studies and transfer these benefits to 19 regions within the Basin. The attributes they consider are in terms of water-based recreation, native vegetation, native fish, colonial water-breeding events, and for waterbirds and other species. Aggregation of the values across Australia over all attributes—except for water-based recreation—yields a present value of $4.3 billion. Thus, the present value of the willingness to pay for key environmental improvements within the MDB—and that are dependent upon environmental flows—are estimated to exceed $4 billion.

A key challenge in using non-market estimates for environmental improvements is to relate willingness-to-pay measures for improved riverine habitats to increased environmental flows. This requires biophysical models that link increased flows or changes in timing of flows to environmental improvements. At present there are prototype models that have been developed for the Narran Lakes and Gwydir Wetlands that predict environmental outcomes in wetlands systems associated with environmental flows (Merritt et al. 2010). If these models were combined with non-market valuation studies they have the potential to provide estimates of the marginal non-market benefits of increased environmental flows.

In addition to non-market benefits, regular flooding of the Basin landscape associated with increased environmental flows provides market benefits. These include benefits in terms of water quality and improvements in soil fertility from the flushing out of salts and other contaminants from soils. There would also be substantial market benefits to floodplain graziers, who would be able to enjoy increased grazing on their properties. Estimates of the grazier benefits associated with regular flooding are some $12.50 per hectare (Arche Consulting 2010:18).

Is 3000 GL/yr Too Little?

The MDBA has established that the minimum volume of increased flows should be, on average, 3000 GL/yr based on the best available science available as of October 2010. Engineering 'works and measures' that use pumps and infrastructure to flood key areas, and possibly adjustments in the timing of environmental flows, can be used to optimise the delivery of water for environmental purposes. There is, however, no credible evidence that such approaches justify a volume of less than 3000 GL/yr in increased environmental flows. Indeed, evidence from the management of the key-icon sites that are part of the 'Living Murray First Step' that delivers approximately 500 GL/yr to these sites suggests the converse. Namely, works and measures without much larger environmental flows are inadequate because 'the inability to direct large volumes of water to sites over the past 5 years...raises real concerns about the ability to maintain or improve the condition of each site' (Swirepik 2009:3).

There is supporting evidence from previous studies that indicates that, for the Murray River alone, an additional 3350 GL/yr in environmental flows, along with improved river operations, would be required to ensure a high probability of a healthy, working Murray River. The Wentworth Group of Concerned Scientists et al. (2010) have estimated that the Basin as a whole would need an additional 4400 GL/yr to avoid a substantial risk that the rivers of the Basin will fail to be in a healthy state. A similar methodology used by the Wentworth Group of Concerned Scientists has also been adopted by the MDBA in defining a target range for desired environmental flows. The MDBA (2010b:108–10) specifies that to ensure key ecosystem functions, environmental flows should be at least 60 per cent of without-development flows. Such volumes, however, would ensure only a *high* degree of uncertainty to achieve desired environmental goals. An environmental flow regime of 80 per cent of without-development flows would give a *low* degree of uncertainty to achieve desired environmental goals.

The 3000 GL/yr increase in environmental flows recommended as a minimum by the MDBA (2010a) represents an even *greater* degree of uncertainty than the case of 60 per cent of without-development flows. The 3000 GL/yr target is also predicted to leave the Condamine–Balonne, Gwydir, Loddon, Lower Darling and the Murray regions of the Basin (MDBA 2010a:74) in 'poor' status. The reason 3000 GL/yr falls outside the MDBA's own target level is because the high uncertainty target for end-of-system flows is actually 3856 GL/yr. Thus, using the MDBA's own confidence intervals of +/– 20 per cent, the lowest boundary of the confidence interval for a high degree of uncertainty consistent with environmental flows at 60 per cent of without-development flows is 3085 GL/yr (MDBA 2010b:114–15). In contrast, the upper-end confidence limit with a high degree of uncertainty is 4615 GL/yr.

Is 4000 GL/yr Too Much?

The MDBA has advised that the maximum desired increase in environmental flows should be no more than 4000 GL/yr based on social and economic information. This upper limit of 4000 GL/yr is an attempt by the MDBA to balance the need for increased water in the landscape against the costs to irrigated agriculture and irrigation-dependent communities of reduced water interceptions and/or watercourse extractions. This is because the *Water Act 2007* obliges the MDBA to prepare a plan such that 'the economic, social and environmental outcomes are optimised and the net economic returns are maximised'. The socioeconomic considerations, however, are subsidiary to the objects of the *Water Act 2007*, which are to ensure levels of extraction 'will not compromise the environmental water requirements of key environmental assets including water-dependent ecosystems; ecosystem services and sites of ecological significance; [and] key ecosystem functions'.

The subsidiarity of socio-economic considerations to meeting the environmental objectives of the *Water Act 2007* is emphasised by legal expert Professor George Williams of the University of New South Wales, who observes 'that these environmental considerations take precedence and that local economic and other concerns must be taken into account "subject to them"' (Williams 2010). Thus, limiting the maximum increase in environmental flows to 4000 GL/yr can be justified only if there is sufficient certainty that such volumes will achieve the environmental goals of the *Water Act 2007*.

Ideally, the desired increase in environmental flows from a societal perspective should be a volume of water that generates environmental outcomes such that the marginal benefit in dollars of an extra gigalitre per year of water for the environment equals the marginal cost in dollars from reduced interceptions or extractions to generate these increased flows. A stylised representation of this outcome is provided in Figure 14.5, where it is supposed that the marginal benefit of additional environmental flows declines with the cumulative amount allocated to the environment.

A possible justification for a downward-sloping marginal benefit curve is that at very low volumes of water allocated to the environment widespread die-offs would be expected in vegetation and species, as well as major problems in water quality, such as from algal blooms and acidity from the prolonged exposure of acid-sulphate soils. If the landscape were to receive successively greater volumes of water, the extra benefits of environmental watering would still remain positive, but would eventually diminish as the minimum volumes needed to avoid crossing minimum critical thresholds were achieved.

In contrast, the marginal cost of reduced interceptions or extractions increases with the amount of environmental flows. The justification for a rising marginal cost is that reductions in extractions can, initially, be achieved at a relatively low cost because there are uses of water for irrigated agriculture that generate low net returns. As more and more water is allocated to the environment, however, progressively more profitable forms of irrigation would have to get by with less water and, thus, the extra costs of reduced extractions would increase.

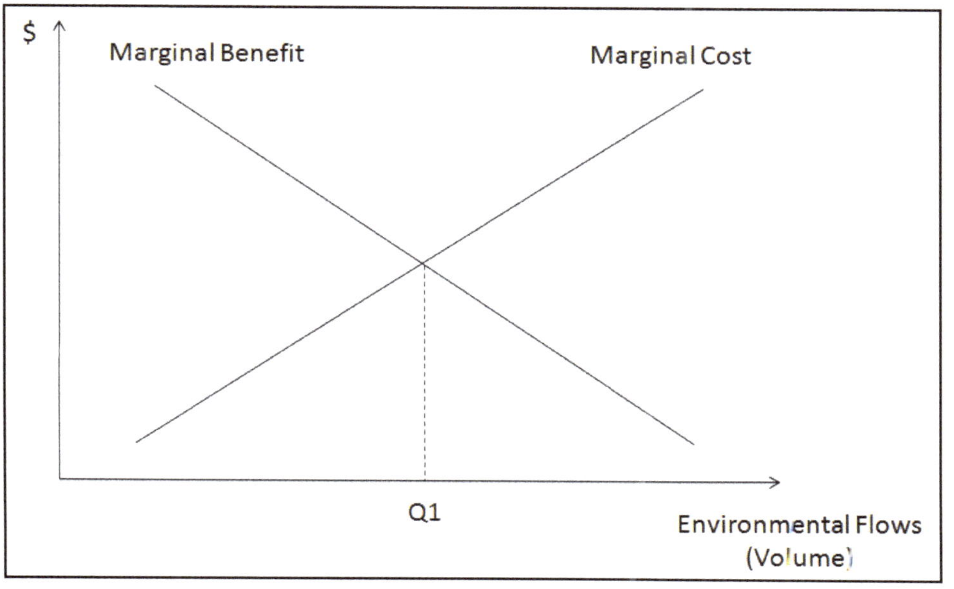

Figure 14.5 Marginal benefits equals marginal costs

Figure 14.5 represents the marginal benefit and marginal cost curves as straight lines that are 'smooth' with no breaks, flat areas or kinks. In reality, we might expect the marginal benefit curve to be steeper, and possibly discontinuous, at low volumes to indicate threshold points beyond which there might be very large losses associated with low environmental flows. Similarly, we might expect marginal cost to rise more steeply as the volumes of water in the environment increase to reflect the fact that higher-value extractive uses might be worth much more than low-value irrigation uses.

The volume of water for the environment that maximises societal benefits is represented by Q1 in Figure 14.5. Volumes of water to the environment *less* than Q1 generate a result that the marginal benefit from environmental flows is greater than the marginal cost of reduced interceptions or extractions. Thus, society, as a whole, would be better off by increasing environmental flows. Volumes of water *greater* than Q1 result in the marginal cost of environmental flows that exceeds the marginal benefit of such flows, from a societal perspective. Consequently, society as a whole would be better off if there were to be lower environmental flows. The desired volume of environmental flows from a societal perspective is, thus, achieved at Q1 where the marginal benefit equals the marginal cost.

At present there is insufficient information to precisely determine the desired increase in environmental volumes such that the marginal benefit of greater environmental flows equals the marginal cost of reduced extractions. There is, however, modelling calibrated to the Murray River that assesses the trade-offs between the environment and surface-water extractions by irrigators that indicates environmental flows should have been much larger over the past decade (Grafton et al. forthcoming). Equally as important, there is also no credible economic evidence to justify the assertion by the MDBA that 4000 GL/yr imposes unacceptably high socioeconomic costs. Indeed, there is evidence that implies the opposite—namely: 1) if large increases in environmental flows are consistent with improvements in environmental assets then increased flows have the potential to deliver multiple billions of dollars of non-market benefits in present value terms, and in excess of $1 billion per year; 2) the annual estimated willingness to pay for improvements in riverine environmental assets in the MDB[2] exceeds the estimated annual reduction in the gross regional product in the Basin associated with a 4000 GL/yr reduction in watercourse diversions, while having virtually no impact on the long-term employment within the Basin (MDBA 2010a:121); 3) reductions in current watercourse diversions to holders of water entitlements will be fully compensated by the $3.1 billion allocated to the purchase of water entitlements in the Water for the Future package, and by the commitment of the Australian Government to acquire water entitlements only from willing sellers; 4) according to the MDBA's own evidence and its own confidence intervals, a decrease in diversions basin wide of *less than* 4615 GL/yr would still achieve only key ecosystem functions consistent with a *high* degree of uncertainty (MDBA 2010b:112).

Mitigating the Costs of Water Reform

To mitigate the costs of water reform, the MDBA has opted to limit the proposed increase in environmental flows. A downside to this approach is that if the proposed environmental flows fail to meet the objects of the *Water Act 2007*, the SDLs on which the Basin Plan is based become vulnerable to litigation. Moreover, the smaller the increase in proposed environmental flows the greater is the risk of incurring at least some of the costs of business as usual.

While it might be possible in the future to achieve the same environmental objectives with less water, or with better timing of water releases, there is at present no credible scientific evidence to support basin-wide reductions in proposed environmental flows on the basis of past 'works and measures'. This is because '[m]ajor wetlands require large-scale inundation events, managed or not' (Kingsford et al. 2010:29). It is possible that, given sufficient time, coupled with appropriate incentives at the site or catchment scale for water environmental managers, and by learning by doing, there will be improved ways to provide water for the environment. If this were to be the case then this would be accounted for in the determination of SDLs in any subsequent Basin Plan.

2 This depends on the discount rate used to generate the present value numbers.

An alternative way to reduce the socioeconomic impacts of lowering current water diversions is to provide adjustment or transitional assistance to irrigation-dependent communities expected to suffer the most from the proposed changes. This addresses the issue that '[a]ny fall-off of economic activity as a result of the Basin Plan will also be felt by those in the irrigated agriculture chain. These people are typically "outside the tent" when it comes to compensation; they need to be included in community adjustment planning' (Marsden Jacob Associates et al. 2010:xxi–xxii).

According to ABARE–BRS, the people most affected and who are currently 'outside the tent' in terms of compensation are individuals who do not own water entitlements and are located in irrigation communities in the Murrumbidgee, Condamine, Murray (New South Wales and Victoria), Gwydir, Barwon–Darling and Goulburn–Broken river regions of the MDB. The Wentworth Group of Concerned Scientists et al. (2010) have proposed a 'reasonable return and community development' option that would provide transitional assistance to affected communities to cope with less water for extraction. Under this approach, the amount of funding available to communities would be predicated on the expected and/or actual economic impacts of reduced water diversions. Communities themselves, with external assistance and advice, would develop proposals for funding and other support and these would be tested for 'value for money' in terms of an expected rate of return. Funds for transitional assistance could come from the Water for the Future program, especially if some of the $5.8 billion currently directed for irrigation infrastructure subsidies in Water for the Future were redirected to buying water entitlements from willing sellers and supporting community adjustment (Grafton 2010).

An approach to structural adjustment proposed by the Wentworth Group of Concerned Scientists is to engage communities in the 'thriving communities' model. This approach includes three steps: 1) facilitation and bringing together of people and leaders in the community; 2) development of plans and ideas for co-investment; and 3) funding and follow through on the initiatives (Miller and Verity 2009). A necessary but not sufficient condition for success, as evidenced by successful adjustment in other resource sectors, is '[t]he absolute necessity for some individuals to be willing to take a leadership role—requiring at times considerable personal courage and a commitment to change—across the interests of all the stakeholder groups in order to progress the necessary change management' (McLoughlin and Rayns 2010:344).

The adjustment process in Australia's Commonwealth fisheries shows that it is possible to undertake structural adjustment and generate better societal outcomes. In this process, the Australian Government spent $220 million in 2006–07 to buy back fishing licences, to provide onshore business assistance, to give three years of relief from government industry levies, and to provide additional funds for scientific research (McLoughlin and Rayns 2010:341). The outcomes of these industry adjustments, along with management rules that enforce reduced harvests when required, have helped to generate higher net returns for harvesters and increased the asset values of the harvesting rights of fishers. In addition, following these changes, between 2006 and 2009, the proportion of fish stocks classified as overfished and uncertain in their stock status fell from 72 per cent to less than half, while the number of species not overfished has more than doubled (Wilson et al. 2010:19).

Concluding Remarks

The *Guide to the proposed Basin Plan* offers a major step forward in water planning in Australia, and worldwide. It performs the difficult task of reconciling the trade-offs between allocating more water for the environment versus water for existing users while complying with the *Water Act 2007*. According to the Murray–Darling Basin Authority, the best available science indicates that the Murray–Darling Basin needs a great deal more water—between 3000 GL/yr and 7600 GL/yr—to meet desired ecosystem functions and the objects of the Act. Using the MDBA's own technical documentation and its own confidence intervals, any reductions in diversions of *less than* 4615 GL/yr would, however, only achieve the goals of the *Water Act 2007* with a *high* degree of uncertainty.

A review of the results of socioeconomic models of the Basin, and also empirical evidence over the past decade-long drought, provides no support for the recommendation by the MDBA that increases in environmental flows should be limited to 4000 GL/yr due to the high socioeconomic impacts of reduced water diversions. Instead, a more effective approach to minimise the socioeconomic impacts of increased environmental flows is to provide transitional assistance to irrigation-dependent communities to help them adjust to a future with lower water extractions. This would require a redirection of existing funding for water reform and/or the allocation of additional funds expressly for this purpose.

Bibliography

Arche Consulting 2010, Socioeconomics of floodplain agriculture in the Murray–Darling Basin: a scoping report, Mimeograph, Arche Consulting, Sydney.

Australian Bureau of Agricultural and Resource Economics (ABARE) and Bureau of Rural Sciences (BRS) 2010a, *Environmentally sustainable diversion limits in the Murray–Darling Basin: socioeconomic analysis*, Report prepared for the Murray–Darling Basin Authority, Canberra.

Australian Bureau of Agricultural and Resource Economics (ABARE) and Bureau of Rural Sciences (BRS) 2010b, *Assessing the regional impact of the Murray–Darling Basin Plan and the Australian Government's Water for the Future Program in the Murray–Darling Basin,* Report prepared for the Murray–Darling Basin Authority, Canberra.

Australian Bureau of Statistics 2010, *Experimental Estimates of the Gross Value of Irrigated Agricultural Production,* 2000-0102007-08, Australian Bureau of Statistics, Canberra.

Bennett, J. 2010, 'Making decisions about environmental water: an economics approach', in *Making decisions about environmental water allocations,* Research Report, Australian Farm Institute, Surry Hills, NSW.

Champ, P. A., Boyle, K. J. and Brown, T. C. (eds) 2003, *A Primer on Nonmarket Valuation,* Kluwer Academic Press, Dordrecht.

CSIRO 2008, *Water availability in the Murray–Darling Basin,* A report to the Australian Government from the CSIRO Murray–Darling Basin Sustainable Yields Project, Canberra.

Davies, P. E., Harris, J. H., Hillman, T. J. and Walker, K. F. 2008, *A report on the ecological health of rivers in the Murray–Darling Basin 2004–2007,* Report by the Independent Sustainable Rivers Audit Group for the Murray–Darling Commission, Canberra.

Grafton, R. Q. 2010, 'How to increase the cost effectiveness of water reform and environmental flows in the Murray–Darling Basin', *Agenda,* vol. 17, no. 2, pp. 17-40.

Grafton, R. Q. and Jiang, Q. (Forthcoming), Economic effects of water recovery on irrigated agriculture in the Murray–Darling Basin, Paper. Australian Journal of Agricultural and Resource Economics.

Grafton, R. Q., Chu, L., Stewardson, M. and Kompas, T. (Forthcoming), '*Optimal dynamic water allocation: irrigation extractions and environmental tradeoffs in the Murray River, Australia',* Water Resources Research.

Hone, S., Foster, A., Hafi, A., Goesch, T., Sanders, O., Mackinnon, D. and Dyak, B. 2010, *Assessing the future impact of the Australian Government environmental water purchase program,* ABARE Research Report 10.03, Australian Bureau of Agricultural and Resource Economics, Canberra.

Jones, G., Hillman, T., Kingsford, R., McMahon, T., Walker, K., Arthington, A., Whittington, J. and Cartwright, S. 2002, *Independent Report of the Expert*

Reference Panel on Environmental Flows and Water Quality Requirements for the River Murray System, Prepared for the Environmental Flows and Water Quality Objectives for the River Murray Project Board, Canberra.

Kingsford, R. T., Brandis, K. J., Jenkins, K. M., Nairn, L. C. and Rayner, T. S. 2010, 'Measuring ecosystem responses to flow across temporal and spatial scales', in N. Saintilan and I. Overton (eds), *Ecosystem Response Modelling in the Murray–Darling Basin*, CSIRO Publishing, Collingwood, Vic.

McLoughlin, R. and Rayns, N. 2010, 'Australia's Commonwealth-managed fisheries', in R. Q. Grafton, R. Hilborn, D. Squires, M. Tait and M. Williams (eds), *Handbook of Marine Fisheries Conservation and Management*, Oxford University Press, New York.

Mallawaarachchi, T., Adamson, D., Chambers, S. and Schrobback, P. 2010, *Economic Analysis of Diversion Options for the Murray–Darling Basin Plan: Returns to irrigation under reduced water availability*, Risk and Sustainable Management Group, University of Queensland, St Lucia.

Marsden Jacob Associates, RMCG, EBC Consultants, DBM Consultants, The Australian National University, McLeod, G. and Cummins, T. 2010, *Synthesis report: economic and social profiles and impact assessments in the Murray–Darling Basin*, A report to the Murray–Darling Basin Authority, Canberra.

Merritt, W., Powell, S., Pollino, C. and Jakeman, T. 2010, 'IBIS: a decision support system for managers of environmental flows into wetlands', in N. Saintilan and I. Overton (eds), *Ecosystem Response Modelling in the Murray–Darling Basin*, CSIRO Publishing, Collingwood, Vic.

Miller, C. and Verity, F. 2009, Thriving communities: addressing and adjusting to the social impacts of water related changes in the Murray–Darling Basin, Mimeograph, Flinders University, Adelaide.

Morrison, M. and Bennett, J. 2004, 'Valuing New South Wales rivers for use in benefit transfer', *Australian Journal of Agricultural and Resource Economics*, vol. 48, no. 4, pp. 591–611.

Morrison, M. and Hatton-MacDonald, D. 2010, *Economic Valuation of Environmental Benefits in the Murray–Darling Basin*, Murray–Darling Basin Authority, Canberra, <http://www.mdba.gov.au/files/bp-kid/1282-MDBA-NMV-Report-Morrison-and-Hatton-MacDonald-20Sep2010.pdf>

Murray-Darling Basin Authority (MDBA) 2010a, *Guide to the proposed Basin Plan. Volume 1: Overview*, Murray–Darling Basin Authority, Canberra.

Murray-Darling Basin Authority (MDBA) 2010b, *Guide to the proposed Basin Plan. Volume 2: Technical background*, Murray–Darling Basin Authority, Canberra.

National Water Commission 2010, *The Impacts of Water Trading in the Southern Murray–Darling Basin: An economic, social and environmental assessment*, National Water Commission, Canberra.

Norris, R. H. 2010, 'How should Australia decide how much water in Australian rivers should be allocated to the environment, and how can the community be sure environmental water is being used efficiently?', *Making decisions about environmental water allocations*, Research Report, Australian Farm Institute, Surry Hills, NSW.

Norris, R. H., Liston, P., Davies, N., Dyer, F., Coysh, J., Dyer, F., Linke, S., Prosser, I. and Young, B. 2001, *Snapshot of the Murray–Darling Basin River Condition*, Murray–Darling Basin Commission, Canberra.

Swirepik, J. L. 2009, Establishing and delivering environmental water allocations in the Murray River, Australia, Paper presented at the International Conference on Implementing Environmental Water Allocations, 23–26 February 2009.

Wentworth Group of Concerned Scientists in association with Grafton, R. Q., Kowalick, I., Miller, C., Stubbs, T., Verity, F. and Walker, K. 2010, *Sustainable Diversions in the Murray–Darling Basin: An analysis of the options for achieving a sustainable diversion limit in the Murray–Darling Basin*, Wentworth Group of Concerned Scientists, Sydney, <http://www.wentworthgroup.org/uploads/Sustainable%20Diversions%20in%20the%20Murray-Darling%20Basin.pdf>

Williams, G. 2010, 'When water pours into legal minefields', *Sydney Morning Herald*, 26 October 2010, <http://www.smh.com.au/opinion/politics/when-water-pours-into-legal-minefields-20101025-170uf.html>

Wilson, D., Curtotti, R. and Vieira, S. 2010, 'Overview', in D. T. Wilson, R. Curtotti and G. A. Begg (eds), *Fishery Status Reports 2009: Status of fish stocks and fisheries managed by the Australian Government*, Bureau of Rural Sciences, Australian Bureau of Agricultural and Resource Economics, Canberra.

Wittwer, G. 2010, *The regional economic impacts of sustainable diversion limits*, 5 November 2010, Report prepared for the Murray–Darling Basin Authority, Canberra.

15. Multiple Benefits through the Life Cycle of the Basin Plan

Darla Hatton MacDonald, Rosalind Bark, Dustin Garrick, Onil Banerjee, Jeff Connor, Mark Morrison

Introduction

The objectives of the new water-sharing plan for the Murray–Darling Basin Plan are set out in the *Water Act 2007* (Cwlth); cardinal among these objectives is the maximisation of economic, social and environmental outcomes for the Australian community (*Water Act 2007*, ss. 3[c] and [d][iii], s. 20[d]). Meeting this objective requires an extensive process of biophysical modelling and economic analysis of the costs and benefits associated with resetting the balance among multiple water users, including the environment. Maximising net benefits applies to plan development where implementation can be thought of as a two-stage process: a) establishing long-term sustainable diversion limits (SDLs) that set the overall balance of water available for the environment versus consumptive uses; b) developing basin-level plans for environmental watering, water quality, accreditation, and water-trading rules. Included in this is the design of mechanisms and incentives to incorporate basin-wide targets and goals into State water-sharing plans and operational delivery mechanisms.

The *Guide to the proposed Basin Plan* (MDBA 2010), released on 8 October 2010, could be interpreted as the first stage in this process of establishing a feasible range of long-term reductions in diversions. The framework for assessing the balance among consumptive users and the environment in the Guide involves weighing the cost to irrigated agriculture in terms of forgone production, including the potential flow-on impacts to employment and irrigation-dependent communities, and a comparison of these with the potential benefits derived from ecological responses to various environmental water-allocation scenarios. Ecological modelling has determined that over the long term the amount of additional surface water required for the environment averages between 3000 and 7600 gigalitres a year. Based on internal analysis, the Murray–Darling Basin Authority (MDBA 2010:XXI) has determined that it would only examine scenarios with reductions of between 3000 GL/yr and 4000 GL/yr. These reductions in diversions have been met with resistance by irrigators (ABC 2010). Meanwhile, environmental-conservation groups and ecologists have reservations that a volume in the lower end of the range will be sufficient to restore ecological character to the Murray–Darling Basin (MDB) (Finlayson 2010). Indeed, the 3000 GL/yr scenario would leave the Condamine–Balonne, Gwydir, Loddon, Lower Darling and Murray regions in poor condition (MDBA 2010:74).

In the next phases of finalising the Plan and implementing State water plans, there are significant opportunities to maximise the benefits associated with reallocating water. Key to achieving this objective is the application of the equi-marginal principle of equating marginal benefits associated with reallocating water from one use to another. By utilising the economic-value estimates associated with different uses and ecological modelling, it is possible to approximate the marginal benefits to multiple users across the Basin. While this is a powerful organising principle, it is complex to implement in practice. Even partial implementation of the principle will deliver considerable gains as the exercise requires a concerted effort to understand the complex, dynamic relationships among water quantity and quality and flow regimes necessary to deliver the outcomes. Investments in institutional arrangements and incentives that are compatible with water being put to highest and best use across the Basin will also be required.

The remainder of this chapter is organised around three central challenges in achieving the objective of maximising the economic, social and environmental benefits of the Plan. First, we consider the importance of improving understanding of the economic benefits and positive effects of improved environmental flows. This section also discusses potential approaches to integrating these benefits into a rigorous and transparent benefit–cost decision-making framework to support the policy process. Second, we discuss how the benefit–cost framework must serve multiple purposes, including the evaluation of long-term average allocation trade-offs, temporal and spatial variations in flow, as well as the systemic risks associated with an uncertain future climate. Finally, we outline how the maximisation of net benefits depends on a robust and efficient set of institutional arrangements that combines top-down and bottom-up processes to reconcile competing and complementary goals of different planning elements at the Basin and State levels.

Environmental Benefits

There have been a number of studies assessing the benefits associated with improving rivers, wetlands, red-gum forests and the Coorong. Hatton MacDonald et al. (2011) and Morrison and Hatton MacDonald (2010) summarise the tourism and recreational benefits, amenity values and the non-use values (bequest, intrinsic and existence values) across the Basin. The body of work on benefits can guide decision making in that it provides insight into the relative value of water for different environmental goals, including providing water to ecosystems (for example, waterbird breeding versus fish populations), providing water to different locations (upstream versus downstream) and to different ecological assets. For the Murray River, as an example, if there were a 10 per cent increase in healthy native vegetation, a 15 per cent increase in fish populations, the frequency of waterbird breeding increased by three years, and the number of waterbird and other species increased by 20 per cent and the quality of the Coorong also improved from poor to good, the total increase in non-use value would be an estimated $7.5 billion on a 10-year net present-value basis (Morrison and Hatton MacDonald 2010:31).

Several messages emerge from the non-market valuation literature completed in recent years with respect to the Murray–Darling Basin. In general, values appear to be increasing over time, with more recent studies producing larger-value estimates for a number of attributes. The evidence is compelling but not conclusive. In general, it appears that the values held for the River Murray are much larger than for other rivers in the Basin or Australia (Morrison and Hatton MacDonald 2010). This likely reflects the greater significance of the Murray, environmental education and media attention in recent years. There is some evidence for the iconic status of Murray-based ecological assets: values for fish populations and waterbird breeding were larger for more significant rivers/wetlands. Similar evidence was found for native vegetation. In contrast, values for providing habitat for waterbirds and other species did not appear to be a function of the type of asset class—that is, bird breeding and bird habitat were important whether the asset was a coastal wetland such as the Coorong or major wetlands such as Barmah–Millewa.

To enable benefit–cost analysis with a more robust consideration of environmental values, a number of challenges remain to be addressed. In particular, there is the choice of appropriate assumptions relating to the geographic extent of preferences of respondents. There might be some spatial variation in willingness to pay when comparing ACT and Victorian responses (Hatton MacDonald et al. In review). ACT survey respondents were willing to pay significantly more, which might be due to differences in environmental preferences. Zander et al. (2010) demonstrate that southern urban residents hold values for less-developed, northern tropical rivers (for example, the Daly, Fitzroy and Mitchell rivers). State-based research of water-based ecological assets reveals intra-state variation in estimated benefits (Bennett et al. 2008; Morrison and Bennett 2004). This might reflect differences in study design, sampling and preferences. There are also valuation differences based on frequency of visitation. Rolfe and Prayaga (2007) find regular visitors have higher values than one-time visitors to freshwater impoundments.

Another segment of water-based and flow-based values is recreational benefits. Some recreational activities—such as boating and fishing—have developed opportunistically in the Murray–Darling system, capitalising on the regulated flow regimes associated with irrigation (Howard 2008). Recreational boating and fishing on Lake Hume during high summer irrigation flows are examples (Crase and Gillespie 2008). Some of these recreational activities might be adversely impacted by changes in flow regimes, however, any negative effects on these recreation pursuits might be offset by improvements in nature-based recreation that might be enhanced by more natural episodic flooding and drying patterns.

Another challenge lies in quantifying 'dose-response' relationships for changes in flow—that is, estimating the marginal improvement in environmental health per unit of additional flow. Also necessary is an understanding of the relationship between a given level of improvement in environmental health and economic value. This relationship is complicated by non-linearities between improvements in environmental quality and value. Further work in advancing ecological response functions and values associated with marginal improvements is required for effectively guiding public policy.

This body of work provides greater confidence and understanding about a suite of environmental benefits. It also reveals a series of gaps for future science and decision support. First, there are only estimates for a subset of environmental and other positive externality benefits relevant to the Plan and its implementation. There are issues of spatial and temporal coverage including the need for dynamic assessments for periodic trade-offs across multiple uses over the full life cycle of water reform. For example, estimated benefits might inform decisions between upstream and downstream environmental watering.[1] Flow regimes for healthy floodplain vegetation (for example, red-gum forests) require episodic flooding and drying versus long periods of inundation that occur with river regulation. Further, peak irrigation and municipal summer demand might not coincide with the watering requirements of iconic wetlands. The management of flow patterns to maximise environmental and socioeconomic values will be complex, requiring the application of the best ecological science with socioeconomic assessment. Furthermore, there are real challenges to translating benefit–cost analysis into a decision-support framework that can support the dynamic implementation and adaptive learning at the appropriate spatial and temporal scales as the Plan moves into the implementation phase.

1 There are, however, gaps and uncertainties that would preclude disconnecting assets from the river if guided by the precautionary principle.

Other Benefits and Multiple Objectives

There is a long history of investment in the River Murray, without a significant reversal in degradation (Lee and Ancev 2009). Setting the long term SDLs and re-establishing a share of the water for the environment present a rare opportunity to move out of the realm of cost effectiveness in water policy to full benefit–cost analysis. In this section, we consider the challenges in providing rigorous hydro-economic modelling, linked with costs and benefits, as a constructive contribution to the implementation of the Plan. A number of non-market valuation studies have been conducted in the Murray–Darling Basin, which can be used as part of the process of setting SDLs and guiding implementation trade-offs. There remain regions of the Basin, however, where no primary data are available. Targeted investment in primary research to fill the gaps will make decision making more robust. It is also recommended that a thorough stocktake and evaluation of costs of the 'do-nothing' alternative be completed. The status quo is one of declining environmental quality. This would serve as a baseline against which potential benefits of the Plan can be compared.

The advantage of the benefit–cost approach is that it involves a systematic consideration of all relevant benefits and costs. As a result, there might be more transparency in the consideration of trade-offs between environmental and economic outcomes. The trade-off between irrigated agriculture and environmental water was highlighted in the *Guide to the proposed Basin Plan* (MDBA 2010). There are, however, also multiple objectives that will benefit from increased flows intrinsic in the proposed SDLs. These objectives include improving ecosystem health, providing adequate conveyance water, irrigation, recreation, cultural flows for Indigenous people across the regions, water quality and critical human needs. For instance, conveyance water requirements will be met more of the time, reducing the need for investment in pumps and weirs. The dilution effects of more flow should improve water quality and the economic benefits of reduced salinity are considerable for irrigators, water utilities, and urban and rural consumers. More water for the environment could also enhance urban water supply reliability and quality, potentially reducing, or delaying, investments in alternative water supplies. There are considerable challenges in estimating the benefits attributable to the overall volume of environmental water and the timing of its delivery. The goal of maximising multiple objectives requires a detailed ecological and socioeconomic understanding of how these objectives interact.

Many of these interactions are positive. For instance, an environmental watering plan could be designed to take into account the ancillary benefits of specific flow regimes that not only achieve ecological goals but also reduce the incidence of blue-green algae blooms, which compromise habitats and water quality. Jackson et al. (2010) make recommendations on how the Basin Plan could be more sensitive to improving Indigenous community benefits. A first step to incorporating Indigenous water values in the implementation of the Basin Plan is to catalogue the spectrum of these benefits in each catchment. A second step is to assess the extent to which they are coincident geographically and temporally with other value streams, such as environmental water. Where these value streams are convergent, there is no trade-off. Where, or when, they diverge, a mechanism—such as a set of rules or incentives for optimising joint benefits or for prioritising one set of values at the catchment or basin-wide level, and for various time slices—needs to be developed. A recommendation of this work is that Indigenous participation in the co-management of environmental water through shared governance arrangements is a potential mechanism to ensure the Basin Plan is 'more inclusive of Indigenous water values, use and priorities' (Jackson et al. 2010:154).

The biophysical modelling underpinning the Plan was based on analysis of flow regimes required for ecological health and water quality across the MDB system. More research is required to understand the full range of impacts on water for conveyance, cultural values, and municipal industrial water issues. A thorough socioeconomic assessment in line with best science and policy practice might provide transparency and accountability in the determination of Plan implementation trade-offs in the pursuit of maximising net total benefits.

Balancing Top-Down and Bottom-Up in Institutional Design of the Plan

The *Water Act 2007* (s. 22, Items 3 and 5) mandates the MDBA to develop strategies, including flexible institutional arrangements, to achieve multiple objectives and address risks associated with interception, climate change and limits to knowledge. These institutional arrangements must 'achieve efficient and cost effective water management and administrative practices' (*Water Act 2007*, s. 21, ss. 4[b]). The Act explicitly provides a directive for 'adaptive management' (ss. 21, 4[c][i]), which implies an active feedback process to link basin-level planning and State implementation. In this context, once SDLs are set, the Plan will involve a set of discrete elements, recorded in Table 15.1. For each element, the responsible level or levels of government are indicated. This tabulation highlights the interaction among the Commonwealth, the MDBA and the States.

Table 15.1 Plan implementation elements with responsible party

Plan element	Responsible party
Transitional arrangements	Commonwealth
Environmental water plan	MDBA and States
Water-quality and salinity plans	MDBA and States
Critical human needs	MDBA and States
Development of surface and groundwater water-sharing plans	States and MDBA accreditation
Conveyance water	MDBA objectives and operator compliance
Water-trading rules	MDBA and Australian Competition and Consumer Commission
Program monitoring and evaluation	MDBA

The optimisation of multiple objectives poses significant coordination and governance challenges to the implementation of the Basin Plan. Implementation will require efficiency in institutional and management arrangements to deliver the required regime (as guided by the best science combined with socioeconomic assessment) and to respond to changing circumstances. Three key questions are: first, at what level of government is it best to implement the different aspects of the Plan? Second, is there sufficient institutional capacity to achieve the Basin Plan's targets and goals? Third, how should linkages be structured across different levels of government to maximise net benefits at the local level to the maximum degree consistent with basin-level objectives?

The *Water Act 2007* addressed a need for coordination among the States for a whole-of-basin approach. The Guide presents the implementation process as a two-tier system of basin-wide planning followed by State-level integration in water sharing (MDBA 2010:Figure 1.3). Although the MDB Plan has involved significant top-down elements, there are considerable opportunities within the Plan for involving localised institutions. This could add significant value by providing the necessary local knowledge to maximise environmental, cultural and socioeconomic values as well as the flexibility to respond to climate-change impacts on the Basin's resources. Conceptually, as the Plan moves into implementation, coordination will involve more or less of three principal approaches (Garrick et al. 2011).

1. Top-down: centralised implementation of central requirements of the Plan such as SDLs—for example, by detailed directive to infrastructure operators by the MDBA-chaired Basin Environmental Watering Committee.
2. Bottom-up: vest decisions at the State and catchment levels with basin-wide oversight limited to accreditation of State plans. For example, the Plan could be implemented allowing significant autonomy to States in decisions about

environmental water scheduling with regard to recommendations from the MDBA Basin Environmental Watering Committee.

3. Nested arrangements: this would involve setting the institutional and management arrangements such that the MDBA, the Commonwealth Environmental Water Holder (CEWH), and States have the incentive to plan for and manage the resource adaptively for the environment, while ensuring: a) appropriate capacity for local, field-based expertise; and b) accountability to both local and basin-wide objectives through environmental, water-quality and water resource-sharing plans.

The relative balance of top-down and bottom-up approaches varies by phase of implementation (see Table 15.1); the critical challenge lies in achieving a balance between basin-wide accountability and proper incentives for local ingenuity. This challenge involves balancing the advantages and disadvantages of central versus local (catchment or State) action in each element of the Plan tabulated in Table 15.1.

The *transitional arrangements* require the development of policy and mechanisms to acquire water efficiently, based on the benefits for the environment and costs to communities detrimentally affected, and the physical constraints on sourcing water to achieve local versus downstream benefits. There might, however, be a role for local communities to have input into the design of transitional arrangements to achieve adjustment in irrigation-based communities with greater efficiency and lower third-party impacts. The advantage of local action lies in knowledge about regional economic productivity, interdependencies and third-party impacts; the MDBA and Commonwealth can provide central oversight to ensure due diligence and accountability in the investments in voluntary buybacks and infrastructure.

The *Environmental Water Plan* and delivery arrangements require a combination of centrally established criteria and local expert knowledge of time and place-specific information. This could be achieved by enabling and accrediting a set of decentralised environmental water trusts and providing rules and incentives for them to work in a coordinated way with the CEWH (Young 2010). This model allocates a portion of the Commonwealth environmental water holdings to regional trusts and reserves the balance for central management. This approach establishes a nested-governance arrangement that provides incentives for entrepreneurial management and delivery decisions to maximise local environmental outcomes in concert with basin-wide and cross-catchment objectives. The Environmental Water Plan established as a component of the Basin Plan would ensure consistency of State and regional environmental watering decisions by 'specifying overall environmental objectives for the Murray Darling Basin's water dependent ecosystems' (MDBA 2010:162). It would also guide development of strategic environmental water plans for each surface-water resource plan area. Because environmental water delivery will remain with the States, mechanisms for pairing local decisions and priorities with those of the Basin-wide Environmental Watering Advisory Committee will need to be carefully structured. Nested-governance arrangements take advantage of knowledge and capacity to maximise local benefits of the pool of environmental holdings, as well as the ability to adapt to real-time opportunities to adjust environmental water-delivery schedules to achieve desired flow regimes. The MDBA offers the central coordination to ensure proper trade-offs between local and downstream benefits across the States, including the timing and delivery of environmental holdings across State boundaries.

Water quality and salinity plans identify roles for both basin-wide and State-level action. Basin salinity management is an area of past success in nested governance (Connor 2008). The Water Quality and Salinity Management Plan will set targets for the Basin-wide water quality for aquatic ecosystems as well as for drinking, recreation and irrigation water. Water-quality issues will require integration into State Water Resource Plans that will be required to incorporate Water Quality Plans to achieve the overarching objectives. Nested arrangements have been successful through technical assistance provided to the States regarding the relative effectiveness and costs of salinity-control options. These information flows link Commonwealth guidelines and capacity building at the State level to achieve water-quality objectives and promote a feedback loop for adaptive learning as States implement plans and projects under the Basin Salinity Management Strategy.

The *accreditation process* is outlined in the Guide as a nested arrangement that involves federal oversight of State water-sharing plans (MDBA 2010:Figure 12.2). The MDBA will set criteria for accreditation against which the States' surface and groundwater management plans will be evaluated. Approved plans are accredited for 10 years. The MDBA may recommend that a plan not be accredited, in which case the minister may overrule this recommendation or recommend that the MDBA prepare the relevant plan. If the MDBA is directed to prepare a water-resource plan, the minister may approve this plan or direct the MDBA to revise the plan. The accreditation process could be guided explicitly by a full socioeconomic cost–benefit assessment of environmental and cultural benefits.

From an institutional perspective, the objective of maximising net benefits applies in the development of the Plan and its implementation. Implementing the Plan requires basin-wide planning with State-level water-plan integration. The relative balance of top-down and bottom-up approaches could vary by phase of implementation to achieve a balance between basin-wide accountability and incentives to realise and maximise multiple benefits by means of capitalising on local incentives and knowledge. Substantial opportunities exist for developing nested institutional arrangements and governance structures that capitalise on local knowledge and expertise, whilst assuring consistency, enforcement and accountability to higher-order institutions.

Conclusions

This chapter reviewed challenges in achieving the objective of maximising the economic, social and environmental benefits of the Plan. First, we consider the importance of understanding of the economic benefits and positive externalities associated with improved environmental flows. We outlined how to do this by building on existing environmental-benefits economics and by addressing some key gaps, support for plan implementation can be improved.

Second, we outlined how realising maximum net benefits to society—at least implicitly—involves a benefit–cost framework that accounts for multiple objectives and trade-offs, temporal and spatial variations in flow, as well as the systemic risks associated with an uncertain future climate. There are multiple objectives, such as water quality, which improve with the dilution effects of more flow, with benefits of reduced salinity for irrigators, water utilities, and urban and rural consumers. More water for the environment will also enhance urban water supply reliability, reducing, or delaying, investments in alternative water supplies. The goal of maximising multiple objectives of ecosystem health, irrigation, recreation and cultural flows for Indigenous people across the regions, water quality and critical human needs is inherently complex. Continued effort to improve detailed ecological and socioeconomic understanding of how these objectives interact will allow better realisation of improved outcomes in Basin Plan implementation.

Finally, we discussed how the maximisation of net benefits depends on a robust and efficient set of institutional arrangements that combines top-down and bottom-up processes to reconcile competing and complementary goals of different planning elements at the Basin and State levels. From an institutional perspective, the objective of maximising net benefits applies in the development of the Plan and its implementation, including scope for adaptive learning. Substantial opportunities exist for developing nested institutional arrangements and governance structures, which capitalise on local knowledge and expertise, whilst assuring consistency, enforcement and accountability to higher-order institutions. The institutions must be flexible in order to adapt to climate change and the inevitable surprises that result from limits to knowledge.

Ultimately, the process of fully understanding all costs, benefits and trade-offs will never be complete. Interactive rounds, however, which improve the understanding of the costs and benefits, can support and potentially improve decision making. Thus, there is a need for built-in, periodic updates of the evaluation of the trade-offs across multiple uses over the full life cycle of water reform. For example, under reduced rainfall and inflow scenarios, estimated benefits might be required to inform decisions between upstream and downstream environmental watering. The management of flow patterns to maximise environmental and socioeconomic values will be complex, requiring the application of the best ecological science with socioeconomic assessment. Further, the benefit–cost analysis must be slotted into a decision support framework that is dynamic and allows for adaptive learning as the Plan moves into the implementation phase.

The advantage of the benefit–cost approach is that it requires the systematic consideration of all relevant benefits and costs. Even then there will always be limits to our understanding of costs, benefits and trade-offs. Nested-governance arrangements provide coordinated institutional capacity and information flows as a response to an integrated assessment of benefits and costs at a whole-of-basin scale, over the life cycle of the Plan and with proper feedback between Commonwealth and local levels to promote adaptive learning. Implementing this framework will require updating of the shared understanding of local communities, States and the Commonwealth Government as the Plan is rolled out. This approach could also lead to more transparency in consideration of trade-offs between environmental and economic outcomes and better realisation of multiple objective outcomes.

Bibliography

Australian Broadcasting Corporation (ABC) 2010, 'Brutal, or bare minimum?: mixed response to Basin Plan', *ABC News*, 11 October 2010, <http://www.abc.net.au/news/stories/2010/10/11/3034582.htm>

Bennett, J., Dumsday, R., Howell, G., Lloyd, C., Sturgess, N. and Van Raalte, L. 2008, 'The economic value of improved environmental health in Victorian rivers', *Australasian Journal of Environmental Management*, vol. 15, pp. 138–48.

Commonwealth of Australia 2004, *Intergovernmental Agreement on a National Water Initiative*, Commonwealth of Australia, Canberra, <http://www.nwc.gov.au/resources/documents/Intergovernmental-Agreement-on-a-national-water-initiative.pdf>

Commonwealth of Australia 2007, *Water Act 2007*, No. 137 (2007), Department of the Attorney-General, Canberra.

Connor, J. 2008, 'The economics of time delayed salinity impact management in the River Murray', *Water Resources*, vol. 44, <doi:10.1029/2006WR005745>

Connor, J. and Young, M. 2003, *Environmental water bank options for the South Australian River Murray*, Final Report to SA Department of Water, Land and Biodiversity Conservation, CSIRO Land and Water, Adelaide.

Crase, L. and Gillespie, R. 2008, 'The impact of water quality and water level on the recreation values of Lake Hume', *Australasian Journal of Environmental Management*, vol. 15, pp. 21–9.

Finlayson, M. 2010, 'Experts respond', *Rapid Roundup*, Murray–Darling Basin Authority Guide release, Murray–Darling Basin Authority, Canberra, <http://www.aussmc.org/2010/10/rapid-roundup-murray-darling-basin-authority-guide-release-%e2%80%93-experts-respond/>

Garrick, D., Lane-Miller, C. and McCoy, A. (2011), 'Institutional innovations for governing environmental water in the western US', *Economic Papers*, vol. 30, pp. 1-18.

Hatton MacDonald, D., Morrison, M., Rose, J. and Boyle, K. (In review), 'Valuing a multi-State river: the case of the River Murray', *Australian Journal of Agricultural and Resource Economics*.

Hatton MacDonald, D., Tapsuwan, S., Albouy, S. and Rimbaud, A. 2011, 'The value of tourism and recreation in the Murray Darling Basin', in L. Crase and S. O'Keefe (eds), *Boom and Bust: Water, tourism and recreation in Australia*, Earthscan Publications, London.

Howard, J. 2008, 'The future of the Murray River: amenity re-considered?', *Geographical Research*, vol. 46, pp. 291–302.

Jackson S., B. Moggridge and C. Robinson. 2010. Effects of changes in water availability on Indigenous people of the Murray-Darling Basin: a scoping study. Report to the Murray Darling Basin Authority. June 2010.

Lee, L. and Ancev, T. 2009, 'Two decades of Murray–Darling water management. A river of funding, a trickle of achievement', *Agenda*, vol. 16, pp. 5–23.

Morrison, M. and Bennett, J. 2004, 'Valuing New South Wales rivers for use in benefit transfer', *Australian Journal of Agricultural and Resource Economics*, vol. 48, pp. 591–611.

Morrison, M. and Hatton MacDonald, D. 2010, *Economic valuation of environmental benefits in the Murray–Darling Basin*, A report to the Murray–Darling Basin Authority, Canberra.

Murray–Darling Basin Authority (MDBA) 2010, *Guide to the proposed Basin Plan. Volume 1: Overview*, Murray–Darling Basin Authority, Canberra, <http://www.thebasinplan.mdba.gov.au/>

Rolfe, J. and Prayaga, P. 2007, 'Estimating values for recreational fishing at freshwater dams in Queensland', *Australian Journal of Agricultural and Resource Economics*, vol. 51, pp. 157–74.

Young, M. 2010, *Managing Environmental Water. Making decisions about environmental water allocations*, Farm Policy Research Institute, Sydney.

Zander, K., Garnett, S. and Straton, A. 2010, 'Trade-offs between development, culture and conservation—willingness to pay for tropical river management among urban Australians', *Journal of Environmental Management*, <doi:10.1016/j.jenvman.2010.07.012>

16. Modelling Challenges

Qiang Jiang

Introduction

This chapter discusses the controversy surrounding the modelling work that has been undertaken in the Murray–Darling Basin Authority's *Guide to the proposed Basin Plan*. Furthermore, the future modelling challenges of the social and economic impacts of the sustainable diversion limits (SDLs)—such as data availability, and the methods and objectives of modelling—will be identified, together with a review of existing modelling work within the Basin. To meet these challenges, the Australian Government should systematically facilitate researchers to accumulate the relevant basin data, or allow them access to their current basin data.

In October 2010, the *Guide to the proposed Basin Plan* was released by the Murray–Darling Basin Authority (MDBA). The calculation of SDLs is specified within the Guide. The Authority proposed that the amount of additional surface water needed for environmental flows is between 3000 gigalitres a year and 4000 GL/yr, which represents a cut of between 27 per cent and 37 per cent in watercourse diversions (MDBA 2010a). As the majority of water-use reductions are in irrigated industries, the Australian Bureau of Agricultural and Resource Economics–Bureau of Rural Sciences (ABARE–BRS 2010a) have predicted that these water-use reduction scenarios might result in losses of between $800 million and $1.1 billion a year for irrigated agricultural production (ABARE–BRS 2010a). Furthermore, if a 3000 GL/yr additional environmental flow target is adopted, 800 jobs will be lost in the long term (ABARE–BRS 2010a).

The irrigators in the Basin reacted angrily to the proposed water-use cuts in the Guide, and disagreed with the MDBA's social and economic impact findings for these additional scenarios. For many farmers in the Basin, the Authority underestimated the social and economic losses of additional environmental flows.

Although most of the ABARE–BRS results are similar to those from previous studies of the Basin (Adamson et al. 2007; Jiang 2010), there are rising concerns about the report's findings and the modelling that underpins the Guide. These have led to a parliamentary inquiry that is set to examine the social and economic impacts of SDLs on regional communities. This inquiry might have a strong focus on understanding the human impact of any proposed changes. The MDBA has also commissioned a future study to deliver a better understanding of the social and economic impacts of the SDLs. The modelling of social and economic impacts will be a major issue in these two future studies.

ABARE–BRS Modelling

In this section, we examine the details of ABARE–BRS modelling including the method, models, implementation, limitations and results. The general findings from ABARE–BRS are similar to the conclusions from previous studies in the Basin (Adamson et al. 2007; Jiang 2010). Not all models in the ABARE–BRS modelling are, however, designed to simulate the 19 Basin Plan regions in the Guide. Therefore, this modelling can reveal only an overall picture of the impacts and predictions of SDLs in some regions, which might not necessarily be accurate. ABARE–BRS stated their model limitations: 'While large-scale economic models are suitable for analysis at broad regional (for example, catchment) levels, they are limited in their ability to provide accurate estimates at smaller geographical scales' (ABARE–BRS 2010a, p.15). Despite the limitations, the models used in the ABARE–BRS study are the best available at this time. Future research might reduce these limitations.

In the method of the ABARE–BRS study, a two-stage approach has been applied (Figure 16.1). The first stage involves using ABARE–BRS's Water Trade Model to estimate the direct effects of changes in SDLs on the gross value of irrigated agricultural production (GVIAP) by the CSIRO's Sustainable Yield regions. In the second stage, these GVIAP estimates have been imported into a computable general equilibrium (CGE) model of the Australian economy, AusRegion, to estimate flow-on effects to regional, State and national economies.

In the first stage, the ABARE–BRS Water Trade Model consists of two parts: hydrologic modelling based on the CSIRO's Sustainable Yields project, and social and economic modelling based on Australian Bureau of Statistics (ABS) data. The hydrologic study modelled the water availability and environmental demands of the Basin. The various simulated water usages under proposed additional environmental-water scenarios in the hydrologic study have been used as the inputs for the social and economic study. This approach has been widely used in previous integrated hydrologic and economic modelling.

In the second stage of the study, these gross value loss estimates from the Water Trade Model have been exported into the AusRegion model. The flow-on effects of SDLs, and the long-term employment change, have been simulated by the AusRegion model.

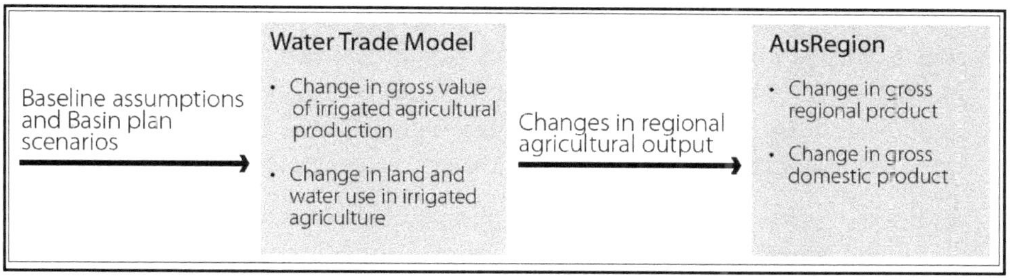

Figure 16.1 Two-stage approach in the ABARE–BRS study

Source: ABARE–BRS (2010a).

In the baseline scenario of the ABARE–BRS Water Trade Model, the 10 375 GL reported in 2000–01 is 2000 GL less than the 12 300 GL of water use under the long-term condition reported by the MDBA. Although this water-use difference was between simulated baseline water use and historical water use, the 2000–01 data are the best available at the moment, and were therefore used by ABARE–BRS to simulate the 3000 GL, 3500 GL and 4000 GL additional environmental flow scenarios. One important assumption in the ABARE–BRS study is that the water market in the Basin is efficient, and allows water to be traded away from relatively lower-value crops to relatively higher-value horticultural crops.

According to the simulated results of the ABARE–BRS study, it is estimated that the gross value of irrigated agricultural production would decline by about 13 per cent under a 3000 GL/yr scenario, 15 per cent under a 3500 GL/yr scenario and 17 per cent under a 4000 GL/yr scenario. As stated earlier, basin-wide employment would fall by approximately 800 full-time jobs, or about 0.1 per cent of current employment levels. These forecasted job loss figures are for the long run and mitigated by the government investment in water saving infrastructure. Without these assumptions, the job loss number could be higher. In a following study by ABARE–BRS, if without the government investment in water saving infrastructure, the short term job loss number could be around 5000 (ABARE–BRS 2011).

Irrigators are suspicious of the modelled social and economic impacts put forward by ABARE–BRS. They have been viewed as inadequate, and the estimated economic and employment losses described as optimistic. In general, the findings from ABARE–BRS are consistent with those from the literature (Adamson et al. 2007; Jiang 2010). In addition, ABARE–BRS state the assumptions and limitations in their report. These statements and results have, however, been largely ignored and misunderstood by the public.

Despite some limitations in the ABARE–BRS modelling, the used models are appropriate. More research is required, however, to improve them. This could include the following.

1. Developing a better regional boundary design in models, consistent with the Basin Plan regions in the MDBA's Guide. Neither the 22 regions in the ABARE–BRS Water Trade Model nor the seven regions in the AusRegion model can fully match the 19 Basin Plan regions in the Guide. Therefore, the prediction qualities in some regions are not as good as the overall results.
2. Continuing to accumulate the relevant basin data in a way that is consistent with the Basin Plan's regions. The best available data for the Basin are currently 10 years out of date, relating to 2000–01.
3. Water saving from irrigation infrastructure has not been examined in the ABARE–BRS Water Trade Model. Irrigators claim it is very important in Australia's water reforms.
4. Some findings, such as employment losses, are focused on the long term. Some short-term impact studies are also required.
5. Water used in the baseline of the ABARE–BRS modelling is 2000 GL less than the MDBA's long-term water-use data. This is the equivalent of an additional 2000 GL water buyback.
6. Water-use data for some regions—for example, 79 GL in eastern Mt Lofty Ranges—are 10 times greater than the water-use data from the CSIRO.

Other Water-Management Modelling in the Murray–Darling Basin

A large amount of literature exists on the subject of water management in the MDB. The Risk and Sustainable Management Group developed their model (RSMG) at the University of Queensland (Adamson et al. 2007). Two similar models were developed with different regional boundaries: the Integrated Irrigated Agriculture Water Model (IIAWM) was developed by the Centre for Water Economics, Environment and Policy at The Australian National University (Jiang 2010), and the ABARE–BRS Water Trade Model was developed by ABARE–BRS (Hafi et al. 2009).

The common approach in modelling water management is to combine hydrologic and economic models. The hydrologic studies examine water availability, water movement and barriers of the MDB under various scenarios such as climate change and water buybacks. In contrast, the economic studies simulate and optimise water use to maximise economic returns. These often use the hydrologic model data as the inputs or constraints. The output data of the existing economic models are limited in direct economic impacts such as gross value and profit. The broad economic impacts, such as indirect economic effects and employment changes, are not examined in the existing economic models.

As a typical example of these kinds of integrated hydrologic and economic models, Adamson et al. (2009) have assessed the effects of irrigated agriculture under different climate scenarios and states of nature using the RSMG model. They used inflow projections by Jones et al. (2007) for 2030, and found under their global solution (optimal adaptation to reduced water availability) that the social value in the Basin declines by at least $178 million a year, and up to $444 million a year (Adamson et al. 2009). The losses occur because of reduced revenues from lower yields due to deficit irrigation, smaller areas under irrigation and because of increased costs from accessing water. A recent SDL impact study using the RSMG model estimated water use fell by 3746 GL, or 35.5 per cent, and the resulting reduction in GVIAP was $1.445 billion, or 16 per cent (relative to the baseline). In the number of economic loss, the result from the RSMG model is 50 per cent higher than the ABARE–BRS Water Trade Model.

The Centre for Water Economics, Environment and Policy at The Australian National University, on behalf of the Wentworth Group of Concerned Scientists, recently developed a combined hydrologic and economic model (IIAWM) to study the economic implications of SDLs in the Basin. IIAWM assumed unrestricted water trade in the Basin and used the CSIRO's regional boundaries. One important feature of IIAWM is that an employment-change forecast function has been developed within simulated land use.

In IIAWM, under a basin-wide diversion of 30 per cent, agricultural diversions were estimated to require a reduction of 65 per cent in the Murrumbidgee River and 39 per cent in the Murray River, while other regions were relatively unaffected. For a 40 per cent basin-wide reduction, the regional reductions were more evenly spread, with the Campaspe and Loddon–Avoca rivers requiring reductions in agricultural diversions of 98 per cent and 84 per cent, respectively. The resulting reduction in annual net returns, averaged across the Basin, was 9.5 per cent for a diversion of 30 per cent and 16.3 per cent for a 40 per cent reduction.

The general findings from these models show that foregone profits resulting from additional environmental flows can be modest with a free water market. Particular regions, however—the Murrumbidgee, for example—might suffer substantial reductions in profits with water trade.

The Challenges of Existing Models

The findings and results from existing models should be treated with caution and considered in the context of model assumptions. Social and economic impacts in these studies, such as flow-on impacts and employment change, are insufficient. In the future social and economic study commissioned by the MDBA, it demands that the wider social and cultural implications of the proposed Plan should be scrutinised. The existing models are, however, limited in these broader impact studies, and the relevant science and social data are insufficient or not available to researchers.

The first challenge is from the free water market assumption. In most models, it is assumed there is a free water market at the Basin scale, and that water can be transferred between regions without barriers. The purpose of this water market is to transfer water from low economic value crops to those with higher economic value. Although some regions might suffer greater economic losses than others, with a free water market, the losses can be minimised at a basin-wide level. Under this free water market assumption, it is inevitable that transaction costs—such as water-purchase costs between farmers—and the uncertainty of the farmers' water-purchase mechanism might diminish the contribution of water trade. Therefore, economic returns from existing models represent long-term results of a best-outcome scenario. In the short term, the economic loss will be higher than the simulated results found in the literature. To overcome this problem, we need to measure and simulate water-trade costs and agricultural production under some water market failure scenarios. As there are no consolidated and consistent water-trade records for the Basin, these issues cannot, however, be fully examined.

The second challenge arises from inadequate employment-change data for irrigation industries—another controversial issue in terms of employment data availability. In 2006, the MDB accounted for approximately 10 per cent of total national employment—about 920 000 people, with about 96 000 engaged in agriculture. We do not, however, have any records for the farmers who worked only in the irrigated industries in the Basin. This poses a further data availability problem for modellers. Similarly, no historical data or models can estimate how many workers will find jobs in other industries.

The third challenge is the assumption in predicting employment change caused by water-use reductions. Using the SDL scenarios, relatively small reductions in employment have been reported by the AusRegion model, with a 0.1 per cent loss of employment across the MDB as a result of the Basin Plan, relative to the baseline. This figure was, however, based on two important assumptions.

1. Employment-change simulations focus on the long term and job losses in irrigated industries can be absorbed by new employment opportunities generated by other industries, such as dryland agriculture and mining. In the short term, the employment losses will be greater than the results for the long term.

2. The AusRegion model assumed Australia is close to full employment and able to adjust and send displaced agricultural labour to other industries. The flexibility and efficiency of transferring irrigators to other industries in the future Australian labour market are uncertain. For instance, the current skills of farmers might not be sufficient for them to find new jobs.

The fourth challenge is how to simulate employment change caused by water reductions. In current employment-change studies, the number of farmers in work is linked with GVIAP; however, commodity prices might change in the future, and employment change forecasted by GVIAP change is unable to consider this. Recently, IIAWM developed a new employment-change forecast function by considering land use and the average farm-worker numbers by crop.

The final challenge is data availability. In 2010, during the design of the *Guide to the proposed Basin Plan*, the Australian Government provided ABARE–BRS with a range of assumptions on irrigation water for the future, including projected total regional expenditures, volumes of water recovered and the distribution of these expenditures and water recovery over time, for both water-entitlement purchases and infrastructure investment. The infrastructure data are not, however, available to the public. Without the water-saving data from infrastructure investment, water-saving studies are not sufficient in the existing models.

A further example involves the Basin Plan's regional boundary data, which were not released by the MDBA. The regional boundaries play a vital role in our model development, yet some models cover only some parts of the Basin (Dixon et al. 2009). These models fail to take a holistic approach to simulate the Basin's water management and are therefore not suitable to model the whole Basin as required.

There are different ways to divide the Basin into regions, but adopting different regional boundaries results in a data-sharing problem between different government agencies. The first way to divide the Basin is based on the catchment management authority areas. Under these rules, the Basin has been divided into

The second way to divide the Basin has been developed in the CSIRO Sustainable Yield report (Figure 16.3). In hydrologic modelling, this method involves linking 24 existing water models from New South Wales, Victoria, Queensland, South Australia and the Australian Capital Territory to construct an integrated model with surface and groundwater intersections. This project modelled the water availability in the Basin for the past 100 years, and under several climate-change scenarios. A third way to adapt for the specific needs of the Basin Plan has seen the MDBA update the methods and tools underpinning the CSIRO study with new regional boundaries and hydrologic features (Figure 16.4).

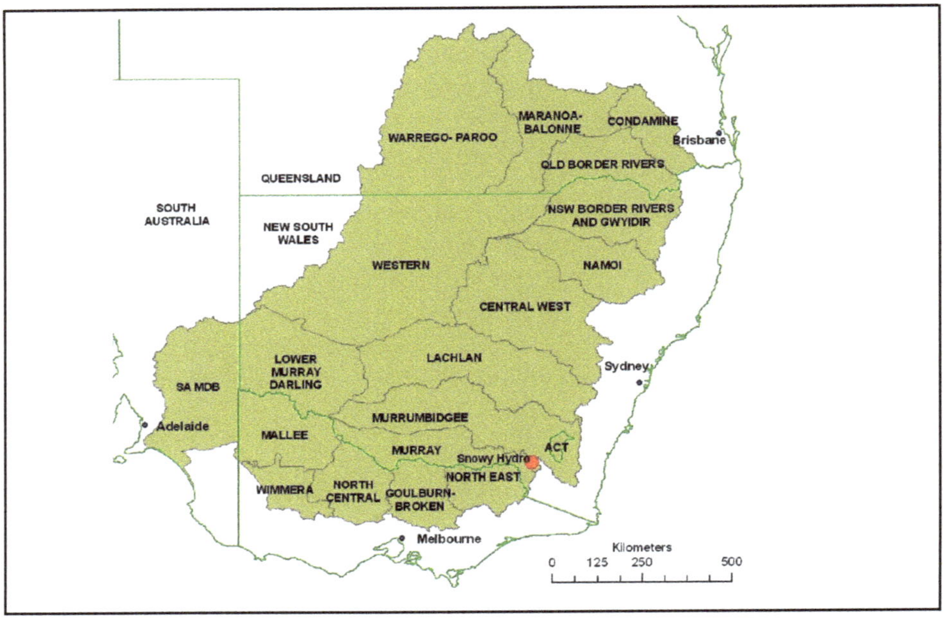

Figure 16.2 Catchment management authority areas

Source: Quiggin et al. (2008).

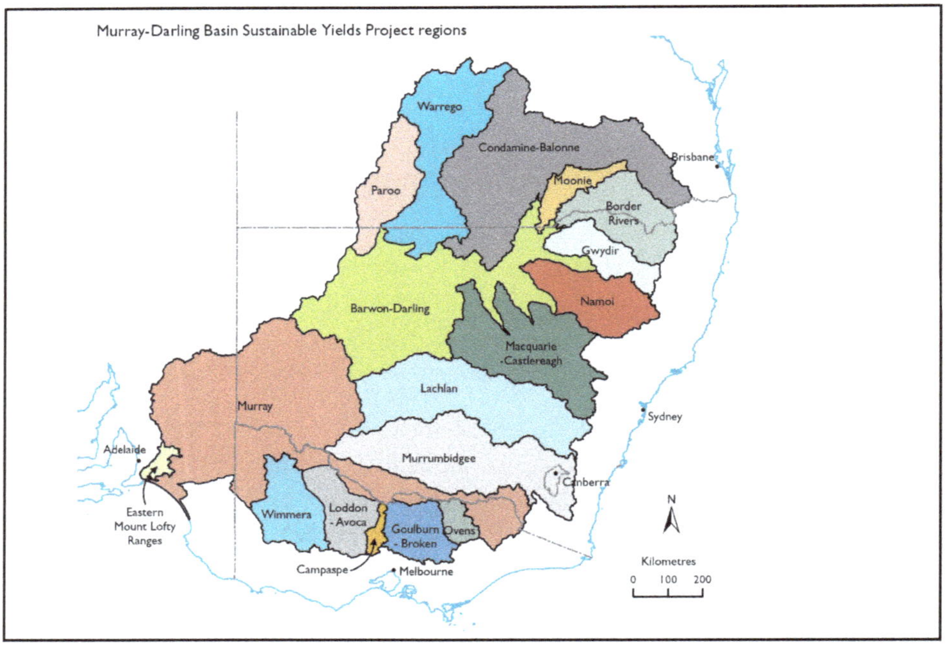

Figure 16.3 CSIRO Sustainable Yield regions

Source: CSIRO (2008a).

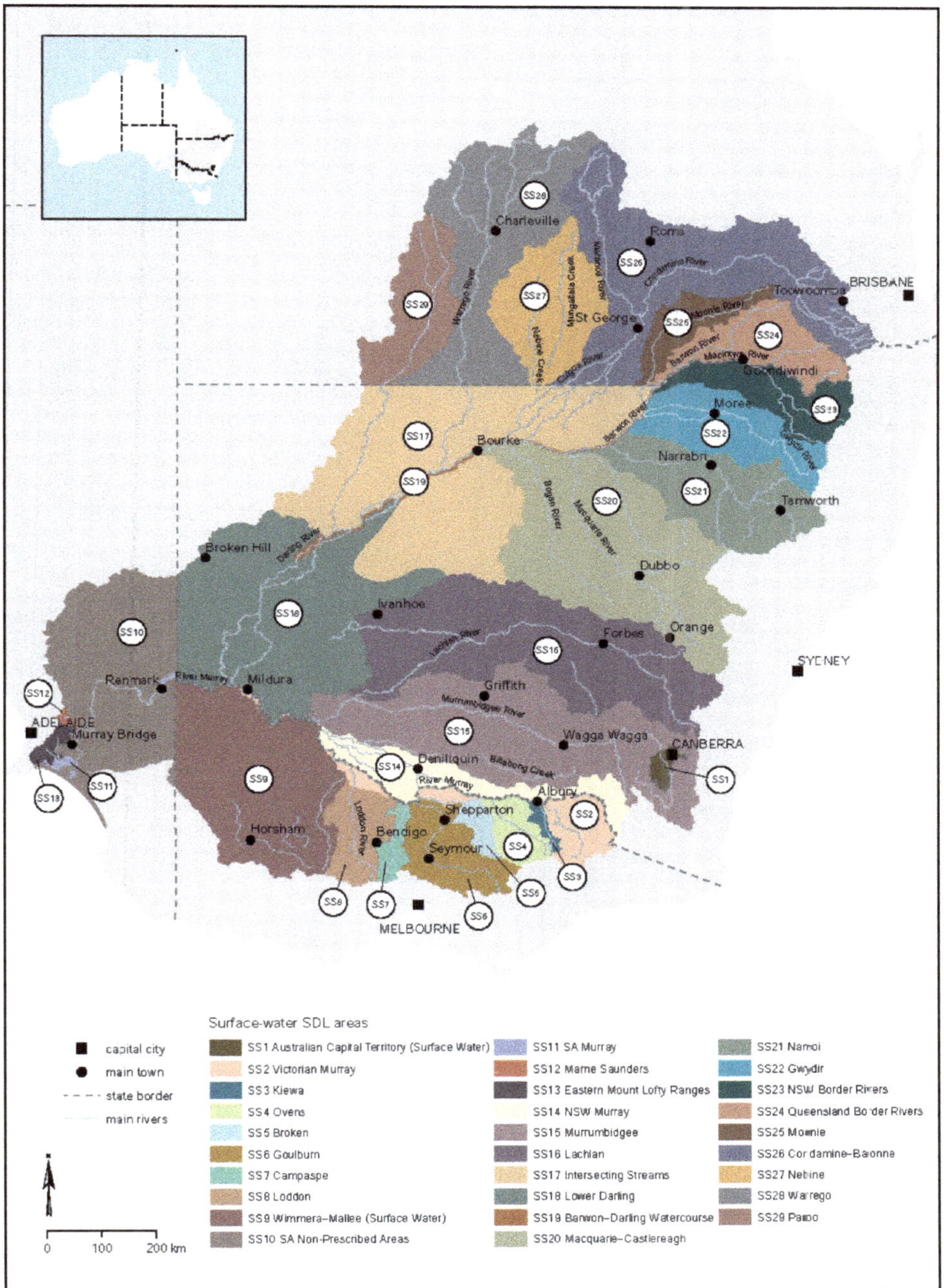

Figure 8.2 Surface-water SDL areas: Murray–Darling Basin

Figure 16.4 Basin Plan regions

Source: MDBA (2010a).

Figure 16.5 Regions in the ABARE–BRS Water Trade Model

Source: ABARE–BRS (2010a).

The regions in the ABARE–BRS Water Trade Model are similar to the Basin Plan regions (Figure 16.5). Some regions, however, such as the Barwon–Darling in the ABARE–BRS Water Trade Model (Figure 16.6), are totally different from their equivalent region in the Basin Plan (Figure 16.7). Without relevant spatial data made publicly available, researchers cannot modify existing models to match the Basin Plan regions.

Clearly, the new Basin Plan regions are different from previous regional definitions in the Basin. Therefore, much of the existing regional data or economic models are spatially incompatible with the new Basin Plan regions, and require significant effort in manipulating and scaling for a comprehensive economic analysis to be possible.

Figure 16.6 Barwon–Darling region in the ABARE–BRS Water Trade Model

Source: CSIRO (2008b).

Conclusions

In 2010, ABARE–BRS modelled the social and economic impacts of the SDLs and predicted that the impacts of additional environmental water are modest. Although the results from ABARE–BRS are similar to previous studies, the Basin communities have challenged their reliability.

With the different regional boundaries and model assumptions between the ABARE–BRS models and the Basin Plan, these SDL impact findings are the results of the first stage of analysis. We need a new model that can fully match the Basin Plan's regions and simulate the impacts under model assumptions.

Furthermore, new methods to analyse employment change need to be investigated, and some simulation scenarios, such as possible water-market failure, should be examined.

After the release of the *Guide to the proposed Basin Plan*, we need to improve our social and economic models of the regions of the Basin. As a substantial amount of data are not available to the public, however, researchers are unable to develop a full model using the Basin Plan's regions. To address this problem, the Australian Government should provide more resources relating to basin data, and allow researchers to access and share them.

Figure 16.7 Barwon–Darling region in the proposed Basin Plan

Source: MDBA (2010b)

References

Adamson, D., Mallawaarachchi, T. and Quiggin, J. 2007, 'Water use and salinity in the Murray–Darling Basin: a state-contingent model', *Australian Journal of Agricultural and Resource Economics*, vol. 51, no. 3, pp. 263–81.

Adamson, D., Mallawaarachchi, T. and Quiggin, J. 2009, 'Declining inflows and more frequent droughts in the Murray–Darling Basin: climate change, impacts and adaptation', *Australian Journal of Agricultural and Resource Economics*, vol. 53, no. 3, pp. 345–66.

Australian Bureau of Agricultural and Resource Economics (ABARE) and Bureau of Rural Sciences (BRS) 2010a, *Environmentally Sustainable Diversion Limits in the Murray–Darling Basin: Socioeconomic analysis*, Australian Bureau of Agricultural and Resource Economics and Bureau of Rural Sciences, Canberra.

Australian Bureau of Agricultural and Resource Economics (ABARE) and Bureau of Rural Sciences (BRS) 2010b, *Assessing the Regional Impact of the Murray–Darling Basin Plan and the Australian Government's Water for the Future Program in the Murray–Darling Basin*, Australian Bureau of Agricultural and Resource Economics and Bureau of Rural Sciences, Canberra.

Australian Bureau of Agricultural and Resource Economics (ABARE) and Bureau of Rural Sciences (BRS) 2011, *The economic and social effects of the Murray–Darling Basin Plan: recent research and next steps*. ABARE Conference Paper 11.12, March, Presented at the 41st Outlook Conference.

Commonwealth Scientific and Industrial Research Organisation (CSIRO) 2008a, *Water availability in the Murray–Darling Basin*, A report to the Australian Government from the CSIRO Murray–Darling Basin Sustainable Yields Project, CSIRO, Canberra.

Commonwealth Scientific and Industrial Research Organisation (CSIRO) 2008b, *Water availability in the Barwon–Darling*, A report to the Australian Government from the CSIRO Murray–Darling Basin Sustainable Yields Project, CSIRO, Canberra.

Dixon, P. B., Rimmer, M. T. and Wittwer, G. 2009, *Modelling the Australian Government's Buyback Scheme with a Dynamic Multi-Regional CGE Model*, Monash University, Clayton, Vic.

Hafi, A., Thorpe, S. and Foster, C. 2009, *The impact of climate change on the irrigated agricultural industries in the Murray–Darling Basin*. ABARE Conference Paper 09.3, February, Presented at the Australian Agricultural and Resource Economics Society (AARES).

Jiang, Q. 2010, *An overview of the Integrated Irrigated Agriculture Water Model (IIAWM) of the Murray–Darling Basin*, CWEEP Research Paper 10-03, Centre for Water Economics, Environment and Policy, The Australian National University, Canberra.

Jones, R., Preston, B., Brooke, C., Aryal, S., Benyon, R., Blackmore, J., Chiew, F., Kirby, M., Maheepala, S., Oliver, R., Polglase, P., Prosser, I., Walker, G., Young, B. and Young, M. 2007, *Climate Change and Australian Water Resources: First risk assessment and gap analysis*, Australian Greenhouse Office and the National Water Commission, Canberra.

Murray–Darling Basin Authority (MDBA) 2010a, *Guide to the proposed Basin Plan. Volume 1: Overview*, Murray–Darling Basin Authority, Canberra.

Murray–Darling Basin Authority (MDBA) 2010b, 'Summary of Barwon–Darling region', in *Guide to the proposed Basin Plan*, Murray–Darling Basin Authority, Canberra.

Quiggin, J., Adamson, D., Schrobback, P. and Chambers, S. 2008, *Garnaut Climate Change Review: The implications for irrigation in the Murray–Darling Basin*, University of Queensland, St Lucia.

17. Irrigators, Water Trading, the Environment and Debt: Buying water entitlements for the environment

Henning Bjornlund, Sarah Wheeler, Jeremy Cheesman

Since January 2008, the Australian Government's Water for the Future initiative has secured at least 705 gigalitres (long-term cap equivalent, or LTCE) of surface-water entitlements through the Restoring the Balance (RTB) environmental water buyback. The Government has $1.8 billion remaining in RTB for future entitlement purchases, and to close the 2300–3300 GL environmental watering gap defined in the *Guide to the proposed Basin Plan*. The Government has indicated a willingness to draw deeper into the public purse to close the environmental watering gap if required. The Government has emphasised that environmental water buybacks have been made, and will continue to be made, from willing sellers. A counterargument is that by conducting environmental water buybacks during the worst drought sequence in 100 years, the Government is predatorily acquiring water from 'forced' rather than 'willing' sellers.

The human, social and economic impacts of selling water to the Government have been considered at the individual, farm and regional community scales. While irrigators who choose to sell some or all of their water entitlements will be financially compensated at some level, they might experience adverse human and social effects from their decision. Moreover, communities that depend on irrigation might experience impacts of water entitlements leaving their region, for example via declining populations and loss of jobs and services. Community-level impacts are likely to be more significant in those communities whose economies have a greater reliance on irrigated agriculture, and that produce agricultural commodities with lower marginal value products of water, such as irrigated broadacre. These and other impacts have already been discussed and reported in the context of trading of water entitlements out of individual irrigation districts in the southern Murray-Darling Basin (Edwards et al. 2007; Fenton 2006).

Qualitative analyses of interviews with irrigators indicate that few regard themselves to be willing sellers of water entitlements. Many irrigators claim that they have been 'forced' to sell as a last resort due to financial stress, not least as a result of the current prolonged drought (Kuehne et al. 2010). They suggest that post drought the number of willing sellers may fall off significantly of the price paid for water entitlements by the Australian government does not rise.

In the context of future environmental water recovery through buyback, the issue is that the impacts of the reduction in total entitlements available for irrigation will depend on how the tenders to purchase water are designed, the extent to which 'willing' sellers will continue to sell entitlements, and how innovative policy makers are going to be to hone the capacity of markets to maximise agricultural output from the available water. Thus, understanding the characteristics of irrigators who sell water entitlements, and the factors they consider in this sale, are core to effective and efficient environmental water recovery by the Government.

Towards this understanding, this chapter provides some insight into the motivations of irrigators who sell water entitlements; what irrigators have historically done with the money they receive from entitlement sales; whether irrigators themselves perceive benefits from having an environmental water market in place; what irrigators think about irrigators who trade water to the Government; and whether there is a widespread belief amongst irrigators that greater environmental flows are necessary in the Murray–Darling Basin (MDB).

The chapter is organised in six parts. The first two parts briefly discuss irrigators' views on water markets and the environment, and how views have changed over time. The next part provides a broad overview of the relationship between debt, selling water entitlements, and what irrigators do with the proceeds. The following part discusses the issue of whether there will be sufficient willing sellers in the future for the Government to meet its environmental water-recovery objectives. The final part concludes and discusses some alternatives to the current environmental water recovery program.

Irrigator Trade Attitudes

Various sources (for example, National Water Commission 2010) show that water markets have been important in allowing irrigators to cope with the drought, by either selling or buying water allocations and/or water entitlements. In particular, the allocation market has been instrumental in allowing farmers to manage cash flow and farm risk. Bjornlund has been surveying water traders in key irrigation districts of the MDB over more than two decades. This chapter details irrigator answers to questions asked in surveys in the Goulburn–Murray Irrigation District (GMID) of Victoria in 1998–99, 2003–06, 2008–09 and 2010–11, the Riverland region of South Australia in 2008–09 and 2010–11, and the NSW Murray in 1998–99 and 2010–11. Irrigator answers provide an overview of irrigators' perceptions of trade issues in the southern Murray-Darling Basin. Tables 17.1 to 17.4 illustrate some key trading questions that have been asked over time, from 1998–99 to 2010–11.

Since water trading was first introduced in Australia in the mid 1980s, irrigators and their communities have held diverse and shifting opinions about the need for and impact of water markets. Over time, irrigator views towards trading have become more accepting. There has generally been more concern about: 1) entitlement trading than allocation trading; 2) trading out of districts than within districts; and 3) what has been termed speculative trading than trading between irrigators. Table 17.1 suggests that NSW Murray irrigators view the market very differently. Bjornlund (2005) found that: i) sellers in the allocation market agree significantly more with allocation trading than buyers; ii) sellers in the entitlement market agree most with the need for entitlement trading than do buyers; and iii) approximately 85 per cent of those who have traded agreed that water markets are a good idea, while only 48 per cent of those who had never traded as of July 1999 agree.

Table 17.1 Trade attitudes by GMID and NSW Murray in 1998–99 (per cent)

	GMID (allocation buyers and sellers and non-traders)					NSW Murray (allocation and entitlement buyers and sellers)				
	SA	*A*	*N*	*D*	*SD*	*SA*	*A*	*N*	*D*	*SD*
Water trade is a very good idea	46	27	13	5	10	48	24	10	7	11
I only agree with temporary transfers since the water stays on the property	35	27	12	13	13	31	20	12	18	19
It has to be possible to transfer water permanently otherwise it is not possible to make long-term commitments	27	28	19	13	13	37	27	16	11	9
Water trade should not be allowed because it activates otherwise unused water and reduces annual sales water	9	15	15	28	32	12	13	15	27	33
Water trade is a good way for some farmers to get out of irrigation	25	39	11	11	14	27	34	15	19	9
It is essential to make allocations to the environment otherwise irrigation will not be long-term viable	27	33	21	10	8	17	38	19	15	11
I am willing to reduce annual sales water allocations in order to ensure sufficient allocations for the environment	7	20	20	22	31	4	15	17	20	44

SA = strongly agree; A = agree; N = neutral, D = disagree; SD = strongly disagree. Notes:

N = 300 (GMID); N = 311 (Murray).

Source: Based on steering committee reports from LWRDC project USA3 available from <www.waterresearch.net>

Table 17.2 Trade attitudes in GMID in 2003–06 (per cent)

	SA	*A*	*N*	*D*	*SD*
Water trading should be allowed only within districts	52	10	14	4	20
Export out of districts imposes additional cost on remaining irrigators	65	15	11	3	5
Export out of districts has significant social flow-on effects within the district, with loss of businesses, development, jobs and population	66	15	10	4	5
If water and land are separated, I fear that big corporations will purchase large volumes of water and control who gets it and at what price	68	14	8	3	6
Trading in water should be allowed only between people who own land on which the water can be used	79	8	5	3	5
I am sceptical about the Living Murray Process—it looks as if the objective is to trade-off country voters for city voters	57	17	14	3	8
It is essential to make allocations to the environment otherwise irrigation will not be long-term sustainable	19	15	30	12	24
We would be willing to reduce our seasonal allocations in order to ensure sufficient allocations for the environment	9	8	16	14	54
We would be willing to give water to the environment as a reasonable contribution to improve conditions and as part of a wider community effort	8	9	20	12	51

Note: N = 1068. Source: Based on data from Bjornlund (2007).

Table 17.3 Trade attitudes in GMID and Riverland in 2008–09 (per cent)

	GMID					**Riverland**				
	SA	*A*	*N*	*D*	*SD*	*SA*	*A*	*N*	*D*	*SD*
People who buy and sell water regularly are just greedy for money	3	38	17	35	7	2	37	16	39	7
I would seriously consider selling my water if I was offered substantially more than market price	5	37	12	40	6	4	28	8	47	14
If I had unused water I would rather use it to expand my irrigation than sell it at an attractive price	2	29	16	46	7	2	39	20	36	2
If I had unused water I would probably not sell it because I would expect the price to increase in the future	2	32	23	37	2	2	28	20	47	4
When I retire I will move to an area that has more to offer	1	58	17	23	1	2	56	14	25	3
If I sold my water I would be letting my community down	2	24	7	53	14	2	38	14	42	5

Notes: N = 300 in GMID; N = 324 in Riverland.

Source: Based on data from RIRDC project by Bjornlund and Cheers; data not previously published.

Table 17.4 Trade attitudes in New South Wales, Victoria and South Australia, 2010–11 (per cent)

	NSW					Victoria—GMID					SA				
	SA	*A*	*N*	*D*	*SD*	*SA*	*A*	*N*	*D*	*SD*	*SA*	*A*	*N*	*D*	*SD*
I believe water trading has been a good thing for farming	7	38	12	23	17	8	32	15	26	18	7	49	14	21	7
Trading water allows me to cope with seasonal uncertainty	14	58	8	13	5	16	57	8	13	4	11	58	17	9	1
We would be willing to have our seasonal allocations reduced to ensure sufficient water for the environment	1	5	9	42	41	0	9	7	41	41	0	17	10	42	28
Most irrigators think increasing environmental water flows is a good thing	1	21	10	48	19	1	20	11	47	16	5	64	11	17	3
It is essential to make allocations to the environment otherwise irrigation will not be long-term sustainable	6	38	12	28	12	4	39	14	32	8	8	71	7	10	1
I am well informed about the trading rules in my district	16	68	6	6	2	15	68	6	9	2	8	73	4	7	1

Notes: N = 274 (SA); N = 358 (Vic.); N = 313 (NSW).

Source: Based on data from ARC Linkage grant by Wheeler, Bjornlund and Shanahan; data not previously published.

During more than a decade of surveying, four concerns about water trading have been consistently raised by irrigators: 1) the sale of entitlements from irrigation regions will result in irrigation costs rising for the remaining irrigators; 2) if entitlements are traded away from farmland it might be left uncultivated and become infested with weed and pests, which is a nuisance for neighbouring farmers; 3) if entitlements are traded away from farmland then the value of the land could decline—as a result, the council rate base will be reduced leading either to higher rates for remaining irrigators or to less services; and 4) if entitlements are sold out of a local area then farm production might decline. This could result in a loss of both on and off-farm jobs, which may in turn lead to regional decline in population and services.

Despite these concerns, irrigators have widely adopted water trading over time, in particular the trading of allocations. When water trading commenced in the mid 1980s, only a few per cent of farm businesses traded each season. In the late 1990s and 2000s, as water scarcity increased and farmers became more familiar with water trading, market participation rates increased markedly. Research within the GMID indicates that in the early 2000s up to 60 per cent of farm businesses traded water in some form each season, while less than 10 per cent had never engaged in water trading (Wheeler et al. 2009).

There is increasing evidence that the allocation market has become more competitive over time, with more than 30 per cent of the water applied in the field within the GMID being purchased in that market (Bjornlund and Rossini 2008; Wheeler et al. 2010a). There has also been an acceleration in entitlement trading, with the percentage of the total entitlement base traded within the GMID increasing from less than 1 per cent prior to 1998–99 to 4.5 per cent by 2007–08 (Bjornlund and Rossini 2010). In the GMID the 4 per cent cap on entitlement trade out of districts has been reached in successive years between 2007 and 2010, which suggests that total entitlement trade would have exceeded 4.5 per cent if this cap was not in place.

Water trading is thus having an increasing influence on who use and own water. There is clear evidence to suggest that irrigators have used water markets to manage the adjustment process and to manage water scarcity and production risk with respect to water supply (Bjornlund 2002, 2004, 2006). Despite the early reluctant acceptance of water trading, there is now a clear understanding within the irrigation community that without water trading the socioeconomic impact of the current drought would have been much harsher.

Irrigator Attitudes to the Environment

In general, irrigators agree with statements about the importance of, and need for, greater environmental flows in the Murray (Bjornlund 2005; Tisdell and Ward 2003). Over time however there have been significant declines in irrigators' acceptance of environmental needs. Tables 17.1, 17.2 and 17.4 summarise irrigators' responses to two questions—one about the need for more environmental flows in the River Murray and one about their preparedness to contribute to environmental flows by reducing their seasonal allocations. In the GMID, the percentage of irrigators agreeing with both statements was highest in 1998–99. Between the 1998–99 and 2010–11 surveys, irrigators' stated willingness to contribute towards environmental flows dropped from 30 per cent to 10 per cent amongst NSW and GMID irrigators. Over the same period, the percentage of irrigators agreeing with the need for more environmental flows declined from 60 per cent to 43 per cent in the GMID, and from 55 per cent to 44 per cent in New South Wales. Although no significant difference was found between NSW and GMID irrigators' attitudes to environmental flows, it seems that SA irrigators hold significantly different views to their upriver counterparts. In 2010–11, nearly 80 per cent of SA irrigators believed in the need to increase environmental flows, while about 17 per cent agreed with some form of voluntary uncompensated reduction in seasonal allocations.

These findings suggest that irrigators' willingness to sacrifice their own water allocations for the environment has been falling sharply over the past decade as water scarcity increased and the national water-reform policy process intensified from a point of discussion to reality. Drought and the real extent of the sustainable diversion limit (SDL) 'cuts' have fostered strong resentment amongst irrigators for what is regarded as a Basin Plan Guide that aims solely to meet the environmental watering requirements defined by federal bureaucrats. However, it may be a real surprise to many that over 10 per cent of southern MDB irrigators surveyed from August to the first week of October 2010 agreed that they were willing to see some of their water allocations sacrificed (without compensation) to ensure enough water for the environment.

The Relationship between Debt and Selling Water

Qualitative research suggests that farm debt is a main reason for irrigators selling water to the Government for environmental flows (ABARE 2009; Kuehne et al. 2010). There is no doubt that some farmers do sell water as a measure of last resort; however, the sale of water to finance or retire farm debt is not a new phenomenon—nor is it associated exclusively with environmental water purchases by the Government. Surveys of irrigators trading during the 1987–94 period within the GMID and Riverland found that cash-flow management was the main reason for 57 per cent of GMID irrigators selling water, and 71 per cent within the Riverland. Consistent results were observed in 1994–96, when 61 per cent of GMID and 58 per cent of Riverland irrigators gave this reason for selling (Bjornlund 1999). A similar survey of water sellers within Murray Irrigation Limited from 1990 to 1999 found that 52 per cent of sellers cited financial reasons as their main motivation for selling their entitlements (Bjornlund 2000).

The relationship between farm debt and selling water is complex. To fully understand the issue one needs to consider all the possible influences on why farmers sell water back to the Government. Several studies have looked at this relationship in detail. In 2008–09, 624 irrigators in the Riverland and the GMID were surveyed and asked about their willingness to sell water to the Government. Wheeler et al. (2010b) sought to understand what influenced irrigators in the two regions to sell their water. Farm debt was a strong, significant influence on GMID irrigators' willingness to sell. It was found that the higher the total farm debt, the higher was the likelihood that they were thinking of selling. Indeed, the higher the debt level, the higher was the percentage of their total water entitlements they were considering selling. This was not the case, however, for Riverland irrigators, where no significant association between debt and willingness to sell was found.

Marsden Jacob Associates (2010) surveyed 1021 farmers (808 irrigators and 213 dryland farmers) in 11 regions of the Basin between March and April 2010. This project sought to explore how irrigators may exit the industry under proposed water-entitlement cuts of up to 40 per cent of LTCE allocations. Overall, irrigation farms were found to be conservatively geared, with farm equity ratios generally near 80 per cent or greater. Over the past five years, however, average annual returns from farming have been low to negative due to drought. The average gross margin return on farm assets for horticulture, broadacre, livestock, dairy and mixed farms interviewed is in the range of –2 per cent to 2 per cent. Adding debt and interest costs will mean the average annual return on assets during the past five years has been negative for the majority of farms surveyed. This is an unsustainable cash-flow situation, and places a significant constraint on the ability of farmers to service farm debt and grow their operations. Cash flow—not assets and gearing—now appears to be the fundamental challenge for most farmers.

Marsden Jacob Associates (2010) also found that younger farmers have progressively higher debt than their older counterparts, and that—holding age and farm type constant—increasing farm debt was systematically linked to lower subjectively assessed personal wellbeing and future optimism scores. Moreover, holding all other factors (such as age, farm type, farm size, and so on) constant, in response to permanent reductions in irrigation allocations, irrigators with higher debt stated they would be more likely to: 1) seek to exit irrigation altogether; [1] and 2) restructure their farming operations. [2]

One of the main findings from these studies is that there is a wide range of influences on selling water. Farm debt does play a part, but irrigators' attitudes, values and wellbeing appear to be the most significant and strongest influences on decisions to sell water or exit the industry. For example, the survey undertaken by Marsden Jacob Associates found that the strongest predictor of the likelihood of an irrigator stating they would exit farming in response to permanent reductions to LTCE allocations was their Deakin wellbeing score. [3]

These studies show that how irrigators feel about their relationship with their community, their family, their health, and their standard of living counts more when they plan what to do with their farm and their water than many other

1 Significant at the 1 per cent level.

2 Significant at the 1 per cent level.

3 The Deakin Wellbeing Index measures subjective wellbeing. It comprises seven questions relating to satisfaction with life domains, such as 'health' and 'standard of living'. A considerable body of research has demonstrated that most people are satisfied with their own life. In Western nations, the average value for population samples is about 75 percentage points of satisfaction. That is, on a standardised scale from 0 (completely dissatisfied) to 100 (completely satisfied), the average person rates their level of life satisfaction as 75. The normative range of wellbeing is tightly grouped around the average, between from about 73 to 76 points. Scores of below 65 signal individuals whose personal wellbeing might be at risk.

farm and farmer variables, and this result is significant across all basin regions studied (unlike findings for other influences, which are often region/industry specific). Although other factors—such as farmer age, years farmed, type of farm, water entitlements, irrigated farm size, and technology adopted—do play a significant part in farm decisions, the influences of these variables differ from region to region.

The 2008–09 surveys in the GMID and Riverland asked irrigators how and where they planned to invest after selling their water entitlement. One-third of GMID irrigators and 40 per cent of Riverland irrigators planned to reduce debt, while the next largest response was off-farm investment (Figure 17.1) (Wheeler et al. 2010b). Only a small number of irrigators planned to use the funds for on-farm investment. Set in the context of the Basin drought, these results make intuitive sense. About one-third of irrigators in both communities suggested that they would spend the funds locally if they did sell water.

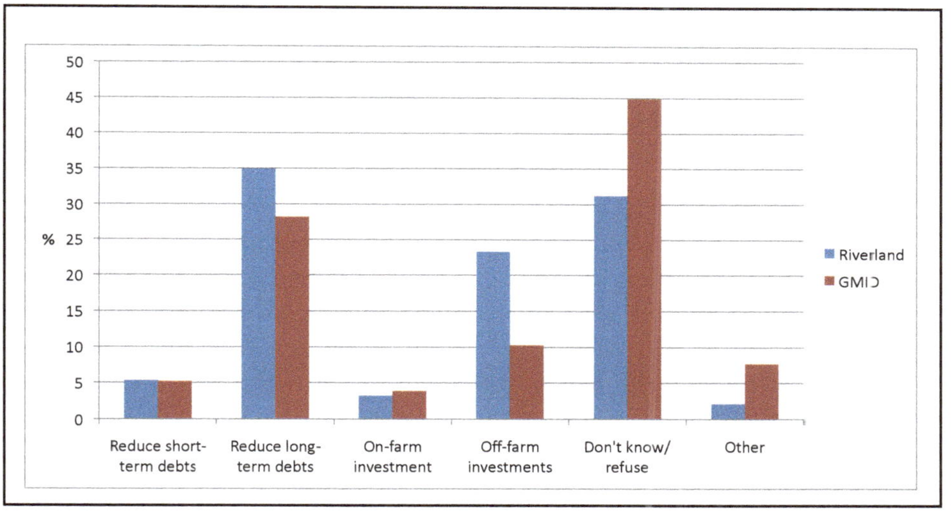

Figure 17.1 Plans of where to invest after selling water, 2008–09

What irrigators intend to do with the proceeds of water sales has shifted significantly over time. Bjornlund (1999) asked entitlement sellers in the GMID and the Riverland during the 1994–97 period how they had spent the proceeds from their water sales. Within the GMID it was found that 63 per cent put the money towards general revenue, 26 per cent towards debt reduction, 20 per cent to improve their irrigation and drainage system, and 11 per cent to purchase other assets such as machinery and land. In the Riverland, 27 per cent used the proceeds to improve their irrigation system, 17 per cent for debt reduction, 12 per cent to buy other farm assets and 44 per cent put the proceeds towards general revenue.

Will Sufficient Willing Sellers Continue to Come Forward?

The future success of the *Restoring the Balance* program depends on securing enough water sales from willing irrigators in targeted regions to meet the environmental watering requirements of the Guide. Surveys of irrigators in 2008–09 suggest that between 40 and 60 per cent of irrigators in the Riverland and GMID stated they have not thought about selling water to the Government at all. Table 17.3 indicates that about 30–40 per cent of irrigators would consider selling their water entitlements, but only if they were offered 'substantially' more than market prices. In the Riverland in 2008–09, the minimum average price suggested by irrigators for a water entitlement per megalitre was just less than $2700, while GMID irrigators' minimum average was a little more than $1900 (Wheeler et al. 2010b).

Wheeler et al. (2010b) estimated the total water volume that irrigators were planning to sell to the Government in 2008–09. It is of some concern that the estimated volume was only about half of what the Government plans to buy back If right, these results suggest that the Australian government will need to pay a higher price for water entitlements into the future than it has paid in the past if it wishes to secure the volumes of water it needs for the environment.

A caveat to our findings is that these estimates have been based on static one-year responses in a period of extreme drought and low commodity prices. Irrigators' past stated willingness to sell could change considerably if farm conditions improve (or worsen). The dynamic aspect of changing preferences for selling water is an area deserving further investigation. If willingness to sell water entitlements has fallen as the drought eases then the Government might need to pursue different approaches or restructure the buyback in order to encourage future participation and reasonable bids from irrigators in the future.

Conclusions

The release of the Guide to the proposed Basin Plan has aggravated a long-term distrust held within some irrigation communities about the permanent transfer of water entitlements from their regions. Irrigation communities have historically been wary of selling water entitlements, although this wariness appears to be easing over time. While trading in allocation water has grown considerably, the growth of water-entitlement trading has been considerably slower. Studies of irrigators' attitudes towards trading show a growing acceptance of the benefits of water markets over time, although often vocal dissenters persist within communities. Irrigators selling water (or exiting farming) are frequently motivated by debt considerations.

It is important to realise, however, that debt management is not the sole (or perhaps the main) rationale for irrigators selling water permanently. Values, attitudes and wellbeing of irrigators appear to play a more central role. While increasing farm debt does appear to be correlated with lower personal wellbeing, financial stress is only one attribute of personal wellbeing.

References

Australian Bureau of Agricultural and Resource Economics (ABARE) 2009, *Irrigated Agriculture in the Murray–Darling Basin: A farm level analysis by region and industry*, Australian Bureau of Agricultural and Resource Economics, Canberra.

Bjornlund, H. 1999, Water trade policies as a component of environmentally, socially and economically sustainable water use in rural southeastern Australia, D.Phil. Thesis, University of South Australia, Adelaide.

Bjornlund, H. 2000, *Water market activities, outcomes and perceptions; section 4: permanent trade along the Murray and Murrumbidgee rivers—New South Wales*, Second Steering Committee Report for LWRRDC Project USA 3, <www.prres.net>

Bjornlund, H. 2002, 'The socio-economic structure of irrigation communities—water markets and the structural adjustment process', *Journal of Rural Society*, vol. 12, no. 2, pp. 123–45.

Bjornlund, H. 2004, 'Formal and informal water markets—drivers of sustainable rural communities?', *Water Resources Research*, vol. 40, no. W09S07.

Bjornlund, H. 2005, 'Irrigators and the new policy paradigm—an Australian case study', *Water Policy*, vol. 7, no. 6, pp. 581–96.

Bjornlund, H. 2006, 'Can water markets assist irrigators managing increased supply risk? Some Australian experiences', *Water International*, vol. 31, no. 2, pp. 221–32.

Bjornlund, H. 2007, *The monitoring of, and reporting on, water trading within the Goulburn–Murray Irrigation District (2)*, Industry Partner Report No. 5, ARC Linkage Grant Water Scarcity and Rural Social Hardship—Can water markets alleviate the problems?, University of South Australia, Adelaide, <www.prres.net>

Bjornlund, H. and Rossini, P. 2008, Are the fundamentals emerging for more sophisticated water market instruments?, Fourteenth Annual Conference of the Pacific Rim Real Estate Society, Kuala Lumpur, January, <www.prres.net>

Bjornlund, H. and Rossini, P. 2010, Climate change, water scarcity and water markets—implications for farmers' wealth and farm succession, Sixteenth Pacific Rim Real Estate Society Conference, Wellington, <www.prres.net>

Edwards, J., Bjornlund, H. and Cheers, B. 2007, *The impact of trading of water out of districts: a case study of the Kerang region in Victoria*, Industry Partner Report No. 5, ARC Linkage Grant Water Scarcity and Rural Social Hardship—Can water markets alleviate the problems?, University of South Australia, Adelaide, <www.prres.net>

Fenton, M. 2006, *The social implications of permanent water trading in the Loddon–Campaspe Irrigation Region of northern Victoria*, May, Report Prepared for North Central Catchment Management Authority, Huntly, Vic.

Kuehne, G., Bjornlund, H. and Loch, A. 2010, *Why do farmers make non-profit decisions—investigating decisions made during drought*, RIRDC Publication No. 10/075, Rural Industries Research and Development Corporation, Canberra.

Marsden Jacob Associates 2010, *Delivering the Basin Plan: Economic and social profiles of Murray Darling Basin communities: Appendix 3b: Irrigator survey technical report*, July, Report for Murray–Darling Basin Authority, Canberra.

National Water Commission 2010, *The Impacts of Water Trading in the Southern Murray–Darling Basin: An economic, social and environmental assessment*, National Water Commission, Canberra.

Tisdell, J. G. and Ward, J. R. 2003, 'Attitudes toward water markets: an Australian case study', *Society and Natural Resources*, vol. 16, pp. 61–75.

Wheeler, S., Bjornlund, H., Shanahan, M. and Zuo, A. 2009, 'Who trades water? Evidence of the characteristics of early adopters in the Goulburn–Murray Irrigation District, Australia 1998–99', *Agricultural Economics*, vol. 40, pp. 631–43.

Wheeler, S., Bjornlund, H., Zuo, A. and Shanahan, M. 2010a, 'The changing profile of water traders in the Goulburn–Murray Irrigation District, Australia between 1998–99 and 2003–06', *Journal of Agricultural Water Management*, vol. 97, no. 9, pp. 1333–43.

Wheeler, S., Lane Miller, C., Zuo, A. and Bjornlund, H. 2010b, *Estimating southern Murray Darling Basin irrigators' willingness to sell water entitlements to the Federal Government*, CRMA Working Paper, November, University of South Australia, Adelaide.

18. Economic Perspectives of the Proposed Basin Plan for the Southern Connected Murray

Donna Brennan

Introduction

The *Guide to the proposed Basin Plan* is discussed in the context of the proposed sustainable diversion limits (SDLs) for the southern connected Murray River. The emphasis is on whether the Basin Plan can deliver on the economic efficiency goals set out in the *Water Act 2007*. While the discussion is by no means exhaustive, a number of issues of concern are raised. First, the tendency for a market-price premium in the Goulburn Valley in some years means that the expected opportunity cost of water in sourcing Murray water from the Goulburn is higher than it is elsewhere. This has not been considered in determining the amount of contribution the tributaries and the main channel should make to the Murray flow requirements—contravening economic-efficiency criteria. The question of the impact of environmental water plans on irrigation reliability is also raised. An example is presented in which it is shown that the economic impact of a spring-flow regime prescribed for the Goulburn by Cottingham et al. (2003) is 75 per cent higher when the impact of reliability is accounted for, compared with an analysis conducted on the mean loss in diversions. That reliability has been deemed to be 'off the agenda' is a serious shortcoming of the Guide. There are many unanswered questions regarding how the operation of environmental watering plans will affect delivery of water for irrigation and how these might affect the water market. Thorough economic analysis of these issues should be undertaken before the Environmental Water Plan is finalised.

The *Guide to the proposed Basin Plan* (MDBA 2010) sets out the principles that will be followed in developing the Basin Plan and proposes a series of sustainable diversion limits (SDLs) associated with basin-wide environmental-flow requirements of 3000 GL to 4000 GL per year. These SDLs take account of consumption from all sources, including diversions and interception activities such as farm dams and forestry. Sustainable diversion limits represent a maximum level of consumption that is allowed on average, but will be implemented like the Cap mechanism, allowing for the effect of climatic variation on consumption. It is likely that the cuts in consumptive use of water will be borne by irrigation diverters.

The development of the Basin Plan is one of the duties of the Murray–Darling Basin Authority (MDBA) under the *Commonwealth Water Act 2007*. This Act charges the Authority with the task of considering the use and management of the Basin's water resources in a way that optimises economic, social and environmental outcomes. It is also required to ensure that water reaches its most productive use through the development of an efficient water-trading regime.

In this chapter, elements of the *Guide to the proposed Basin Plan* are discussed in the context of the proposed SDLs for the southern connected Murray. The outline of the chapter is as follows. In the next section, a brief description of the SDLs is presented. This is followed by discussion of differences in the opportunity cost of sourcing water for environmental needs in the Murray River. The question of the economic impact of reliability and the potential impact of the environmental watering plan on the seasonal market is then discussed.

The Sustainable Diversion Limits

Sustainable diversion limits for catchments in the southern Murray–Darling Basin are shown in Table 18.1, for the 3000 GL and 4000 GL flow scenarios. The proposed cuts in consumptive use are uniform across the major tributaries and the Murray River—being 26 per cent for the 3000 GL scenario and 35 per cent for the 4000 GL scenario. Cuts in consumptive use are lower for the smaller Victorian catchments—ranging from 10 per cent in the Broken River to 21 per cent in the Loddon River for the 3000 GL case. Also shown is the share of consumptive use that comes from interception activities, which is very high in some of the smaller Victorian catchments—as high as 75 per cent in the Broken River. The MDBA expects that it will be too difficult for the States to demonstrate that they can cut the consumptive use for interception activities. Rather, it is likely that consumption from interception activities will be capped but not cut in the water-resource plans. In order to achieve the required cuts in consumption, there will need to be a greater cut in irrigation diversions, with a much larger impact on diversions in those catchments where interception activities are significant. For example, in the Ovens River, to achieve the overall consumptive-use cut of 12 per cent, it will be necessary to cut irrigation diversions by 40 per cent for the 3000 GL scenario. Cuts of 40 per cent are also required in the Broken, Loddon and Kiewa rivers. The Murrumbidgee River is also affected by interception activities, with diversion cuts increasing to 32 per cent over the 26 per cent consumptive-use cut required.

Table 18.1 Share of interception activities in consumptive water use and its impact on the cuts to diversions required for SDLs (per cent)

Catchment	Interception share of consumption	3000 GL per year		4000 GL per year	
		Consumptive-use cut	Diversions cut	Consumptive-use cut	Diversions cut
Ovens	70	12	40	13	45
Goulburn	6	26	28	35	37
Broken	75	10	40	11	45
Loddon	49	21	40	23	45
Campaspe	26	26	35	33	45
Murrumbidgee	20	26	32	35	43
Murray (NSW)	6	26	28	35	37
Murray (Vic.)	3	26	27	35	36
Murray (SA)	0	26	26	35	35
Kiewa	56	18	40	20	45

The Guide provides details on the contribution of diversion cuts to environmental flows, according to whether the water is contributed to local or to downstream environmental needs. In the case of the Ovens and the Kiewa rivers, all of the environmental water goes to downstream uses, and irrigators incur a 40 per cent cut in diversions. This is considerably higher than the diversion cuts borne by irrigators on the Murray. The Goulburn and the Murrumbidgee rivers make a contribution to the Murray, with 93 GL, or 21 per cent, going to the Murray from the Goulburn River and 191 GL, or 29 per cent, from the Murrumbidgee River under the 3000 GL scenario.

Table 18.2 Local environmental water needs and the contribution to downstream uses

	Local	Downstream	Downstream (%)	Diversions cut (%)
	3000 GL per year			
Ovens	0	10	100	40
Broken	3	3	50	40
Goulburn	349	93	21	28
Loddon	28	10	26	40
Campaspe	28	12	30	35
Murrumbidgee	474	191	29	32
Kiewa	0	4	100	40
	4000 GL per year			
Ovens	0	11	100	45
Broken	4	2	33	45
Goulburn	504	89	15	37
Loddon	37	6	14	45
Campaspe	39	13	25	45
Murrumbidgee	675	217	24	43
Kiewa	0	5	100	45

Opportunity Cost of Sourcing Water for the Murray River

If the contribution of water to downstream uses in Table 18.2 is to be considered efficient, it is necessary for the opportunity cost of water to be the same in each of the catchments from which water for the Murray is sourced. Presumably, the MDBA has made the assumption that water trade will equalise any divergence in the opportunity cost of water—but this is incorrect.

The presence of physical trade constraints leads to a persistent price premium in the Greater Goulburn trading region, as evidenced by market data. Goulburn irrigators may sell water to other trading regions, but they may not buy water because it is impossible to deliver. The rules governing trade between the major trading regions are summarised in Table 18.3. All regions may sell to the Murray-above-Barmah trading region; the Murray below Barmah may buy from anyone except the Murray above Barmah because of the Barmah Choke. The Murrumbidgee and the Goulburn rivers may only be sellers; buying water in is prohibited. That Murrumbidgee irrigators may not buy water in has no material impact on prices, because the level of water entitlement held there is high relative to the amount of irrigation investment. Murrumbidgee irrigators tend to be net sellers. In the case of the Goulburn River, however, there are some seasonal situations when Goulburn irrigators would be willing to buy water from other trading regions, but they may not.

Table 18.3 Rules governing inter-regional trade in the southern Murray

		Buyer			
		Murrumbidgee	Goulburn	Murray above Barmah	Murray below Barmah
Seller	Murrumbidgee	Allowed	Not allowed	Allowed	Allowed
	Goulburn	Not allowed	Allowed	Allowed	Allowed
	Murray above Barmah	Not allowed	Not allowed	Allowed	Not allowed
	Murray below Barmah	Not allowed	Not allowed	Allowed	Allowed

The seasonal price observed in the Goulburn trading region since 1999 is shown in Figure 18.1, measured as the difference in the Goulburn and Murray-below-Barmah prices. In 2003 and in 2007, the price premium was more than $100 per ML. In 1999 and 2002 it was at least $40/ML. In those cases when the Goulburn trading region is a net seller, there is no price premium. The average premium over the period 1999–2009 was $34/ML.

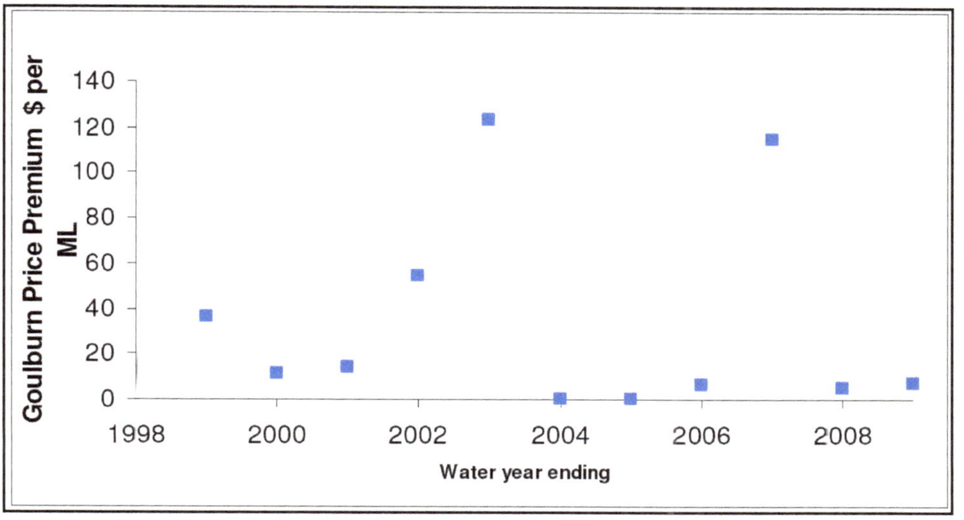

Figure 18.1 Seasonal price premiums in the Goulburn trading region, 1999–2009

Using a market-simulation model developed by the author, Brennan (2010b) demonstrated that this price premium is not simply a phenomenon of the past decade, but would be expected to persist over a 100-year simulation using water-allocation data derived from the CSIRO sustainable yields project (CSIRO 2008). The simulation model determines trade volumes and prices in each trading region for each year of the simulation, maximising the gains from trade given water allocations, using econometrically estimated water-demand curves, and subject to the trading rules of the market. Prices predicted by the model over the period 1999–2009 are compared with actual market data in Table 18.4. The model predicts very well for the Goulburn and the Murray below Barmah. The divergence in predicted prices is greatest for the Murrumbidgee (modelled prices 8 per cent higher), but this could be due to the presence of ad-hoc changes to trading rules over the period, such as the prohibition of trading out of the Murrumbidgee in 2007, which would have depressed actual market prices.

Table 18.4 Modelled and actual market prices over the period 1999–2009

	Actual mean price $/ML	Modelled mean price $/ML	Difference %
Murray above Barmah	146	156	6
Murray below Barmah	162	157	-3
Murrumbidgee	144	157	8
Greater Goulburn	195	200	2

Expected market prices were calculated using allocations from the CSIRO's future climate scenario, using entitlements held in 2008, and entitlement reductions required to meet the sustainable yield cuts. Results are shown in Table 18.5. The model predicts that the price premium—measured using 2008 entitlements—would be $15/ML. The effect of reducing entitlements by the diversion cut for 3000 GL/yr is to raise prices by $58 in the Goulburn and $44 elsewhere, increasing the Goulburn premium to $27.90/ML. The 4000 GL scenario raises prices by about 150 per cent over the baseline (2008 entitlements), and increases the Goulburn premium to $31/ML.

Table 18.5 Expected market prices for water allocations, 2008 baseline, 3000 GL and 4000 GL SDL scenarios

Entitlements:	2008	3000 GL SDL	4000 GL SDL
Goulburn	61.2	118.9	154.2
Murray above Barmah	44.8	88.8	117.3
Murray below Barmah	46.5	91.0	122.9
Murrumbidgee	46.5	91.0	122.9
Goulburn premium	14.8 (32%)	27.9 (31%)	31.3 (25%)

These results, along with the evidence from the actual market over the past decade, suggest that there is a price premium for water in the Goulburn and this is likely to be exacerbated by the proposed SDLs. This higher opportunity cost should be considered when sourcing water for the Murray. Failure to do so contravenes the economic-efficiency criteria set out under the *Water Act*.

Economic–Environment Trade-Offs Regarding Reliability

At the time of writing, the technical documentation supporting the Guide had not been released, and the main criterion provided regarding environmental flows is the end-of-systems flows that would be achieved for each catchment. The economic analysis has been limited to an assessment of the economic cost of an average reduction in water allocated in an average year.

There are a number of operational conflicts between managing water for environmental flows and managing water for irrigation that have implications for the economic returns from water use. If the MDBA is to meet the requirement of trading off economic and environmental objectives, it should provide an assessment of the economic impact of different watering regimes and their environmental outcomes.

Consider spring flows, for example. An increase in the amount of water used to meet medium-scale floods will improve outcomes for bird breeding and other ecosystem functions, but will have implications for irrigation reliability. The extent to which irrigators can mitigate these reliability impacts depends on the rules governing storage and how capacity is divided between the irrigation industry and the environmental water-holder. Different rules of storage management might result in different outcomes for the environment as well as for irrigation reliability. The effect of these rules, and the effect of different overall levels of spring flooding on the achievement of environmental outcomes and the net economic returns to irrigation, should be assessed as part of the decision-making process. It could be that marginal gains in environmental outcomes have a very high economic cost. Or they might not. It is difficult to know when nobody has asked the question, or provided the necessary information to conduct economic analysis of the trade-offs.

If the MDBA followed a rigorous process of quantifying the economic trade-offs of environmental watering it could then make an informed judgment as to what is acceptable. If it chooses to impose higher costs on the irrigation community in order to provide greater security of outcomes for the environment then at least the issue of compensation for the irrigation industry will be clearly articulated.

An Example of the Possible Cost of the Reliability Impact of Spring Flows

A simple example is provided in the potential impacts of spring flows on irrigation reliability and the cost to irrigators. Spring-flow requirements for the Goulburn River were quantified using formulae provided in Cottingham et al. (2003), which express the quantity of environmental water required for spring flows as a function of September and October inflows. These spring-flow requirements were then deducted from inflows available for irrigators, and, using the historical storage rules, the amount of water that would be allocated to irrigators in each season was calculated. This is referred to as the allocation sequence that accounts for spring flows. The mean reduction in diversions was 15 per cent. The mean annual loss in economic value associated with these spring-flow allocations was calculated using irrigation-consumption demand curves reported in Brennan (2008, 2010a). As shown in Table 18.6, the expected annual loss in value in the Goulburn Valley is $20 million per year. Also provided in the table is an estimate of the value of a 15 per cent cut in consumption measured at the mean level of diversions. (This latter approach is the basis on which all the calculations have been done for the Basin Plan to date, as reported in ABARE–

BRS 2010). The estimated cost of a reduction in consumption of 15 per cent valued at the mean level of diversions is only $11.4 million. Properly accounting for reliability increases the impact of spring flows by 75 per cent.

It is possible that irrigators might be able to reduce the impacts of the loss in reliability by using carry-over. The effectiveness of carry-over in improving reliability will depend on rules governing storage-capacity ownership and spill-accounting procedures. Because of non-negativity of storage, however, it is never possible to smooth consumption perfectly and therefore there will be a reliability impact even if storage arrangements are favourable to irrigators. This reliability impact will be a cost imposed on those remaining in the industry, not those selling water under the buyback scheme. The question of reliability (and rules governing how irrigators might manage reliability better through carry-over) should be addressed before the Basin Plan is finalised.

Table 18.6 An example analysis of the economic impact of spring flows if reliability is accounted for

Scenario	Expected annual cost $ m
Average year approach, with allocations cut by 15 per cent	11.4
Properly accounting for spring-flow impact on reliability (old storage regime)	20.0
Difference	75%

Potential Jeopardy in the Water Market

There are a number of other operational conflicts that will have implications not only for the economic cost of environmental flows but also for the functioning of the seasonal water market. This is because of the strategy of adaptive management, which could result in rules changing during the irrigation season, which will cause windfall commercial gains and losses to irrigators who have taken an early position on the water market.

For example, the requirement for low summer flows could conflict with demands for channel capacity for delivery of irrigation water at the peak of the irrigation season. SKM (2006) suggests that this potential conflict will be binding for the Goulburn River, and that investment in alternative delivery infrastructure such as pipelines might be a way of overcoming the delivery constraint. The economic cost of such investment should be assessed and compared with the economic and environmental impacts of alternative summer flow/delivery constraint regimes. But even if the optimal flow/investment regime is determined, there remains the question of how the summer flow/delivery constraint will be implemented. Will it be seasonally dependent? Will adaptive management require that the decision about the summer flow required for that season be adjusted during the season? If adaptation is desirable, this will have the effect of increasing uncertainty in the water market. The economic impact of this should be quantified before the environmental watering plan is finalised.

The *Water Act* permits the environmental water-holder to trade on the seasonal market, if it is beneficial for achieving environmental outcomes. An example would be the selling of water in dry years when prices are high to create revenue for buying substantially more water in wet years when prices are low. But the presence of a large player in the market might have a significant impact on prices. Rules governing the participation of the environmental water-holder in the seasonal market require serious consideration. It might be necessary, for example, to require that the environmental water-holder declare what quantity it intends to buy or sell in the market prior to the commencement of trading. This would avoid the problem of windfall gains and losses to irrigators using the water market.

Conclusion

This chapter has raised questions regarding the adequacy of the Basin planning process with respect to the requirement that water planning be economically efficient. The list of issues raised might have barely scratched the surface of potential conflicts between environmental-flow requirements and irrigation demands, but demonstrates the need for more in-depth economic analysis using output from hydrological models.

From the information reported on market-price premiums, it is reasonable to conclude that the SDLs presented in the Guide do not meet economic-efficiency criteria. The example presented on spring flows demonstrates the folly of using an average approach to estimating economic impact, and the potential for high costs associated with the reliability impact of environmental-flow regimes. That reliability has been deemed to be 'off the agenda' is a serious shortcoming of the Guide. There is a whole range of unanswered questions regarding how the operation of environmental watering plans will affect delivery of water for irrigation and how these might affect the water market. Thorough economic analysis of these issues should be undertaken before the Environmental Water Plan is finalised.

References

Australian Bureau of Agricultural and Resource Economics (ABARE) and Bureau of Rural Sciences (BRS) 2010, *Environmentally sustainable diversion limits in the Murray–Darling Basin: socioeconomic analysis*, Report prepared for the Murray–Darling Basin Authority, Canberra.

Brennan, D. 2008, 'Missing markets for storage and the potential economic cost of expanding the spatial scope of water trade', *Australian Journal of Agricultural and Resource Economics*, vol. 52, no. 4, pp. 471–85.

Brennnan, D. 2010a, 'The economic potential of market oriented water storage decisions: evidence from Australia', *Water Resources Research*, vol. 46, no. W08537, <doi:10.1029/2008WR007407>

Brennan, D. 2010b, Water prices in the lower Murray market: the effect of physical trading constraints, climate change and the spatial pattern of buyback, Unpublished manuscript available from the author.

Commonwealth Scientific and Industrial Research Organisation (CSIRO) 2008, *Water availability in the Murray Darling Basin*, A report from CSIRO to the Australian Government, Canberra, <http://www.csiro.au/files/files/po0n.pdf>

Cottingham, P., Stewardson, M., Crook, D., Hillman. T., Roberts, J. and Rutherfurd, I. 2003, *Environmental flow recommendations for the Goulburn River below Lake Eildon*, Technical Report 01/2003, Cooperative Research Centre for Freshwater Ecology, Canberra.

Murray–Darling Basin Authority (MDBA) 2010, *Guide to the proposed Basin Plan. Volume 1: Overview*, Murray–Darling Basin Authority, Canberra.

Sinclair Knight Merz (SKM) 2006, *Goulburn Campaspe Loddon Environmental Flow Delivery Constraints Study*, November, Sinclair Knight Merz, Melbourne.

19. Managing Risk in the Murray-Darling Basin

John Quiggin

For much of the twentieth century, the expansion of irrigated agriculture in the Murray–Darling Basin (MDB) was treated as a self-evidently desirable objective, to be pursued without excessive regard to questions of economic costs and benefits. Irrigation seemed to offer a 'drought-proofing' solution to the risks and uncertainties that plague dryland agriculture in Australia.

By the late 1980s, however, the capacity of the Basin to support additional diversions was close to exhaustion. Analysis at the time suggested—in the terminology of Randall (1981)—that a move from an 'expansion' phase, in which resource constraints were relatively unimportant, to a 'mature' phase, characterised by increasingly sharp conflicts over access to the resource, was under way. It was hoped that these conflicts could be resolved at low cost through the introduction of market mechanisms.

In reality, however, as noted by Quiggin (2008), the actual outcome was a 'crisis' phase, in which the possibility of a systemic collapse loomed ever larger. The only feasible response, it has become evident, is a 'contraction' phase, in which claims to the resource are scaled back.

Attempts to deal with the problems of the Basin through the creation of markets in water rights, minimising the role of governments, began with the communiqué of the 1994 Council of Australian Governments (COAG) meeting and was developed more fully in the National Water Initiative (NWI) announced in 2004 (COAG 1994, 2004). The NWI was described by the National Water Commission (2009) as Australia's enduring blueprint for water reform', through which 'governments across Australia have agreed on actions to achieve a more cohesive national approach to the way Australia manages, measures, plans for, prices, and trades water.

In practice, however, the NWI failed to resolve many of the key conflicts associated with the mature water economy. Some conflicts between States arose from the need to deal with different systems of water entitlements. Conflicts also emerged between States and within the Commonwealth over the extent to which trade in water entitlements should (or should not be) restricted and over the possibility of transfers of water from rural to urban use. Most importantly, the NWI did little to resolve the conflict between demands for extractive water use and the needs of the natural environment.

Only three years after the announcement of the NWI, however—based on cooperation between Commonwealth and State governments—then Prime Minister John Howard unilaterally announced the National Plan for Water Security (Howard 2007). Although the National Plan was described as 'accelerating the implementation of the NWI', it amounted to an abandonment of the cooperative approach in favour of a Commonwealth takeover of water planning throughout the Murray–Darling Basin. The National Plan was a poll-driven exercise produced largely by Howard himself.

The central element of Howard's plan was the *Water Act* 2007, which called for the newly created Murray–Darling Basin Authority (MDBA) to prepare a management plan for the Basin, based on scientifically determined sustainable diversion limits (SDLs) for each catchment. The hope was that this plan would end the uncertainty surrounding water allocations and water rights, and thereby lead to a resolution of the long-running disputes over water use in the Basin. The *Guide to the proposed Basin Plan* (MDBA 2010a) was released by the MDBA in October 2010. Far from producing a resolution, however, the Guide was a source of new conflict.

In this chapter, it is argued that many of the most intractable management problems of the Basin can be understood in terms of the interaction between uncertainty and property rights.

Uncertainty and Property Rights

As Adamson et al. (2009) observe, variability and uncertainty regarding natural flows are central to the analysis of irrigated agriculture. It is useful to distinguish between predictable variation (for example, seasonal patterns) and uncertainty, and to further distinguish two kinds of uncertainty: risk and ambiguity. Risk arises when the probability distribution of a given variable is known. Ambiguity—also sometimes referred to as Knightian uncertainty (Ellsberg 1961; Knight 1921)—arises when probabilities are unknown, or when it is not possible to describe all possible outcomes in advance.

Even under stable long-term climatic conditions, the probability distribution of inflows to the Murray–Darling Basin displays high levels of risk compared with other major river systems. Farmers and other water users do not respond passively to risk, but choose production strategies to manage risk. To represent this appropriately, it is necessary to analyse production under uncertainty in state-contingent terms. A general theory of state-contingent production is developed by Chambers and Quiggin (2000) and applied to the modelling of the Murray–Darling Basin by Adamson et al. (2007).

Climate change has introduced ambiguity arising from the fact that our understanding of changes in climatic patterns remains limited—particularly at regional and catchment levels. Thus, while we know that the probability distribution of climatic variables will change from the historically observed values, we cannot yet determine the probability distribution that will be applicable in the future. When concern about the sustainability of irrigation policy in Australia first emerged in the 1980s, the possibility of climate change was not seriously considered in this or other discussions of public policy. Even as late as 1994, the water-reform program agreed by COAG (1994) took little account of climate change.

This is a classic case of ambiguity (Ellsberg 1961). More fundamentally, the example of climate change shows that one cannot consider all the possible states of nature that might affect the outcomes of policy decisions or production choices. This problem of 'unknown unknowns'—made famous by Rumsfeld (2002)—has been discussed in detail by Grant and Quiggin (2010).

Property Rights under Uncertainty

Any property right can be considered as a bundle of state-contingent claims. In some cases—such as those of property rights over consumption goods—the element of contingency is relatively unimportant, while in others, such as the rights associated with the purchase of an insurance policy, they are critical. Rights to water for irrigation in Australia are in the latter category. Different entitlements covering the same volume of water might involve radically different state-contingent allocations.

In theoretical discussions of property rights, it is commonly assumed that a right is specified so that, for every relevant contingency, the right is associated with a given claim. This assumption fits naturally with standard assumptions of unbounded rationality and zero transaction costs. As shown by the first and second fundamental theorems of welfare economics, any Pareto-optimal outcome can be implemented as a competitive equilibrium based on an appropriately chosen initial allocation of wealth.

In reality, this can never happen, because it is impossible to identify, in advance, all possible states of nature. As is shown by Grant and Quiggin (2010), a decision maker who proceeds as if all possible states of nature have been considered is vulnerable to adverse surprises and to manipulation by more aware participants in the policy process. Moreover, even if participants in the policy process seek to agree on a set of responses to particular contingencies, disagreement will inevitably arise *ex post* as to whether the necessary conditions have been fulfilled. This issue is discussed by Grant et al. (2009).

Exceptional Circumstances

One example relevant in the current context is that of 'exceptional circumstances' under drought policy. The criteria adopted in 1999 stated that exceptional circumstances assistance should be available only in the event of 'rare and severe events', where a rare event is defined as one that occurs on average only once in every 20–25 years (DAFF 2011).

The severe drought conditions that prevailed in much of Australia in the decade following the adoption of the criteria resulted, however, in exceptional circumstances assistance being made widely available, often for periods of several years. Various responses to this outcome are possible.

One view is that the drought conditions of the early 2000s were indeed exceptional, and that the drought policy worked broadly as intended. Doubtless, those who formulated the policy would have hoped that the first occurrence of exceptional circumstances would not have been so early or so widespread, but the nature of uncertainty is that extreme events occur with positive probability.

A second view is that, as a result of climate change, droughts are likely to be more frequent and more severe in future. On this view, it might be argued that it is necessary to recalibrate the definition of exceptional circumstances to the 'new normal'.

A third view is that much of the exceptional circumstances assistance provided during the drought was not in fact justified under the stated criteria. On this view, the widespread provision of exceptional circumstances assistance represented a breakdown of the central theme of drought policy—namely, that farmers should normally be expected to manage climatic variation.

The incomplete specification of terms such as 'exceptional circumstances' has proved critical in drought policy, and has substantially affected policy responses to drought in the Basin. Even more significant—for the purposes of the Basin Plan—is the question of whether a new determination about the sustainable volume of extractions from a given catchment represents new knowledge or a change in policy. This issue—discussed in detail below—is central to the way in which the risk-sharing principles of the NWI are applied.

The National Water Initiative and the Assignment of Risk

The trading system set up under the 1994 water-reform process provided irrigators with a range of tools for managing farm-level risks arising from uncertain water supply, as well as a marketable asset that could be used to manage financial risk. Thus, in periods of water shortage, farmers facing cash-flow problems could sell water entitlements (either temporarily or permanently) to those with a high demand for reliable water supplies.

By converting revocable licences into property rights, however, the process reduced the capacity of governments and system managers to deal with aggregate uncertainty.

The NWI (COAG 2004) set out principles regarding the sharing of risk arising from changes in the aggregate availability of water.

Two major principles were announced. The first was that, in future, water allocations should be stated as shares of available water, rather than as specific volumes. This approach deals with fluctuations in water availability by sharing the total amount available among users in proportion to their share. It raises the question of whether it will continue to be possible, as at present, to distinguish between high-security and low-security rights. The difficulties with this approach are discussed by Freebairn and Quiggin (2006).

The second principle—and one particularly pertinent in the present context—concerned an approach to the sharing of risk arising from changes in the aggregate availability of water. Under this principle, the risk of changes in water availability due to new knowledge about the hydrological capacity of the system will be borne by users. The risk of reduction in water availability arising from changes in public policy, such as changes in environmental policy, will be borne by the public, and water users will receive compensation for such reductions.

The principles of the NWI were elaborated in more detail in a statement issued by the 2004 COAG meeting. The communiqué specified a framework that assigns the risk of future reductions in water availability as follows.

- Reductions arising from natural events such as climate change, drought or bushfire to be borne by water users.
- Reductions arising from *bona fide* improvements in knowledge about water systems' capacity to sustain particular extraction levels to be borne by water users up to 2014. After 2014, water users to bear this risk for the first 3 per cent reduction in water allocation; the relevant State or Territory government and the Australian Government would share (one-third and two-third shares

respectively) the risk of reductions of between 3 per cent and 6 per cent; State/Territory and the Australian governments would share equally the risk of reductions above 6 per cent.

- Reductions arising from changes in government policy not previously provided for would be borne by governments.
- Where there is voluntary agreement between relevant State or Territory governments and key stakeholders, a different risk-assignment model to the above may be implemented.

Risk and the Environment

The NWI principles contained a crucial ambiguity regarding the allocation of water to the environment. At the time the NWI was adopted, it was clear—in general terms—that the existing allocations of water for irrigation use were environmentally unsustainable. On the other hand, there had been no detailed assessment of environmentally sustainable levels of water extraction for the catchments in the Basin. Governments were committed to undertaking such an assessment—a commitment that was formalised by the *Water Act2007*.

The outcome of the assessments, not surprisingly, was that SDLs in all catchments of the Basin should be set below the current Cap. The key question, in terms of the NWI, was whether these assessments represented new scientific knowledge or were simply a consequence of a change in policy to require diversions to be restricted to environmentally sustainable levels.

On the first interpretation, water users were to bear the costs of reductions. This interpretation appears consistent with the NWI principles, noting in particular the reference to '*bona fide* improvements in knowledge about water systems' capacity to sustain particular extraction levels'.

On the second interpretation, the cost was to be borne by governments. It could be argued that over-allocation of water rights—or, at least, reckless disregard of sustainability constraints—was conscious public policy in the decades leading up to the imposition of the Cap in 2004, and that the Cap froze existing over-allocation in place. On this view, any move towards sustainability would constitute a change in policy, and therefore the costs of any reduction in aggregate diversions should be borne by government. It is hard to see, on this reasoning, why principles of risk allocation were needed at all.

Nevertheless, in the years following the adoption of the NWI, it became apparent that any attempt to make irrigators bear the risk associated with the determination of limits on sustainable levels of extractive water use would be untenable. A number of factors contributed to this outcome.

First, the severe drought conditions that prevailed for most of the first decade of the twenty-first century left large numbers of farmers in severe financial difficulty. An uncompensated reduction in their water entitlements would have forced many to leave agriculture. This was unlikely to be a politically acceptable outcome.

Second, it became apparent that any reduction in water entitlements would entail substantial political difficulties. Even voluntary transfers of entitlements between irrigators faced substantial opposition. The Victorian Government, in particular, imposed limits on the volume of entitlements that could be sold from a given irrigation district.

Finally, the adoption and success of the Restoring the Balance in the Murray–Darling Basin program (Wong 2008) by the Rudd Labor Government demonstrated that purchase of water rights from willing sellers provided a fiscally affordable method of securing large volumes of water for environmental flows. As at 31 December 2009, the Restoring the Balance program had secured the purchase of 766 gigalitres of water entitlements worth more than $1.2 billion (DEWHA 2010).

Risk and the *Guide to the proposed Basin Plan*

The MDBA released the *Guide to the proposed Basin Plan* in October 2010. The release of the Guide was an opportunity to reach a broad agreement on a sustainable allocation of water rights, taking account of all the prevailing uncertainties.

These hopes have so far not been fulfilled. Rather, the Guide has met with a strong, and largely hostile, reaction. The document was publicly burned at meetings of farmers. The Chairman of the MDBA, Mike Taylor, resigned and has been replaced by a former minister in the NSW Labor Government—presumably more attuned to the political realities.

In large measure the failure of the Guide was due to poor communications. The Guide represented, in effect, an abandonment of the principles of risk sharing set out in the NWI, and a massive transfer of wealth to irrigators. The failure to communicate this fact was reflected in the hostile response of those who stood to benefit most from the policies put forward in the Guide.

The key proposal is that the entire reduction in diversions proposed in the Basin Plan should be treated as arising from a change in government policy. This proposal appears inconsistent with the risk principles. Proposals for reduced diversions were based on new scientific evidence about the sustainable supply capacity of the different catchments—the risk of which was supposed to be borne by irrigators.

It could, perhaps, be argued that the unsustainability of existing policies was known at the time the NWI was agreed, and arguably since the imposition of the Cap in the early 1990s. On this view, the requirement to achieve SDLs was itself a change in policy and the associated risk should be borne by governments. Given such a view, it seems hard, however, to conceive of any risk that would be borne by irrigators, and therefore hard to understand the rationale for the elaborate principles of the NWI.

A more plausible view is that the proposals in the Guide reflect political realities that have changed substantially since 2004. On the one hand, the political feasibility of uncompensated cuts in allocations—always limited—now appears non-existent. On the other hand, the 2007 National Action Plan for Water provided (or at least promised) a Commonwealth 'bucket of money' totalling $10 billion, which should be more than sufficient to cover the cost of a reduction in diversions of 3–4000 GL.

Risk and Climate Change

The most important unresolved uncertainty in estimates of the capacity of the Basin to sustain diversions for human use relates to climate change. The estimates of SDLs used in the Guide were based primarily on historical observations over the 114 years since monitoring of the system began. On the assumption of a stable climate, a data set of this length would permit reasonably accurate estimates of the distribution of inflows.

The climate is changing, however, as a result of greenhouse gas emissions, superimposed on natural cycles such as the Indian Ocean Dipole and the El Niño Southern Oscillation. The Guide (MDBA 2010b:4.2) discusses the problems of adjusting the Basin Plan to take account of the uncertain impacts of climate change. The solution adopted—which seems reasonable in the circumstances—is to defer long-term changes until the next 10-year plan, which is due to be developed about 2020. For the next 10 years, the Guide imposes a 3 per cent reduction in entitlements relative to those that would be derived on the basis of historical data. This, it can be noted, is the only policy in the Guide that could accurately be described as a 'cut'.

The Guide (2010b:s. 5.2) maintains the NWI approach under which the risk of climate change is borne primarily by water users. It remains to be seen whether this allocation will prove politically feasible.

Balancing Priorities

The debate over the Guide—already confused and confusing—was derailed further by arguments over whether, and how, social and economic considerations should be balanced against the needs of the environment. These arguments had their genesis in the decision of the Howard Government to pass the *Water Act2007* over the objections of State governments, which necessitated reliance on the Commonwealth's treaty-based power to protect Ramsar Convention-listed wetlands. As a result, the MDBA made public statements that it was required to give primary priority to environmental protection.

These statements were inconsistent with the central policy decision of the Guide—namely, that the volume of water to be restored to the environment should be between 3000 and 4000 GL. The figure of 3000 GL was estimated to be the minimum consistent with environmental sustainability, while 4000 GL was estimated to be the maximum that would not entail unacceptable economic impacts.

Obviously, the treatment of environmental and social/economic objectives in the Guide was symmetrical, with each being treated as a binding constraint. Given this symmetry, the insistence of the MDBA, and its Chairman, Mike Taylor, on the claim that environmental requirements had priority under the *Water Act* is difficult to understand.

On-Farm Water Saving

The main remaining problem is that the great bulk of National Water Allocation Plan funding has been notionally allocated to finance on-farm water-saving measures. It seems highly unlikely—based on the experience of such ventures as the Victorian Food Bowl Modernisation Project—that there exists sufficient cost-effective on-farm options to generate the proposed savings. Quiggin et al. (2011) examine the Food Bowl Modernisation Project and conclude that the cost of water released to Melbourne and the environment could be as much as $10 000/ML—five to 10 times the likely market price.

It seems likely that the infrastructure investments proposed under the National Plan for Water Security will replicate this disastrous outcome on an even larger scale. Given a willingness to purchase water rights at a market price, there is, quite simply, no need to provide infrastructure subsidies. If infrastructure investments can realise cost-effective reductions in water losses, the water saved as a result can be sold at the market price—either for environmental or for irrigation use—to pay for the investment.

In equity terms, proposals to subsidise investment in irrigation infrastructure make little sense. As noted above, under a policy of repurchasing rights from willing sellers, irrigators are virtually guaranteed to be gainers. Their water assets increase in value, and, if they choose not to sell, the value of farm output is likely to increase as a result of the withdrawal of other irrigators.

Assuming that the Commonwealth is willing to bear the full cost of reductions in diversions—contrary to the risk-allocation principles in the NWI—it will be necessary to reorient some funding from the National Water Allocation Plan/ Water for the Future. The central principle should be that of cost effectiveness. Water-saving projects should be funded only if they can deliver savings at a lower cost than the market price of repurchasing water rights.

A crucial requirement for progress is to minimise the dissipation of scarce public funds on cost-ineffective infrastructure investments. Such funds could be better allocated either to the purchase of entitlements from willing sellers or to social infrastructure.

The Way Forward

Given sufficient political will, and more skilful communications, it might still be possible to salvage a sustainable policy from the wreckage of the Guide. At this stage, the optimal political strategy is probably to defer any final plan while proceeding with an interim strategy that is, in essence, a continuation of the voluntary purchase strategy of Water for the Future, with additional funding to ensure that purchase programs have a positive net economic and social effect on the communities concerned. Both the setting of specific targets for reductions in diversions and the funding of substantial investments in irrigation infrastructure should be deferred until the finalisation of the Basin Plan.

Using this incremental strategy, it should be possible to restore substantial volumes of water to the environment at relatively low cost, while addressing many of the adjustment concerns that have effectively derailed the Guide. A reconsideration of SDLs, taking account of environmental, economic and social objectives, could then be conducted.

Such reconsideration must allow sufficient flexibility to respond to new information and unforeseen contingencies, as well as to the seasonal and annual fluctuations in inflows that have always characterised the Basin. The ultimate solution must be a system of property rights, specified in terms of state-contingent allocations to water users and the environment, along with a continued role for government as the ultimate risk manager.

References

Adamson, D., Mallawaarachchi, T. and Quiggin, J. 2007, 'Modelling basin level allocation of water in the Murray Darling Basin in a world of uncertainty', *Australian Journal of Agricultural and Resource Economics*, vol. 51, no. 3, pp. 263–81.

Adamson, D., Mallawaarachchi, T. and Quiggin, J. 2009, 'Declining inflows and more frequent droughts in the Murray Darling Basin: climate change, impact and adaptation', *Australian Journal of Agricultural and Resource Economics*, vol. 53, no. 3, pp. 345–66.

Chambers, R. G. and Quiggin, J. 2000, *Uncertainty, Production, Choice and Agency: The state-contingent approach*, Cambridge University Press, UK.

Council of Australian Governments (COAG) 1994, *Report of the Working Group on Water Resources Policy: Communique*, February, Council of Australian Governments, Canberra.

Council of Australian Governments (COAG) 2004, *Intergovernmental Agreement on a National Water Initiative*, Council of Australian Governments, Canberra, <http://www.coag.gov.au/meetings/250604/iga_national_water_initiative.pdf>

Department of Agriculture, Fisheries and Forestry (DAFF) 2011, *Exceptional Circumstances Criteria*, Department of Agriculture, Fisheries and Forestry, Commonwealth of Australia, Canberra, <http://www.daff.gov.au/agriculture-food/drought/ec/ec_handbook#criteria>

Department of the Environment, Water, Heritage and the Arts (DEWHA) 2010, *Restoring the Balance in the Murray–Darling Basin*, Department of the Environment, Water, Heritage and the Arts, Commonwealth of Australia, Canberra, <http://www.environment.gov.au/water/publications/mdb/restoring-balance.html>

Ellsberg, D. 1961, 'Risk, ambiguity and the Savage axioms', *Quarterly Journal of Economics*, vol. 75, no. 4, pp. 643–69.

Freebairn, J. and Quiggin, J. 2006, 'Water rights for variable supplies', *Australian Journal of Agricultural and Resource Economics*, vol. 50, no. 3, pp. 295–312.

Grant, S. and Quiggin, J. 2010, *Inductive reasoning about unawareness*, Risk and Sustainable Management Group Risk and Uncertainty Program Working Paper R09_1, University of Queensland, St Lucia.

Grant, S., Kline, J. and Quiggin, J. 2009, *A matter of interpretation: bargaining over ambiguous contracts*, Risk and Sustainable Management Group Risk and Uncertainty Program Working Paper R09_3, University of Queensland, St Lucia.

Howard, J. 2007, A national plan for water security, Statement by the Prime Minister, 25 January, Parliament House, Canberra.

Knight, F. 1921, *Risk, Uncertainty and Profit*, Houghton Mifflin, New York.

Murray–Darling Basin Authority (MDBA) 2010a, *Guide to the proposed Basin Plan. Volume 1: Overview*, Murray–Darling Basin Authority, Canberra.

Murray–Darling Basin Authority (MDBA) 2010b, *Guide to the proposed Basin Plan. Volume 2: Technical background*, Murray–Darling Basin Authority, Canberra.

National Water Commission (2009), 'Australian water reform 2009: Second biennial assessment of progress in implementation of the National Water Initiative', http://www.nwc.gov.au/www/html/147-introduction---2009-biennial-assessments.asp

Quiggin, J. 2008, 'Managing the Murray–Darling Basin: some implications for climate change policy', *Economic Papers*, vol. 27, no. 2, pp. 160–66.

Quiggin, J., Chambers, S. and Mallawaarachchi, T. (eds) 2011, *Water Policy Reform: Lessons in sustainability from the Murray Darling Basin*, Edward Elgar, Cheltenham, UK.

Randall, A. 1981, 'Property entitlements and pricing policies for a maturing water economy', *Australian Journal of Agricultural Economics*, vol. 25, no. 3, pp. 195–220.

Rumsfeld, D. 2002, Transcript of press briefing, 12 February, US Department of Defense, Washington, DC, <http://www.defense.gov/transcripts/transcript.aspx?transcriptid=2636>

Wong, P. 2008, Water for the future, Paper presented to the Fourth Annual Australian Water Summit, Sydney Convention and Exhibition Centre, Sydney, 29–30 April.

Part V. Governance

20. The Role of the Commonwealth Environmental Water Holder

Daniel Connell

Introduction

The new governance framework for the Murray–Darling Basin (MDB) created by the *Commonwealth Water Act 2007* is the first attempt to take a comprehensive approach to water management in the region. Previous arrangements applied only to a limited range of issues selected through a decision-making process requiring unanimous agreement by all governments. Realisation that many of the concerns excluded—such as the impacts of climate change—were a serious threat to future water security motivated the recent reforms. The requirement to produce a Basin Plan to be implemented through sub-plans developed by each of the States is the best-known part of the *Water Act*. Since the release of the *Guide to the proposed Basin Plan* in October 2010, it has been a major source of controversy. This chapter argues, however, that another set of provisions in the *Water Act*—those applying to the Commonwealth Environmental Water Holder (CEWH)—will be even more significant in the long term.

As a result of the need to placate the States, the arrangements for the Basin Plan are complex, rigid and subject to many restraints to protect State government priorities. That makes adaptive management in response to new issues difficult and provides many opportunities for opponents to delay and frustrate its implementation. In contrast, the arrangements applying to the CEWH give the Commonwealth autonomous power to shape future water management in the MDB for the first time, particularly in relation to environmental conditions, and are a major break with the past.

The *Water Act 2007* is the eighth major attempt since the 1980s to significantly shift the division of water in the MDB between production and the environment. The seven previous attempts were the Salinity and Drainage Strategy of 1989, the Natural Resources Management Strategy of 1990, the Cap mechanism of the mid 1990s, the 1994 Council of Australian Governments (COAG) rural water-reform package, the Integrated Catchment Management (ICM) Policy Statement of 2000, the Living Murray First Step project of 2003–04, and the National Water Initiative (NWI) in 2004. This chapter will highlight a number of characteristics that have been recurring themes in the history of water reform in the MDB by discussing four of the eight attempts to realign the balance between the environment and production: the Salinity and Drainage Strategy, the Cap, the Living Murray project and the *Water Act 2007*. It will argue that the Basin Plan as proposed in the *Water Act*—like its predecessors—is weakened by its dependence on complex and brittle intergovernmental agreements with weak or unworkable provisions for compliance.

The CEWH is different; the Commonwealth Government controls its budget and operating policies. Through the CEWH, it will be able to achieve the environmental-flow targets of the Basin Plan even if the States do not give their support. The historical record of water reform in the MDB points to a low probability of success for arrangements that depend on close cooperation between governments. It can be argued that new arrangements need to be based on the expectation that cooperation will be variable and rarely sustained. While this might appear regrettable, it is a more realistic perspective about what is possible in a pluralist democracy such as Australia than has been the case in the past. There will always be many competing interests in the MDB. The CEWH—but not the Basin Plan—is better suited to operate in such a rough-house policy and management environment. Like its seven predecessors, the Basin Plan will probably be too delicate a flower to flourish in such conditions.

For more than 90 years—from 1915, when the *River Murray Waters Agreement 1914-1915* was enacted, until the *Water Act 2007*—arrangements for cross-jurisdictional management of the MDB depended on identical legislation enacted in parallel in the four, later six, legislatures with responsibilities in the MDB: the Commonwealth, New South Wales, Victoria, South Australia, and, more recently, Queensland and the Australian Capital Territory. The *Water Act 2007* is Commonwealth legislation based on Commonwealth constitutional powers. A key provision of the earlier legislation was the requirement that all decisions be unanimous. This gave each jurisdiction the power of veto. In addition, it meant that each jurisdiction was able to demand large concessions to protect its interests in the implementation of those few policies that were approved for joint policy making. As a result, MDB policies through the 1990s and 2000s were typically low in compliance capacity and high in transaction costs resulting from provisions to protect State interests. Not surprisingly, there was frequently a very large gap between the aims expressed for a policy and the details put in place for its implementation. These characteristics shaped outcomes in the past and are a serious threat to the success of plans for the future because of their continued presence within the *Water Act 2007*—in particular, in the sections applying to the Basin Plan.

Salinity and Drainage Strategy, 1989

One of the most important policies enacted in the period immediately after the reforms of the 1980s was the Salinity and Drainage (S&D) Strategy. This strategy showed the potential that could be realised through cooperation and also the many ways in which it could be frustrated by State governments if they so wished. The Salinity and Drainage Strategy has been highly successful in containing the impact of salinity in the lower reaches of the River Murray caused by irrigation development. The wider goal of restraining other sources of degradation caused by irrigation development was, however, not achieved. Many of its original designers had thought that capping salinity impacts would also cap the volume of irrigation water that could be applied. Irrigation causes salinisation by bringing dissolved salt and by mobilising salt previously held in the soil profile. It was thought that by capping the level of acceptable salinisation, the volume of water that could be applied would also be capped.

That wider policy goal was frustrated by the way the S&D Strategy was implemented. This involved issues that were relatively inaccessible to the wider public. One was the starting date after which remedial action would have to be undertaken to counter the salinity impacts of any irrigation development. South Australia wanted accountability for impacts under the scheme to start from a time before significant irrigation development began—early in the twentieth century. That would have made the upper States responsible for the very considerable salinity impacts of irrigation development over the previous 60 or 70 years, and substantially increased the size of their required contribution to remedial works. In response, New South Wales insisted on 1 January 1988 as the benchmark and rejected proposals to use existing levels of irrigation activity as the starting point after which all future development would be held accountable, requiring compensatory works to balance the salinity impacts. New South Wales also insisted that any future development based on the use of existing water entitlements, even though they had not yet been activated, should be excluded from the accountability process.

All three states—South Australia, New South Wales and Victoria—had some unused entitlements, but there was little awareness that in New South Wales the total was very large. (Entitlements that had been only partially activated became known as 'dozers'. Those that had not been activated at all were called 'sleepers'.) Subsequently, the Water Audit conducted in 1995 estimated that the average allocations available to diverters during the five years to 1992–93 were 16 902 gigalitres, of which only 63 per cent had been activated—indicating considerable capacity to expand even without the granting of new entitlements (Murray–Darling Basin Ministerial Council 1995:8, Table 2). This additional capacity for expansion contained within the S&D Strategy effectively destroyed its potential to protect the river environment from incremental erosion through increasing extraction. That early episode showed the dependence of basin-wide reform on genuine cooperation by the individual States. The great differences that exist between States makes it relatively easy for special interests to frustrate changes that might not benefit their particular region even if they are beneficial from a wider perspective.

The Cap on Extractions, 1996

Further demonstration of the capacity of the States to frustrate MDB reform by agreeing in principle but opposing in detail is provided by the history of the Cap on extractions. By the early 1990s it had become clear that the S&D Strategy had failed to restrain the growth in extractions, and that riverine conditions in the MDB were still deteriorating. A spectacular algal bloom in the summer of 1991–92 that extended along more than 1000 km of the Darling River gave the issue international prominence. In June 1993, a South Australian member of the Ministerial Council, John Klunder, proposed that 'there should be no further regulation and diversions arrangements which would exacerbate deteriorating flow regimes'. He drew attention in particular to the potential for diversions to increase as a result of the utilisation of existing but previously un-activated entitlements (Klunder 1993)—the sleepers and dozers. In response, the Council commissioned an audit of water use in the MDB that was delivered in June 1995 (Murray–Darling Basin Ministerial Council 1995).

The Water Audit found that conditions were continuing to decline. In response, 'as an essential first step in establishing management systems to achieve healthy rivers and sustainable consumption uses', the Ministerial Council introduced an immediate, temporary Cap on further expansion at 1993–94 levels of development—the irrigation season upon which the Water Audit was based (Murray–Darling Basin Ministerial Council 2000:9). Introduction of the Cap in 1995 was a unanimous joint decision of the Governments of the Commonwealth, New South Wales, Victoria and South Australia (Queensland agreed in principle but never followed through in practice). The process of implementation, however, created the widespread impression that the Cap was an imposition of the Murray–Darling Basin Commission (MDBC) (in the opinion of some people that meant the Commission Office) on reluctant governments. During the 1998 federal election campaign, the Deputy Prime Minister, Tim Fischer, promised to 'zap the cap', although nothing was heard of that promise subsequently. In New South Wales in particular, the benefits of the Cap as a means to protect security of supply were poorly explained.

When implementing its version of the Cap, New South Wales treated un-activated entitlements in the same way as it had at the time of the introduction of the S&D Strategy in 1988. Against advice from a number of quarters, it decided once again to recognise existing entitlements that had not yet been developed. This time this approach created serious problems. In 1988, much of the political pressure that might have been created by implementation of the S&D Strategy was avoided by allowing continued expansion of diversions by permitting activation of existing but unused entitlements.[1]

In 1995, however, the volume of water that New South Wales was allowed to divert as a whole under the Cap formula was fixed at the level that would have occurred under '1993/4 levels of development'. As a result, the total consumptive pool stayed about the same, but now had to be shared among a larger group of diverters, as unused entitlements were increasingly traded and activated.

Under pre-Cap arrangements, NSW Government water managers knew that a certain proportion of irrigators would not use all or any of their water, so they had customarily redistributed the surplus to established irrigators at the same low cost as was charged for the water provided as part of their entitlement. Under the new arrangements for the Cap, established irrigators now lost access to that low-cost water. To maintain their consumption at previous levels, they had to purchase what they had lost at higher prices on the water market. Not surprisingly, many of them felt that their water allocation had been cut because of the introduction of the Cap.

This created great hostility to the new system in New South Wales and generated the 'vast majority' of hostile submissions received by the five-year review of the Cap in 2000 (Murray–Darling Basin Ministerial Council 2000:27). In its final report, the review commented that some of the complainants were not receptive to advice that this was an issue that should be taken up with the relevant State government (New South Wales), not the MDB Ministerial Council or the MDB Commission. The report expressed concern about the potential created by this confusion for negative feedback and publicity that would have a detrimental impact on the Cap. This turned out to be the case and, despite repeated statements in the annual reports of the Independent Audit Group appointed to assess implementation of the Cap that much more needed to be done, there was no progress beyond the original decision to introduce the Cap at 1993–94 levels of development. As a result, the Cap was never extended to include groundwater or tightened further to achieve a more sustainable balance between irrigation and the environment, as envisaged. The approach taken to

1 No new entitlements were granted unless salinity mitigation works were undertaken to counter the salinity effects. The volume of unused existing entitlements, however, was large. During the five-year period to 1992–93, only 63 per cent of existing entitlements were utilised. See Murray–Darling Basin Ministerial Council (1995:8, Table 2).

implementation of the Cap is similar to that planned for the Basin Plan, which will be introduced through sub-plans to be developed by each of the States. This time there are stronger compliance provisions in place but they require extensive consultation before activation and contain great scope for disagreement about what is reasonable. As a result, the potential for the intent of the Basin Plan to be frustrated, as was the Cap, remains high.

Living Murray Project, 2004

Another major issue for the *Water Act* and the Basin Plan is that of transaction costs. An illustrative example of the transaction costs created by the need to respect State sensitivities is the Living Murray project, approved with funding in 2004. (The Living Murray continues as a separate project although its operation is required to be consistent with the requirements of the *Water Act 2007*.) It was a product of the review of the Cap conducted in 2000, which concluded that more needed to be done to arrest the continuing decline in environmental conditions and resource security. Early in the planning phase for the Living Murray, the Ministerial Council established a scientific reference panel to advise about the potential benefits of a range of rehabilitation options. Six were assessed by the panel against the probability that they would restore the River Murray to a condition that could be described as that of 'a healthy working River Murray system'. The first three—do nothing, improved operations only, and improved operations plus 340 GL for new environmental flows—were all considered to have a 'low' probability of success. Improved operations plus 750 GL was given a 'low–moderate' rating. For improved operations and 1630 GL, the probability was 'moderate'. Only 3350 GL plus improved operations was rated 'high'. Eventually, in June 2004, $500 million was approved for use on six sites along the River Murray. At the time, it was hoped that this would result in an additional 500 GL becoming available for the environment. Subsequently, in May 2006, the Commonwealth added another $200 million to the project plus an additional $300 million for works that had been delayed in recent years through lack of funds.

Compared with previous efforts, the Living Murray project represented a much higher level of coordinated inter-jurisdictional activity in the MDB. It brought together a number of projects that have been under way for some time, added new ones and considerably increased the total effort invested in river rehabilitation. Existing activities included fish-ways, restoring fish habitat and a range of engineering works to make better use of environmental water. For some years, the MDBC had been installing fish-ways on its major structures so that fish would be able to move up and down the Murray over more than 2000 km from the Murray mouth to the foot of Hume Dam. After more than a century of de-snagging to improve navigation, thousands of old tree trunks are being placed back in the river to improve fish habitat. On important wetlands—such as those in the Barmah–Millewa forests—environmental flows were being released to extend bird-breeding seasons, and engineering works and pumps were being installed so that important sites could be selectively watered. Efforts were also being made to involve Indigenous people in the planning and management of sites of particular cultural significance to them. These projects are important in themselves and also because of the opportunity they provide for water managers and policy makers to work out in practice what can be done to achieve environmental objectives on a river system subject to intense development pressure. The Living Murray was important because it brought them together as a more coherent, coordinated whole.

But the labyrinthine arrangements put in place to manage the water entitlements assigned to the Living Murray to protect State sensitivities are extraordinary. Satisfying the States' demand to maintain the maximum possible degree of autonomy appears to have taken precedence over the greater efficiency—or at least clarity—that would presumably be made possible by a single accounting system operated by a single accounting body. The structure agreed in the *Intergovernmental Agreement* about the MDB combined the following elements. The six signing governments agreed to commit $500 million over five years with each jurisdiction responsible for a defined proportion. The contributions of the various States were not to be consolidated in a central fund but were to be held in the various State and Commonwealth Treasuries until needed. Working within the parameters of the Living Murray project, each government was to nominate and implement projects that met defined criteria. Jurisdictions were to have the option of investing in each other's projects up to set limits. Each would receive credit for its investment in these projects up to the value of its contribution to the $500 million.

These financial transactions were to be recorded at both the jurisdictional level and by the MDBC—now the Murray–Darling Basin Authority (MDBA)—on behalf of the Ministerial Council, but the central registry was not to be a master registry. (This means that in the event of a dispute about a difference between records there is no final arbiter.) The intergovernmental agreement about the registers to be created to record these transactions stated that they were to be developed independently by each of the six jurisdictions subject to the general oversight of the Ministerial Council (that is, States were not required to use the same approach or criteria but merely to be 'consistent' in the opinion of the Ministerial Council—a body that was empowered only to make unanimous decisions. In other words, a State could be censored only if it agreed with the criticism of it).

As with the plans for water recovery, the use of this saved water was also to be decentralised. Working to the Basin Environmental Watering Plan, States are to arrange for releases of the water listed on their registers at appropriate times. Subsequently, their efforts would be subject to an annual external auditing process to be conducted on behalf of the Ministerial Council. This situation was made more complicated by the fact that States and the Australian Capital Territory are entitled to trade the environmental water they have listed on their registers. As the water is moved around the Basin, it will be necessary to track where the water comes from, where it goes to and the hydrological and environmental implications of changing the point at which it is used. Detailed information about these transactions must be lodged on the registers of all the jurisdictions involved and on the central register to be run by the MDBC, now MDBA (as already noted, not a master register). Although the environmental water documented on the six registers is 'to be clearly assigned in perpetuity for the purposes of the Agreement', its component 'bits' are to be managed through the registers of the various jurisdictions that reclaimed or traded them. For all but the cognoscenti, attempting to trace these transactional trails and determine whether the environment receives the water originally assigned to it is virtually impossible. Not only will the Living Murray arrangements continue in their Byzantine glory under the new regime established by the *Water Act 2007*, they also provide an illuminating example of the types of accounting systems that are still being put in place across the MDB because of the pressure to protect special interests.

The *Water Act 2007*

The most discussed sections of the *Water Act 2007* are those applying to the Basin Plan, which is to be implemented through four State plans and a plan for the Australian Capital Territory. If a State fails to develop a satisfactory plan, the Commonwealth is empowered through the *Water Act 2007* to develop its own plan for that State. Before that happens a lengthy process of consultation is required. If the Commonwealth has to develop a plan, however, it would take a significant amount of additional time to complete. The Commonwealth does not possess the knowledge or capacity required and there would be considerable delay while it acquired it. In addition, there is a range of issues—involving Section 100, for example—that are likely to result in litigation. Consequently, the Basin Plan is likely to face prolonged dispute and delay before implementation.

Fortunately, however, there is a potential alternative mode of implementation contained in the *Water Act*—through the Commonwealth Environmental Water Holder (CEWH). Since 2007, when the *Water Act* was introduced, the future of water management in the Murray–Darling Basin has been transformed by two developments barely mentioned in the guidelines to the Basin Plan. These are the creation of the CEWH and the resolution of an obscure debate about whether the environment should be supplied through so-called rules-based water or through entitlements. In the coming years it is likely that the CEWH will emerge as the most important water-management institution in the Murray–Darling Basin.

The Labor Government has made a commitment that all water reclaimed through the Basin Plan will be purchased at market prices from willing sellers or come from investment in infrastructure improvements. This water will be lodged with the CEWH to be managed as entitlements for the environment. This means that the timetable for developing the State sub-plans will be almost irrelevant. In reality, the driving force behind the program will be the CEWH—limited only by the size of its budget and any limitations that might be imposed by the Commonwealth. When the CEWH's purchasing program is complete, it will hold more than one-quarter of all water entitlements in the MDB. It will operate unfettered by constraints beyond the requirements of the Environmental Watering Plan to be developed by the MDBA—another Commonwealth agency (albeit in consultation with the States).

The long-established approach to managing rivers in the MDB has been to develop plans that allocate a proportion of the flow to entitlement-holders with the balance left in the river. Water left in the river is called rules based because it is the result of applying the management rules. But over the years it has proved difficult to protect rules-based water from encroachment by entitlement-holders. The rules have always favoured entitlements when water supplies are reduced during droughts. The management rules have this bias because the original reason for building dams and regulating rivers was to supply water to agriculture and towns during the summer and in droughts when the natural flow is low. In addition, there is the burden of water theft, which the late Peter Cullen guesstimated was responsible for at least 20 per cent of all extractions. Most of that stolen water comes from the rules-based component of flow.

When they were first introduced in the early twentieth century, water allocations were very loosely defined. In recent years, they have been tightened, sharpened and designated as 'entitlements' and 'rights'. As entitlement-holders increased their certainty and the reliability of their 'rights', the risks of non-supply in over-allocated systems in a highly variable climate became increasingly concentrated within the proportion of flows supposedly reserved for the environment: the rules-based component. In response, some environmental advocates argued for buying the entitlements that were previously seen as the source of the problem. That way they would have water for the environment with the same legal and security characteristics as other entitlements. But in the absence of substantial funds those scheming dreams were little more than fantasies.

The situation was transformed by the then Prime Minister John Howard's $10 billion water-reform package announced in January 2007, which included $3 billion for the purchase of environmental water. Although it might have looked like a radical green gesture from some perspectives, it provided a way to reimburse irrigators for water that they were going to lose without compensation if the National Water Initiative was implemented in its original form (NWI, Paragraphs 48–9). Under the traditional approach to water management, in which the environment is supplied through rules-based water, most of the water purchased from irrigators would have been returned to the entitlement pool during times of drought because of the way the management rules work. To sceptics of the environmentally orientated campaigns to restore the MDB, this might have looked like a win–win situation. The irrigation community would get $3 billion in compensation and then get most of the water back anyway in times of drought. The shift to supplying the environment through the use of entitlements changed that dynamic, however. Under this arrangement, that water will be permanently outside the consumptive pool. It will continue to be available to the environment even in times of severe drought (unless there is political interference with the operation of the CEWH).

Bibliography

Australian Bureau of Statistics, Australian Bureau of Agricultural and Resource Economics and Bureau of Rural Sciences 2009, *Socio-economic context for the Murray–Darling Basin*, Descriptive Report, MDBA Technical Report Series: Basin Plan, BP02, Murray–Darling Basin Authority, Canberra.

Australian Competition and Consumer Commission 2009, *Water Trading Rules: Draft advice*, Australian Competition and Consumer Commission, Canberra.

Blackmore, D. 2002, 'Protecting the future', in D. Connell (ed.), *Uncharted Waters*, Murray–Darling Basin Commission, Canberra.

Clark, S. D. 1971, 'The River Murray question: part II. Federation, agreement and future alternatives', *Melbourne University Law Review*, vol. 8, pp. 215–23.

Commonwealth Parliament 1993, *Murray–Darling Basin Agreement (1992)*, Commonwealth of Australia, Canberra.

Connell, D. 2007a, *Sustainability and the National Water Initiative*, Land & Water Australia, Canberra.

Connell, D. 2007b, *Water Politics in the Murray–Darling Basin*, The Federation Press, Sydney.

Council of Australian Governments (COAG) 2004, *Intergovernmental Agreement on a National Water Initiative*, Council of Australian Governments, Canberra, <http://www.nwc.gov.au/resources/documents/Intergovernmental-Agreement-on-a-national-water-initiative.pdf>

Cullen, P. 2004, Interview with Peter Cullen, December, Interviewer: D. Connell, Oral History Collection, National Library of Australia, Canberra.

Davies, P., Harris J., Hillman T. and Walker K. 2008, *Sustainable rivers audit: a report on the ecological health of rivers in the Murray–Darling Basin*, Prepared by the Independent Sustainable Rivers Audit Group for the Murray–Darling Basin Ministerial Council, Canberra.

Department of the Attorney-General 2007, *Water Act 2007*, Department of the Attorney-General, Commonwealth of Australia, Canberra, <http://www.comlaw.gov.au/ComLaw/Legislation/Act1.nsf/all/search/E511A5A262379705CA25734C001ED679>

Department of Environment, Water, Heritage and the Arts 2009, *A framework for determining Commonwealth environmental water actions*, Discussion Paper, Department of Environment, Water, Heritage and the Arts, Commonwealth of Australia, Canberra, <http://www.environment.gov.au/water/policy-programs/cewh/pubs/cehw-framework-discussion-paper.pdf>

Grafton, R. Q. and Hussey, K. 2007, 'Buying back the Living Murray: at what price?', *Australasian Journal of Environmental Management*, vol. 14, pp. 74–81.

Grafton, R. Q. and Jiang Q. 2010, *Economics of drought, water diversions, water recovery and climate change in the Murray–Darling Basin*, CWEEP Research Paper 10-01, Centre for Water Economics, Environment and Policy, The Australian National University, Canberra, <http://cweep.anu.edu.au/pdf/publications/research_papers/10-01_WaterEconomics.pdf>

Hamstead, M., Baldwin, C. and O'Keefe, V. 2008, *Water Allocation Planning in Australia—Current practices and lessons learned*, National Water Commission, Canberra.

Horridge, M., Madden, J. and Wittwer, G. 2005, 'The impact of the 2002–2003 drought on Australia', *Journal of Policy Modeling*, vol. 27, no. 3, pp. 285–303.

Klunder, J. 1993, The changing demands for surface water in the Murray–Darling Basin, 25 June, Submission to Murray–Darling Basin Ministerial Council Meeting No. 12, Melbourne.

Murray–Darling Basin Authority (MDBA) 2009, *Development of sustainable diversion limits for the Murray–Darling Basin*, Issues Paper no. 49/09, Murray–Darling Basin Authority, Canberra, <http://www.mdba.gov.au/files/publications/sustainable-diversion-limits-issues-paper-12-11-09.pdf>

Murray–Darling Basin Ministerial Council 1995, *An Audit of Water Use in the Murray–Darling Basin*, June, Murray–Darling Basin Ministerial Council, Canberra.

Murray–Darling Basin Ministerial Council 2000, *Review of the Operation of the Cap*, August, Murray–Darling Basin Ministerial Council, Canberra.

National Water Commission 2009, *Australian Water Reform 2009—Second biennial assessment of progress in implementation of the National Water Initiative*, September, National Water Commission, Canberra.

Senate Standing Committee on Rural and Regional Affairs and Transport 2008, *Water Management in the Coorong and the Lower Lakes*, Parliament of Australia, Canberra.

Tabara, D. 2005, *Sustainability learning for river basin management and planning in Europe*, HarmoniCOP Integration Report, HarmoniCOP Project, Work Package 6, European Commission, <http://www.harmonicop.info/_files/_down/WP6%20Integration%20reportFINAL.pdf>

Wong, P. 2008, Rudd Government to invest $12.9 billion in water, Media release no. PW 56/08, Parliament House, Canberra, <http://www.climatechange.gov.au/~/media/Files/minister/wong/2008/Media%20Releases/April/mr20080429.ashx>

21. Improving Water Planning Processes: Priorities for the next five years

Mark Hamstead

Introduction

The Murray–Darling Basin Plan will be an overarching statutory document that relies on a range of subsidiary arrangements for its implementation.

Water-resource plans are to be prepared at a regional level by basin State governments[1] and accredited by the Commonwealth minister. Requirements for accreditation will be set out in the Basin Plan. The timing of the preparation of these varies from 2012 to 2019, depending on when current water-allocation plans come up for review. Regional strategic environmental watering plans will also be developed by basin State governments within one year of the making of the Basin Plan.

The Commonwealth will be investing heavily through buybacks of entitlements or other means to discharge its obligations in regard to funding reductions under the *Water Act 2007* risk-sharing arrangements. This should be done in a manner consistent with best achieving Basin Plan outcomes.

Lastly, collaborative arrangements will be needed to coordinate the actions of multiple parties. Government and non-governmental environmental water-holders will need to collaborate in managing water to achieve Basin Plan outcomes, and various State and Commonwealth agencies will need to collaborate in relation to complementary management of land, and facilitating structural adjustment.

Achieving the objectives of the Plan—and gaining the expected value from major reductions in water extraction—is heavily reliant on the effectiveness of these arrangements.

At the time of writing, only the *Guide to the proposed Basin Plan* (MDBA 2010) had been released, with the Draft Plan to follow after further consultation. The

1 References to 'basin States' in this chapter means each of the Murray–Darling Basin States and the Australian Capital Territory.

Guide sets out in plain text what the Murray–Darling Basin Authority (MDBA) proposes to put into the Draft Plan. It is assumed that the Plan will generally conform to this, though details might change.

Under the Basin Plan, water-resource plan areas encompass whole river valleys. State-based water-resource plans that will be accredited under the Basin Plan are expected to be an amalgamation of one or more instruments prepared under existing State legislation. For example, an accredited water-resource plan could be a package of one or more State water plans covering different water resources within the area, some statutory orders, a natural-resource management plan and a regional water-quality strategy.

This chapter focuses on aspects of implementing the Basin Plan that could make a big difference to its effectiveness. Implementing the diversion limits will clearly be a central aspect of water-resource plan accreditation and compliance, and will depend on the existing Basin Cap compliance arrangements. Likewise, there is a strong and worthwhile emphasis on adaptive learning and building resilience in environmental management. Rather than dwell on these already heavily emphasised matters, in this chapter, other matters that are—in the author's view—less well addressed but still essential for achieving the desired outcomes of the Basin Plan are discussed. These are

- greater integration of groundwater and surface-water management
- greater integration of natural-resource investment plans and water-allocation plans
- supporting increased resilience in the irrigation industry
- improving the rigour of water-resource planning
- governance at a regional level.

Greater Integration of Groundwater and Surface-Water Management

Until recently, water plans for groundwater systems have been prepared separately to those for surface-water systems, and have largely ignored the effects each can have on adjoining connected systems. The Basin Plan notes that across the Basin more than 60 per cent of groundwater systems were assessed as being highly connected to surface-water systems.

Not recognising the cross-connection effects of water extraction has meant that there are unconsidered and unaddressed impacts. Studies over recent years have shown that time-lagged effects of past groundwater extraction on current and future surface-water availability are substantial in many areas. The benefits obtained by the past groundwater users are thus offset in part by the lost future benefits to surface-water users and ecosystems.

The Basin Plan has addressed this by limiting increased extraction of groundwater in systems that are connected to rivers. Integrated management across connected groundwater and surface-water systems can, however, go beyond this to deliver more optimised outcomes.

If there is to be integrated management of connected water systems then either they must be addressed in a single water plan or, if they are in separate plans, they must include appropriate linkages. It is possible to achieve a level of integration of management through linking plans, but it is more difficult than where the water systems are in a combined plan. Where the plans are totally separate, the making and review of the plans are generally unsynchronised, making development of coordinated-management strategies challenging.

The spatial scale of planning is important for integration. In the Basin States, water planning is moving towards a larger scale and inclusion of both surface and groundwater. In Queensland, water plans are being developed mostly at a catchment or multi-catchment scale. While the early plans addressed only major rivers within the catchments, they have now in several areas been amended to include major aquifers, and some recent plans include both ground and surface water. In New South Wales, the most recent 'macro' plan—the Peel—includes all the surface and groundwater in the Peel Valley. Victoria is currently developing its first plan that addresses both surface and groundwater.

There seems to be little reason why connected aquifers should not be included in the same water plan as the rivers, even if in some parts time lags are large. Ideally, all jurisdictional water plans within a basin water-resource plan area should be pulled into a single instrument. Decisions can then be made within the plan development process on the level of integration of management that is optimal. At a minimum, narrow, highly transmissive unconfined alluvial aquifers should be brought into the same plan as the rivers to which they are connected.

With planning for connected systems combined in this way, it opens up the ability to develop and apply coordinated strategies and rules. These will vary depending on the nature of the connection—whether a unit of groundwater extraction results in a unit less water in the river or a lesser amount, and how long the time lag is between groundwater extraction and impact on river flows. Strategies and rules that can be used to optimise cross-system water management include linked caps on water rights or extraction, linked seasonal allocation determinations, linked restrictions on rates of extraction tied to water levels or flow triggers, providing for trade of water entitlement and/or allocation across the connection, constraining the location of groundwater extraction to alter the timing or extent of cross-connection impacts, and provision for measures such as works, purchase of water entitlements and structural adjustment to offset cross-connection impacts.

The Basin Plan does not limit more integrated management occurring, and indeed provides flexibility to adjust for more optimal integrated solutions. For example, the Guide notes that 'some groundwater systems are highly connected to surface-water systems, but may be capable of sustaining further take. In these SDL [sustainable diversion limit] areas, further take could be feasible provided that there is a corresponding reduction in surface water take to offset the resultant impact on streamflow' (MDBA 2010:139). Also, basin water-resource plan areas provide an opportunity for jurisdictional water-allocation plans for groundwater and surface water to be combined.

It would be worthwhile for the Basin Plan to go a step further and require an assessment of possible integration strategies as part of water-resource plan accreditation.

Greater Integration of Natural-Resource Investment Plans and Water-Allocation Plans

All basin States have processes for water-allocation planning and natural-resource management investment planning. While these two separate planning processes serve different purposes, there is an area of overlap. Both include provisions relating to the maintenance or improvement in the condition of freshwater aquatic ecosystems (rivers, aquifers, wetlands, and so on).

Achievement of aquatic ecosystem outcomes is dependent on many factors. For example, the condition of a river can be affected by a combination of altered flow regime, channel modification, poor water quality, introduced exotic flora and fauna, loss of riparian vegetation, stock access, in-stream obstructions to fish passage, and catchment disturbance. Often addressing only one of them will not achieve the intended result, and might be ineffective because of inaction to address the others.

For the Basin Plan, provision of additional flows to preserve water-dependent ecosystems might be ineffective if, for example, those ecosystems are degraded by stock access or poor water quality. Thus, there is a need to consider all the threats together and develop coordinated plans of action.

The Commonwealth Caring for our Country investment program includes investments in riparian vegetation, wetland conservation, catchment condition and river-form protection—through works, partnerships and education. This is prioritised through regional natural-resource management plans. As noted in Hamstead et al. (2008), coordination of this activity with water-allocation planning has varied substantially in effectiveness.

Within any planning area, there are those ecosystems that are more highly valued than others, those that are more under threat than others, and those that have greater potential for protection or recovery from damage. Effective integration requires action to be planned and coordinated in a way that takes these factors into account so as to focus investment and maximise natural-resource outcomes. Recognising that both types of planning processes require considerable resourcing and that government and community resources are limited, improved integration can also increase the return on this investment by minimising duplication and increasing synergy.

Improved integration can be achieved through a range of approaches ranging from establishing linkages between parallel processes to improve alignment, synchronising and linking the making and review of plans in parallel to combining processes into one. Victoria already does this to a large extent through the way it uses regional sustainable water strategies to coordinate regional natural river health strategies with water-allocation management. A pilot of how water-sharing plans and catchment-action plans could be better aligned was recently undertaken in New South Wales with good results (Hamstead forthcoming).

The Guide (MDBA:194) states explicitly that creating this sort of coordination is outside its scope, being ruled out by legislation. This means that it is reliant on collaboration and cooperation rather than statutory mandate. Without changing legislation, the obvious way forward is for the MDBA to work with the Commonwealth agency responsible for the Caring for our Country funding to require coordination of natural-resource investment and water plans as a prerequisite for provision of funding to States. Guidelines for doing so could be prepared.

Supporting Increased Resilience in the Irrigation Industry

Water planning is occurring in an ever-changing environment with considerable uncertainty about what the future holds. Future climate scenarios prepared by the CSIRO and others are just that—scenarios. Each scenario is intended to show a plausible future. They do not represent a probabilistic range or a normal distribution, with the median being most likely. There is substantial uncertainty in these forecasts, arising from assumptions made about emission scenarios and the modelling process. Not only are trends uncertain, the frequency and extent of extreme events are expected to increase.

Bates et al. (2010:11) comments that despite continuing research, 'this uncertainty cannot or will be only partially reduced'. As stated by Aerts and Droogers (in Ludwig et al. 2009:87): 'the new element in adapting to climate change in water management is an unknown future.'

With this in mind, it is clear that responding to climate change is more about dealing with uncertainty than anything else. Future scenarios provide a plausible, rationally derived range to inform thinking, but there remains the need to consider that something different again could be the reality. This requires a fundamental change in thinking and approach. Folke et al. (2002) describe this as follows:

> Paradoxically, management that uses rigid control mechanisms to seek stability can erode resilience and enhance breakdown of socio-ecological systems...In contrast to an efficiency-driven, command-and-control approach, management that accepts uncertainty and seeks to build resilience can sustain social-ecological systems, especially during periods of transformation following disturbance.

The Basin Plan is commendable in placing a focus on ecosystem resilience. The first of its four major outcomes is water-dependent ecosystems in the Basin would be more able to withstand short and long-term changes in watering regimes resulting from a more variable and changing climate (MDBA:7).

It would have been, in the author's view, appropriate to have a similar outcome for water users—something like this: basin entitlement holders and communities will be more able to withstand short and long-term changes in watering regimes resulting from a more variable and changing climate.

Risk assessment and management, as described later, can contribute to achieving this end. The level of uncertainty, however, is such that future threats and events cannot all be identified or catered for. Building resilience and adaptability is the alternative approach to helping ensure the future of our communities.

Beare (in Bates et al. 2010) observes that an important strategy for handling uncertainty is to have a range of options available. In relation to water management, institutional arrangements that increase options for water users include facilitating water trade (permanent and temporary), allowing access to system storage capacity (dams or aquifers) to carry over unused annual allocations, supporting the construction of private storages to carry over unused allocations, facilitating use of aquifers for active storage and recovery of water, unbundling water entitlements, and facilitating conjunctive access to multiple water resources.

A culture of hedging for future uncertainty is needed—meaning that, like insurance, a reduction in short-term benefits is necessary to provide for longer-term security. This might mean, for example, water users retain more in carry-over storage than they have in the past, wearing the risk that if conditions are wet it will spill.

Most farming businesses are familiar with hedging—for example, diversification of crops, consolidation of properties, maintaining financial reserves, insurance, and so on. So this is not generally new territory. What is different is a change in assumptions about water availability. Past signals from government about certainty of supply are no longer appropriate. Businesses will not put in place their own measures to deal with water-supply uncertainty if they have an expectation that the Government has thought things through and that supply will be reliable.

Where additional water-management options are provided, investment is also necessary to increase understanding and capacity to use the options effectively. The importance of capacity building should not be underestimated. Hamstead (2008:13) noted that the

> general shortfall in technical support capacity and a lack of irrigator awareness of the operational aspects of modernised irrigation equipment was a major risk to any on-farm efficiency programme...there is no shortage of cases where high technology irrigation systems have been installed but failed to deliver either the promised water savings or productivity improvements...On site evaluations of irrigation systems noted a high rate of faults in either design, commissioning or operation.

The same will undoubtedly be true of any other facility provided. Capacity building is a critical and often under-resourced strategy that can be built into water planning.

The Basin Plan does provide for some important measures that can help water users build resilience. Enhanced water trading is a major element of the Plan. Greater certainty about rules and planning is another. Other important measures listed above, however—such as improved access to carry over, unbundling of water entitlements, and building capacity to use water-management tools—are not mentioned. They could be addressed through water-resource plan accreditation requirements, and through collaboration with State and Commonwealth agencies.

Improving the Rigour of Water-Resource Planning

Well-Structured Plans

Appropriately, the Basin Plan will establish an adaptive-learning framework for managing water to achieve environmental outcomes. The benefits of this could be lost, however, if water-resource plans have poorly expressed objectives, poor internal logic and low levels of investment in monitoring, reporting and research—all of which are required so that the managers can exercise this flexibility meaningfully.

Many current basin State water plans have general aspirational objectives, which allow for wide interpretation. With such objectives, it is difficult to know whether success is being achieved or not. This ambiguity creates an expectation of a level of ecological health and water-entitlement security that is not consistent with the trade-offs inherent in the body of the plan. Monitoring programs and performance indicators are also frequently general and do not provide a clear indication of whether objectives are being achieved or not. The logic of the plan—the rationale explaining the anticipated cause-and-effect relationships between strategies and objectives—is rarely well documented.

There is a need for the discipline of specific, realistic, measurable objectives and targets, with documentation of the rationale for selecting the strategies and rules for achieving them, and monitoring and reporting aligned to measuring achievement. This applies not just to environmental watering, but to all aspects of the water-resource plan. The Basin Plan could facilitate this through the water-resource plan accreditation requirements.

A Greater Emphasis on Risk Management

Past management has tended to focus planning on typical or average conditions. What is needed is a greater focus on extreme conditions and handling what can go wrong. Risk management is a well-developed approach that is underused in water planning.

Unfortunately, the term 'risk' in the Basin planning context has become synonymous with a few specific matters that could affect future water availability in the longer term—for example, interception by farm dams, stock and domestic bores and pumps, and plantation forestry. This narrow interpretation of risk is not what is intended here.

Risk management requires identification of threatening processes and events that could have negative consequences in relation to the achievement of plan outcomes, and assessment of their likelihood and consequence. Once these risks have been identified and evaluated, risk-mitigation measures can be identified. Risk-mitigation measures can include measures implemented well before the occurrence of events, adaptive response at the onset and during the occurrence of events, and recovery after the occurrence of events.

For example, if there is a risk of water-supply failure to a town, responses could be to do something beforehand to remove the risk (for example, add a reserve water supply), to monitor the situation and have in place a plan to implement should the event occur (for example, truck in water, or buy irrigation allocations on the market if they are available), or to decide to let the town run out (people temporarily move out and come back when the drought breaks—admittedly, an unlikely option).

Similarly, for an endangered fish that would die out in a severe drought, options could be to improve and extend habitat now so the species is less vulnerable, have an emergency plan to buy water and shore up drought refuges should the event occur, or stock up an aquarium or aquaculture pond with the fish so they can be used to restock the river after the drought.

The nature of the response would depend on the cost and feasibility of each option and the level of risk. If the risk is large, multiple mitigation options might be applied.

The Basin Plan mandates some risk-mitigation measures. It includes an emphasis on building ecosystem resilience to reduce vulnerability to future events in environmental water management, and provision for setting aside of water for essential human needs in severe droughts in the River Murray system.

Further, it requires that water-resource plans 'are robust enough to continue to operate during extreme and unprecedented events' for them to be accredited (MDBA 2010:174). To achieve this, risk management will need to be a feature of water-resource plan development. It might then be expected that there will be a rigorous risk-assessment process for each water-resource plan, and public documentation of how each identified risk will be addressed.

Addressing Equity

Equitable sharing of the costs of ecosystem maintenance and the benefits of ecosystem services is a fundamental, yet often overlooked, aspect of environmentally sustainable management. Major international conventions (Ramsar, Agenda 21, Convention on Biodiversity) all recognise that addressing the current needs of people in an equitable manner is essential. They also recognise that future generations must be considered. Australia's National Strategy for Ecologically Sustainable Development has as one of its three objectives to 'provide for equity within and between generations'. Protecting ecosystems is done not because they have intrinsic value, but because they provide important services to society as a whole and to future generations. Equity is the core of sustainability.

Equity and fairness are objectives common to natural-resource legislation across Australia. In its essence, water planning is fundamentally about equity informed by science, not the other way around.

Yet the methods for achieving equity are poorly defined and are not explicitly thought about. It is left largely to the skills of the planners and decision makers, and assumed that standard processes will somehow deliver it. Community and stakeholder involvement in water planning—while important to this—is not of itself sufficient. Weak equity processes drive stakeholders to bypass plan makers and resort to political processes and courts.

The Basin Plan is addressing water-sharing equity at a broad level through the establishment of diversion limits. The interests of the broader, even global, community and future generations in relation to ecosystem services are represented by statutory requirements for preserving 'key' environmental assets and functions. What will be traded off in deciding the diversion limits is the level of risk these assets and functions are exposed to, in return for economic benefits from consumptive use of water. The Basin Plan will address equity at this broad level for each water-resource plan area.

Applying the principles of procedural fairness is an essential part of achieving equity. Plan development processes should ensure that all affected parties have the opportunity to hear and understand the potential implications of the Plan for them, and have their views presented and considered in decision making. Decision making should both be and be seen to be unbiased and informed, and open to cross-examination. Transparency and openness at all stages of the process are essential.

It remains to be seen how well these principles will be applied in finalising the Basin Plan. Certainly, one challenging area is explaining in ways that can be understood by the broader community what is at stake in terms of ecosystem services if no action is taken, and how this might effect the wellbeing of them and future generations.

Whatever the result is, there will still be another level of equity decisions to be made in water-resource plans, in relation to the sharing of the available consumptive water between water users. Some of the major debates in past water plans have been about how available water is shared between different classes of water rights.

The *Guide to the proposed Basin Plan* (MDBA 2010:171) indicates that water-resource plan accreditation will include the requirement to specify the consultative arrangements upon which the water-resource plan will be developed. This ought to be expanded to set out requirements to ensure that procedural fairness principles and other means for addressing distributional equity as described in the literature are applied.

Governance at a Regional Level

Coordination

As was discussed in the introduction, the Basin Plan will be relying on the collaboration of a number of players to achieve its desired outcomes in each valley. In each basin, State responsibilities for water-allocation planning, management of environmental water, water-entitlement management, operation of water infrastructure, natural-resource investment, and various elements of land and water-quality management are split between multiple State agencies. Additionally, Commonwealth agencies responsible for Commonwealth environmental water holdings and allocation of Commonwealth funds under various relevant programs will be involved. Coordination of these agencies around the achievement of common regional outcomes and targets will be very challenging. Current mechanisms are not sufficient.

Funding

It is clear that the Commonwealth has invested heavily in the Basin Plan and has committed to continue to do so. This level of investment is not reflected, however, in all current State planning processes.

Lower-cost approaches are appropriate in situations where the risks and competition for water are low. Applying them to high-risk, high-competition cases, where there are real impacts on water users and substantial risks to the environment, is, however, inviting difficulty and putting at risk the achievement of Basin Plan outcomes and the value of returning water to the environment.

Accreditation processes need to include assessment of resourcing to deliver the necessary rigour to planning and management of water at the regional level.

Agreements about funding commitments and cost sharing between basin States and the Commonwealth are a critical underpinning. In the end, you get what you pay for.

References

Bates, B., Walker, K., Bears, S. and Page, S. 2010, *Incorporating climate change in water allocation planning*, Waterlines Report Series No. 28, May 2010, National Water Commission, Canberra.

Folke, C., Carpenter, S., Elmqvist, T., Gunderson, L., Holling, C. S. and Walker, B. 2002, *Resilience and Sustainable Development: Building adaptive capacity in a world of transformations*, Environment Advisory Council to the Swedish Government, Stockholm.

Hamstead, M. 2008, *The national plan for water—opportunities and risks for the irrigation industry*, Discussion Paper, April 2008, Prepared for Irrigation Australia Limited, Hornsby, NSW.

Hamstead, M. 2010, *Alignment of water planning and catchment planning*, Waterlines Occasional Paper, No. 36, Dec 2010, National Water Commission, Canberra.

Hamstead, M., Baldwin, C. and O'Keefe, V. 2008, *Water allocation planning in Australia—current practices and lessons learned*, Waterlines Occasional Paper No. 6, April 2008, National Water Commission, Canberra.

Ludwig, F., Kabat, P., van Schaik, H. and van der Valk, M. 2009, *Climate Change Adaptation in the Water Sector*, Earthscan, London.

Murray–Darling Basin Authority (MDBA) 2010, *Guide to the proposed Basin Plan. Volume 1: Overview*, Murray–Darling Basin Authority, Canberra.

22. Connecting Science and Engagement: Setting groundwater extraction limits using a stakeholder-led decision-making process

Stuart Richardson, Ray Evans, Glenn Harrington

Introduction

Groundwater is a critical part of Australia's water resources. Though it is generally of lower available volume than surface water, it still provides key supplies for both consumptive and non-consumptive uses for key regions and at specific times. About 6000 gigalitres of groundwater is estimated to be extracted annually from aquifers across Australia, with this water used primarily for consumptive purposes, such as irrigation. Groundwater supplies about 15 per cent of water for irrigation across the Murray–Darling Basin (MDB), but this can increase to more than 70 per cent of the total water available in some catchments during extended dry periods.

At the national level, there are also high-value ecosystems that use groundwater, especially in dry periods when surface-water supplies are short. These ecosystems include terrestrial vegetation, aquatic systems in wetlands and rivers, and marine and sub-surface ecosystems.

Extraction of groundwater creates a change in the water resource, which in turn creates an impact on users or systems dependent on that resource. The question is not whether the impact will occur; rather it is a question of the location, magnitude and timing of impacts. There is a range of impacts from groundwater extraction that need to be managed, including

- depletion of storage (mining) within an aquifer
- increasing groundwater salinity
- reduced availability of groundwater to dependent ecosystems
- loss of production for consumptive users
- impacts to stream flow.

Governments and water-management agencies in Australia are in the process of (re)defining environmentally sustainable levels of groundwater extraction (as required by the National Water Initiative, NWI). In the past, the general approach has been to develop an extraction limit via a technical process and then provide that to the community via a water plan. In a sense, it was a 'DAD' (Decide, Announce and Defend) approach. Although there were successes with this approach, there were also occasions when it led to tensions. Generally, this tension came about through the over-allocation of groundwater resources because of a mind-set that focused on development in concert with the absence of information on what ecosystems occur in specific areas together with the environmental water requirements for these ecosystems.

This chapter outlines a different approach to the estimation of sustainable limits to extraction, where sound science is brought to stakeholders who are engaged in determining proposed extraction limits to inform policy development. The approach has been developed via a number of specific projects dealing with water-allocation planning in areas of significant use and where potential impacts to the groundwater resource can occur—that is, areas where adjustments in entitlements may be needed.

The unique aspects of the approach described in this chapter are

- re-conceptualisation of sustainable yield as acceptable yield
- characterisation of acceptable yield in terms of resource-condition limits
- use of a stakeholder-led process for making decisions.

Policy Setting

The approach—developed as an aid to groundwater management—was designed with the requirements of the NWI and the *Natural Resources Management Act 2004* (SA) in mind. Specifically, the NWI requires that over-allocated and over-extracted systems are returned to environmentally sustainable levels of extraction (substantial progress by the end of 2010). The SA *Natural Resource Management Act 2004* requires that

- the needs of the natural environment and human demands be considered in determining appropriate limits of extraction
- the rate of the use of water is sustainable, including to meet the reasonably foreseeable needs of future generations.

What is Safe, Sustainable and Acceptable?

The movement to sustainable development originated with the findings of the Brundtland Commission, *Our Common Future*, produced in 1987. This was a seminal step in the process of managing the impacts of humanity on the world's resources, and introduced the term 'sustainable development' into our lexicon. This process of sustainable development flowing from Brundtland led to the adoption of an ecologically sustainable development policy within Australia in the early 1990s.

In one sense, the concept of sustainability is a reaction to a perception that we are 'living beyond our means' in terms of the environmental cost of economic development. In the early years of considering this issue, it became apparent that certain groundwater systems were being used at a rate faster than they were being replenished. This led to the use of the term 'safe yield' to denote an upper limit to the total water available for use on an annual basis without depleting storage. Safe yield did not, however, promote the safeguarding of other, predominantly environmental, users of the resource. As can be seen, if all of the resource is used by one part of the user group then all other users will not be supplied.

Following on from this, there was pressure to limit use to lower rates, so that the needs of non-consumptive users could be balanced against broader expectations over resource use, generally for the environment. The standard approach has been to estimate the various components of the water balance and to then apportion a certain percentage of the long-term average recharge to consumptive uses. The percentage allocated usually approached 100 per cent of recharge; any reduction below 100 per cent depended on specific information on the presence of ecosystems. It was rare to find data that quantified the exact water requirements for ecosystem function. This approach generally worked well and addressed the issue of sustainability by limiting use to less than the long-term average recharge. The approach still did not preclude impacts from occurring, however, especially changes that affected key environmental assets. The approach also did not deliver optimum water-use outcomes in that the extraction limit defined was not spatially specific, and ecosystem function could still be impaired, even though the use was below the long-term recharge rate.

The historical approach made good ground in meeting sustainability objectives in a general sense; however, even though it was based on the knowledge of the water balance, it failed to take account of the concept that any extraction from a water source will create impacts somewhere in the hydrologic system.

The corollary to the knowledge that extraction will create an impact somewhere in the hydrologic system is that a decision is required when extracting water from such a system as to the acceptable level of impact that can be accommodated within the system. Obviously, a necessary condition for an acceptable impact is that the extraction cannot exceed the long-term average recharge rate. Apportioning a somewhat arbitrary percentage of the total—based generally on the need to manage current levels of consumption—is not, however, the best solution to the problem. A more rigorous approach is to define a set of limits that governs the agreed sustainability of the hydrologic system, and to manage extractions so that the impacts fall within the limits. This produces an *acceptable level of impact* according to specified *resource condition limits*. In one sense, the term *sustainable yield* should be replaced, at the management level at least, with the term *acceptable yield*.

In Australia, the NWI defines sustainable yield as the level of water extraction that, if exceeded, would compromise key environmental assets, or ecosystem functions, and the productive base of the resource. In addition, the National Groundwater Working Group has defined sustainable yield for groundwater as the groundwater extraction regime, measured over a specified planning time frame, that allows acceptable levels of stress and protects dependent economic, social and environmental values.

Each State and Territory implements this definition in a slightly different manner.

One can see from the National Groundwater Working Group definition of sustainable yield that the concept of acceptability of impact is already embedded and that the impacts are framed in terms of the values of the system. The definition does not prescribe that impacts should be avoided; rather, that they should be either optimised or minimised within a values framework.

Using the Developed Water Balance

In a large number of situations, the water balance considered as the basis for deriving the long-term average recharge rate (as part of normal operational sustainable yield analysis) is based on natural conditions—that is, conditions that existed prior to the development of the water source via extraction. The natural water balance does not, however, always provide the most useful information when considering the water balance of a groundwater source that is developed. This point was reported in the early 1980s by Bredehoeft et al. (1982) and later by Bredehoeft (2002) in what was called the 'water budget myth'. Essentially, the myth states that the natural recharge rate does not inform the sustainable yield that can come from an aquifer. Rather, when an aquifer is pumped, there is a capture of groundwater discharge as well as potential inducement of recharge to form a new water balance: the developed water balance. In this sense, the volumes of the various components of the developed water balance are what will determine the total available water pool from which the consumptive pool of water available for use is derived. A corollary of this consideration is that both the capturing of discharge and the inducement of recharge (generally from surface water) will have an impact somewhere else in the system.

The myth was framed around groundwater considerations only, and it is interesting to consider it within the context of the total water balance. If we call the total volume of water available in a system—inclusive of both surface and groundwater—the available water pool then obviously conservation of mass says that this should remain the same (on average) in a closed system under different pressure states. If, however, the myth also says that the developed groundwater balance can change once pumping commences, logic says that the changed flows are compensated for either out of the surface-water balance or from outside the section of aquifer being considered.

Case Examples

The definition of groundwater sustainable yield clearly establishes an acceptability framework within the context of preserving the key values of the system being considered. If one can identify the values of the groundwater system—in terms of attributes such as its water quality, its reliability, the ecosystems that it sustains and its productive base—then an opportunity exists to set proposed limits on extraction such that the impacts on these values are at a level that is acceptable to all stakeholders. Such an approach has been attempted in three areas of South Australia (Figure 22.1).

Northern Adelaide Plains Prescribed Wells Area

The Northern Adelaide Plains Prescribed Wells Area (NAP PWA) is located on the northern outskirts of the Adelaide metropolitan area. Groundwater is used extensively for irrigation of vegetables, olives and grapes.

The majority of groundwater supplies is taken from the deeper, confined Tertiary aquifers. The upper T1 aquifer and underlying T2 aquifer are generally high yielding and contain low-salinity groundwater.

Historical groundwater use in the NAP PWA has resulted in seasonal groundwater-level declines forming cones of depression in the T1 aquifer to the south and a more permanent cone of depression in the T2 aquifer in the north. Winter extraction of groundwater, together with the usual high summer demand for irrigation water, maintains the cone of depression throughout the year.

The current (2000) water-allocation plan for the region provides for a maximum allocation of 26.5 GL/year, which is greater than what is used. Groundwater use has varied between about 13 GL and 19 GL/yr over the past 10 years.

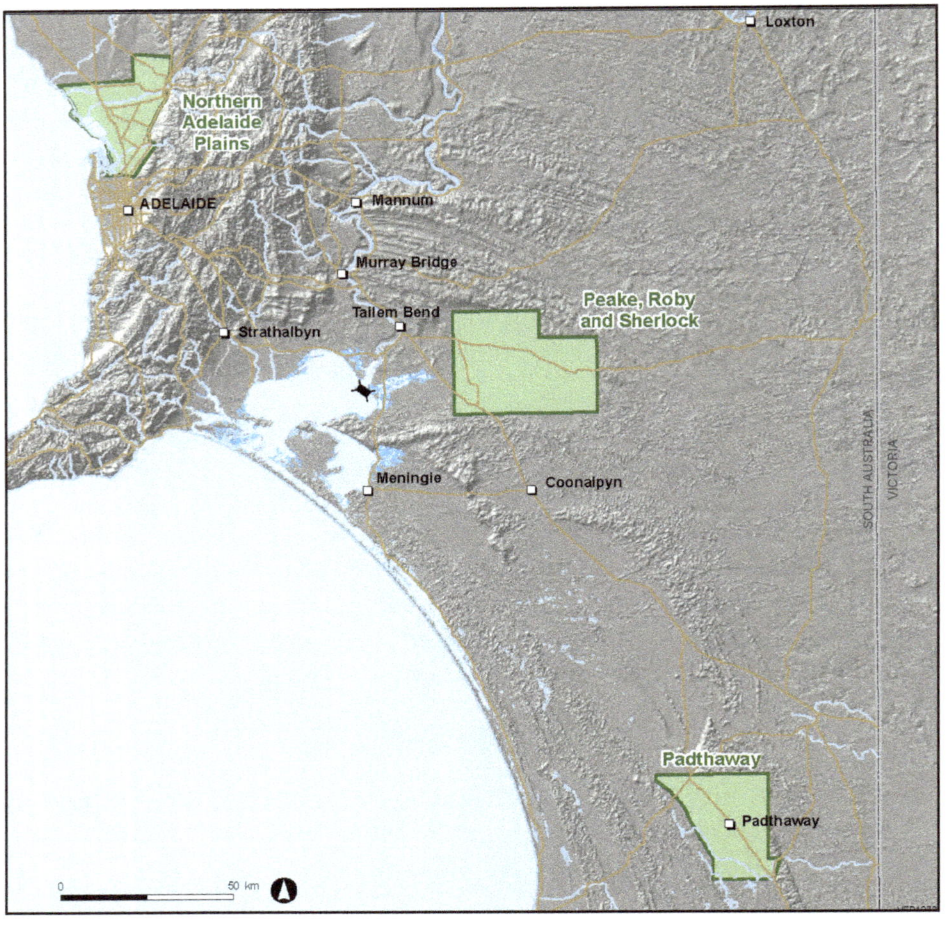

Figure 22.1 Location of the Peake, Roby and Sherlock Prescribed Wells Areas

This aquifer system has been characterised as potentially over-allocated but not necessarily overused. A new water-allocation plan is currently being prepared by the Adelaide and Mount Lofty Natural Resource Management (NRM) Board, which will set new extraction limits.

The Adelaide and Mount Lofty NRM Board has constructively engaged with stakeholders through a community–government committee and implemented the approach described in this chapter to derive a proposed extraction limit.

Padthaway Prescribed Wells Area

The Padthaway PWA is located in the south-eastern region of South Australia and covers an area of about 700 sq km. Most groundwater is extracted from the shallow Padthaway Formation aquifer, which is less than 10 m below the ground surface. Groundwater is also extracted from the Bridgewater Formation beneath the adjacent hills.

The volumetric allocation (following conversion from crop area to volumetric licences) was 89 GL/yr, and use in 2005–06 was nearly 41 GL/yr.

The State Government carried out a review of the condition of the groundwater resources in the area in early 2005. This review confirmed previous observations that although the Padthaway region has experienced minimal long-term change in groundwater levels, the rate of groundwater salinity increase is alarming (often 20–30 mg/L/yr), particularly for wine-grape growers (REM 2007). A great deal of investment in technical studies has shown the increasing salinity is a result of salt mobilisation following clearance of native vegetation over the adjacent hills and recycling of salty water via irrigation.

Analysis also suggested that four of the five groundwater-management sub-areas in the PWA will have licensed allocations exceeding the total available recharge to the aquifer following volumetric conversion. In response, the local NRM Board, which is charged with developing a water-allocation plan, proposed a default approach to address the perceived over-allocation. The proposed approach consisted of a series of five, equal-sized annual reductions in allocation to bring it below the total available recharge volumes, unless significant improvements in resource condition were observed after the third reduction. This approach would have meant that irrigators faced a significant reduction in their allocations in some areas. Whilst other sub-areas would require lesser reductions, many of the areas that are apparently over-allocated have not experienced long-term water-table decline.

The irrigators considered this default option unacceptable and developed an alternative strategy in response. The alternative strategy involved the quantification of proposed extraction limits using the approach described in this chapter.

Peake, Roby and Sherlock Prescribed Wells Area

The Peake, Roby and Sherlock PWA is located within the Mallee region of South Australia, approximately 140 km east of Adelaide. Groundwater is extracted from the unconfined Murray Group Limestone aquifer in the eastern part of the PWA and from the deep, confined aquifer within the Renmark Group under the coastal plain in the west. Groundwater is used for irrigation, stock, domestic and town-water supplies.

Metered groundwater extraction was 1.597 GL/yr in 2007–08.

A Water Allocation Plan (WAP) was recently completed, which sets allocation limits for this groundwater resource. The WAP was prepared in response to a decline in groundwater levels (of the order of 10–15 m) following development of irrigated areas. The decline in groundwater levels caused pumping infrastructure to become stranded in stock and domestic wells and increased the potential for lateral ingress of more saline groundwater from surrounding areas.

The plan sets an allocation limit of 5.383 GL/yr across six management sub-areas. The allocation limits were set based on an analysis of draw down of groundwater levels and flow of saline groundwater from the west (SAMDBNRMB 2010a).

Acceptable Yield and Resource Condition Limits

Whilst the primary aim of managing any natural resource (including water) is to ensure the long-term sustainability of that resource (preferably in terms of both resource abundance and resource quality), defining short-term management objectives and/or targets can be less obvious. Ultimately, this task requires differentiation between what is 'sustainable' and what is 'acceptable'. The former has connotations that the resource will last forever, although recent definitions of sustainability for water-resource management purposes have tried to incorporate social and economic benefits as well as ensuring environmental longevity. When these three factors are considered simultaneously (providing a 'triple bottom line'), it might be more appropriate to use the term 'acceptable' rather than sustainable. It is therefore critical that all stakeholders are involved in the development of the adaptive-management approach.

Values of acceptable yield are determined through a process that links an understanding of how an aquifer system behaves when stressed with an analysis (via stakeholder engagement) of water-management objectives and resource condition limits.

Resource condition limits (RCLs) are upper limits to the levels of impact on groundwater resource condition that cannot be exceeded due to the extraction of groundwater. The RCLs are measured using resource condition indicators. In some groundwater-management areas, the RCL could relate to an indicator, such as the relationship between groundwater level and the saturated thickness of the aquifer (for example, Figure 22.2); it could be a specified groundwater level, it could be an indicator such as a rate of draw down of groundwater level over an annual cycle with a maximum rate specified as the limit, or it could be a groundwater quality indicator with the limit being the beneficial-use category.

In any case, the RCL is derived from a conceptualisation of how the aquifer system responds to stress, either based on historical time-series data (an evidenced-based approach) or from a numerical model (predictive analysis).

Resource condition limits are set through discussion with key stakeholders. Stakeholders in the groundwater resource—whether they are licensed irrigators, unlicensed stock or domestic users, dryland farmers, the regulators or representatives for the environment—will each have different views on what aspects of the resource are important to their industry and therefore which aspects should be protected.

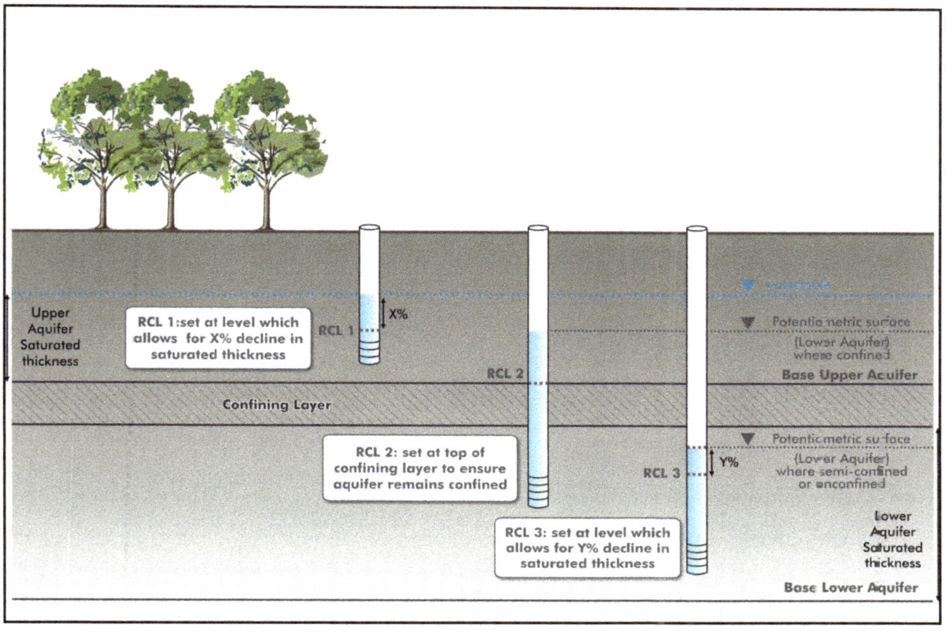

Figure 22.2 Examples of RCLs based on groundwater levels and confined aquifer pressure as percentage of aquifer thickness

A Process to Determine Acceptable Groundwater Yield

The process taken to determine the acceptable impacts is to identify the key values for the water source, the limits to the impacts such that the resource is protected to the desired level of all stakeholders, technical activities that translate the limits into an extraction volume, and a decision process that obtains agreement amongst stakeholders that a suitable solution has been reached. The process is shown schematically in Figure 22.3.

The flow of tasks shows that there is a mixture of both technical and general tasks that allows a decision-making process to proceed on the basis of informed discussion. It is critical that this technical linkage occurs, particularly in periods when the decision makers want to view the implications of various management scenarios. Equally, the technical support must be robust enough that scenario analysis has levels of certainty appropriate to the importance of the decision being made.

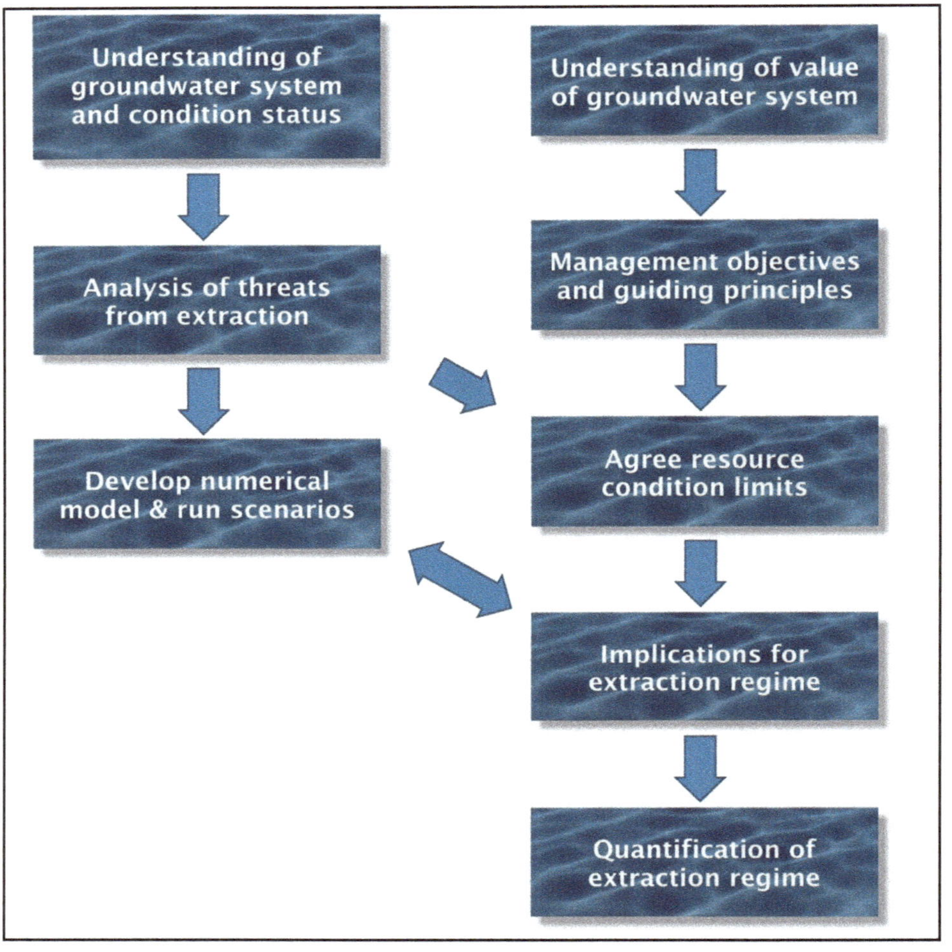

Figure 22.3 A flow of tasks that represents the technical process coupled with stakeholder engagement

The Importance of Shared Objectives and Principles

The strength of the process described here lies in the collaborative approach taken to decision making, and, specifically, to gaining agreement to groundwater-management objectives and principles between all stakeholders.

The engagement process provides the opportunity for all stakeholders to enter a debate that is backed by science and allows all interests to be heard. The objective of the discussions is to get all the viewpoints and issues disclosed, create a common understanding of the way the physical system operates, understand the limits to use and come to an agreement regarding an acceptable condition of the groundwater resource. The engagement process is supported by modelling analysis of a series of future scenarios.

It is important that all stakeholders are represented. It is implicit that if all stakeholders are well represented, and those stakeholders are able to represent their constituency in the decision process, there is consideration of the balance between environmental, social and economic issues. Where stakeholders do not have representative status from their constituency, some time is required to enable a combined stakeholder view to be input to the decision process.

In the case of the Northern Adelaide Plains and Padthaway, the stakeholder groups were comprehensive and included irrigator groups (with Padthaway involving different subgroups based on crop and type of irrigation system), regulators (with State representation of the NWI Agreement, policy, licensing and technical), planners, stock and domestic users, sleeper-licence holders (water entitlement-holders who do not use the water) and those representing the needs of the environment.

The conversations in the engagement process focus on perceptions of management issues, water-management objectives, values placed on the groundwater resource and principles that will guide decision making. In the case of Padthaway, the value of the resource was characterised as

- a high-security water resource (that is, high yield, good quality and reasonably low cost to access)
- having a capacity to generate high revenue per megalitre.

The agreed values were significantly weighted to socioeconomic outcomes. This was similar to the Northern Adelaide Plains, where the 'bankable' value of water entitlements was also recognised, as was the connection between the availability of water and the social value of the region.

A discussion of threats to values provides a mechanism for understanding the 'pressure points' on all stakeholder perspectives. In Padthaway, the key threats to values were related to the trends in the condition of the resource (for example, falling groundwater levels and rising salinity). This further emphasised the need to provide sound science and monitoring data in the discussion.

Our experience indicates that stakeholders are very capable of describing the management issues and highlighting the priority issues. This information provides the focus for technical investigations and further discussion around management responses. It is important that all issues are recognised and impacts quantified (that is, where, when and by how much?).

Guiding principles provide the operating rules that govern the way groundwater resources can be managed. The common themes to these principles are

- equity
- equal share of benefits and costs
- sustainability (leave resource in no worse condition than the start)
- precautionary (will make a cautious decision if not enough data)
- manage for the long term across climate variability
- continuous improvement in knowledge base
- adaptive management
- security.

The discussion with stakeholders leads to the development of water-management objectives. These objectives reflect broadly what condition the groundwater resource should be in and what access users should have to the groundwater resource.

Box 22.1

The stakeholders in the Padthaway region developed the following objectives.

Irrigators

- seek to maintain groundwater levels (especially on flat) at least at 2004 levels (or some average)
- at break-in-slope seek to maintain groundwater salinity at acceptable levels (800 mg/L or current)
- improve groundwater salinity on the flat
- flush stored salt from the soil
- create opportunities for recharge of water from surface-water sources.

Regulator

- needs to be NWI compliant
- share irrigators' objectives.

Environment

- maintain water levels under wetlands to west, although not sure of dependence of these wetlands on certain water levels
- level of salinity in water—too high for red gums at the moment, particularly on flats to the west.

NRM Board

- consider the social, environmental and economic impacts of WAP policy decisions
- continuation of a thriving viticultural and small seeds industry/enterprise in the Padthaway area in the next 100 years
- sustainability
- conditions of resource at least maintained if not improved
- no triggers are exceeded
- provide adequate water for the environment (for example, 10 per cent of recharge for environmental purposes—including through-flow)
- approach to groundwater management needs to be NWI compliant.

Resource Condition Limits

The continuum of impacts due to resource use can be subdivided into three categories of acceptable, tolerable and intolerable, where intolerable impacts are those that must never occur, tolerable impacts are those that may occur but must be avoided by management action, and acceptable impacts are those that can be 'lived with' over a long time. Resource condition limits represent a 'sustainability position' or the boundary to an acceptable condition of groundwater. They define when an impact moves the resource to a tolerable or intolerable condition.

There are a number of key aspects to the agreement on RCLs to consider. Setting RCLs must be based on a sound understanding of the dynamics of the groundwater system and implications for threats to water-management objectives. For example, in the Padthaway region, the technical analysis indicated that the thickness of the productive zone of the aquifer was relatively small (about 5 m) and excessive draw down of groundwater levels resulted in less water coming from the aquifer, affecting irrigators' capacity to obtain enough water for their crops.

It was necessary to support the stakeholder conversations related to this point with the time series of historical groundwater levels and connect that information with stakeholders' experiences. The time-series data had to be evaluated to understand the relationship between extraction and climate and geological variability. The issue of geological variability led to the conclusion that blanket approaches applied to the region as a whole were not appropriate and management targets needed to be spatially variable.

In other regions, RCLs were set to protect water access for stock and domestic users where declining groundwater levels were predicted to fall below the base of stock and domestic wells.

To be effective, measures of acceptable impacts and RCLs (and their related indicators) must

- be explicit in space and time
- relate to an agreed water-management objective
- be measurable.

Being explicit and measurable allows the RCLs to be operationalised.

Examples of proposed RCLs from the Northern Adelaide Plains case are that the groundwater resource will be no worse off (and therefore the impacts will be acceptable) if

- the water level in the T1 aquifer is greater than the specified level in September each year as measured at agreed bores
- the water level in the T2 aquifer is greater than the specified level in September each year as measured at agreed bores
- the salinity in all aquifers does not exceed the beneficial use where it is below that level.

Our experience is that few RCLs can be constructed that relate to the protection of the environment. Environmental objectives are often provided but these broader statements cannot often be supported by the more detailed RCL. This occurs mainly because there is often a lack of monitoring infrastructure near ecological assets dependent on groundwater and because there is little quantified information regarding the groundwater needs of ecological assets. This obviously is a flaw and is being rectified.

Determining Acceptable Yield

The acceptable yield volume is generated through the evaluation of a series of extraction scenarios modelled using a groundwater-flow model. The groundwater-flow model estimates future groundwater levels (and in some cases salinity), making it an ideal tool to connect extraction scenarios with the agreed condition of groundwater (via RCLs).

In the case of Padthaway, the results from a series of model scenarios were compared with RCLs and it was noted that extraction at 48 GL/yr (about 45 per cent of allocation) maintained groundwater levels, through-flow and groundwater salinity within acceptable limits (Table 22.1).

Table 22.1 Summary of whether the model scenario results meet the RCL requirement for the Padthaway PWA. The results suggest scenario 2 most clearly complies with the RCLs.

RCL	Scenario 01 100% Allocations	Scenario 02 45% Allocations	Scenario 03 60% Allocations	Scenario 04 TARd
June 2004 WL	N	Y	Y/N	Y (mainly)
50% Saturation	N	Y	Y/N	Y (mainly)
Maintain N-S flow	N	Y	Y	Y

Similarly in the Northern Adelaide Plains, a proposed extraction limit was agreed that would meet the requirements of the proposed resource condition limit..

In the case of the Peake, Roby and Sherlock PWA, the proposed extraction limit was chosen based on maintaining groundwater salinity values within a suitable limit (salinity value of 5000 mg/L) over the next 200 years (SAMDBNRMB, 2010b).

Summary

- The involvement of stakeholders in the decision-making process can be made to work and will generally lead to better outcomes. Meaningful engagement with stakeholders, and the sharing of key knowledge with them, allowed joint decisions to be made by regulators, irrigators and representatives of the environment. The maxim was 'start the discussion early and start with science and the *basics*'. The *basics* include the resource values, guiding principles and a shared view of resource-management objectives. The examples of Northern Adelaide Plains and Padthaway were successful partly because the stakeholders (irrigators) who were wearing the cost of changes to water entitlements also accrued the benefits.
- Couch extraction limits in terms of acceptable yield. Extraction will always have an impact. The question is: what impact is acceptable? If this is accepted then it is clear science does not provide all the answers and that value judgments are needed. A trade-off decision needs to be negotiated.
- Trade-off decisions need to be supported by reliable science and technical understanding. The decision process should be informed by the science, not deliberated entirely by technical specialists and revealed to stakeholders at the end of the process. A robust and reliable technical assessment requires comprehensive data as input.
- Ecosystems are not strongly represented in discussions. Our experience is that there is little quantitative science regarding ecosystem-water requirements brought to the discussion because of lack of data and limited understanding of the response of ecosystems to incremental changes in groundwater conditions. There are some exceptions to this where site-specific data are available.
- There is a need to develop tools that explicitly link changes to the condition of groundwater to indicators of social and economic value. These links were dealt with in a qualitative manner in the work described in this chapter, but quantitative analysis allows for a more rigorous evaluation of scenarios.

- The examples used in this chapter cover local to subregional scales. Further analysis is needed to understand whether the approach described can be used as a template for application at the broader basin scale.

Acknowledgments

This work is supported by examples from a series of case studies where the approach has been applied. The process has been applied in the Northern Adelaide Plains Prescribed Wells Area (NAP PWA), Peake, Roby and Sherlock Prescribed Wells Area (PRS PWA) and the Padthaway Prescribed Wells Area (PPWA).

The authors wish to thank the following organisations for support and funding to undertake these projects: Adelaide and Mount Lofty Ranges Natural Resource Management Board; South Australian Murray–Darling Basin Natural Resource Management Board; and Department for Water and South East Natural Resource Management Board.

The authors would also like to acknowledge RPS Aquaterra for the groundwater-flow and solute-transport modelling undertaken for Padthaway.

References

Bredehoeft, J. D. 2002, 'The water budget myth revisited: why hydrogeologists model', *Groundwater*, vol. 40, no. 4, pp. 340–5.

Bredehoeft, J. D., Papadopolus, S. S. and Cooper, H. H. jr 1982, 'The water budget myth', in *Scientific Basis for Water Resource Management, Studies in Geophysics*, National Academy Press, Washington, DC, pp. 51–7.

Resource & Environmental Management Proprietary Limited (REM) 2007, *Padthaway water allocation planning*, Discussion Paper, Prepared for the Department for Water, Government of South Australia, Adelaide.

South Australian Murray–Darling Basin Natural Resource Management Board (SAMDBNRMB) 2010a, *Water Allocation Plan for the Peake, Roby and Sherlock Prescribed Wells Area*, South Australian Murray–Darling Basin Natural Resource Management Board, Murray Bridge, SA.

South Australian Murray–Darling Basin Natural Resource Management Board (SAMDBNRMB) 2010b, Guide to the Draft *Water Allocation Plan for the Peake, Roby and Sherlock Prescribed Wells Area*, South Australian Murray–Darling Basin Natural Resource Management Board, Murray Bridge, SA.

23. Comparative Perspectives on Basin Governance in the Murray–Darling Basin: Insights from the western United States

Dustin Garrick, Rosalind Bark

Introduction

The Murray–Darling Basin (MDB) is internationally recognised for innovation in basin governance to achieve more sustainable outcomes. This chapter examines water-management experiences from the western United States relevant to the emerging public dialogue about the Murray–Darling Basin Plan. We argue that a comparative perspective offers more scope for exchanging policy and practical lessons than previously recognised. Australia imparts more lessons than it stands to learn from the exchange, but the western United States has both success stories and failures that are instructive in four elements of sustainable water reforms, including: a) balancing multiple objectives in water reform; b) defining environmental-flow requirements; c) basin planning across jurisdictions; and d) adaptive learning in environmental-water recovery. The Columbia and Colorado river basins bracket the diverse range of management challenges and institutional innovations in the western United States. Both have engaged with diverse stakeholders to reconcile multiple objectives in long-range planning, particularly with indigenous communities. The Colorado River has responded to water-supply variability by building technical capacity for engagement in State and tribal stakeholder groups. The Columbia has pioneered adaptive learning in market-based environmental-water recovery through nested governance arrangements that combine bottom-up planning and implementation with basin-wide financing and accountability. This preliminary comparison highlights the broad scope for mutual learning and the need for sustained and systematic cross-case comparisons to realise this potential.

Australia's experiences with integrated basin governance are the envy of much of the world and have attracted increasing interest about lessons learned (Pilz 2010). The development of the Murray–Darling Basin Plan is the most recent milestone in 'Australia's internationally recognized record of water reform' (Langford et al. 2010). The Basin Plan aims to fulfil the mandate of the *Commonwealth Water Act 2007* to establish environmental-water requirements and sustainable diversions limits (SDLs) that optimise economic, social and ecological outcomes of water use. The public reception of the Plan—particularly in regional Australia—underscores the enduring implementation challenges of a whole-of-basin approach.

Intensified competition for water and concern about future trends in hydroclimatology have forced other countries to join Australia in responding to water scarcity. The western United States is uniquely positioned to offer experience relevant during the ongoing public dialogue about the Basin Plan. It shares many of the Murray–Darling Basin's defining challenges—aridity, highly variable stream flow, over-allocation and coordination challenges across irrigation districts, States and the Federal Government—and the western United States has both success stories and failures that are instructive.

James Wescoat jr (2005), North American Water Resource Geographer, summarised his observations from decades of experience comparing water-policy lessons in the western United States and internationally: 'I found much in common between water management in the western United States and other parts of the world, and I became convinced that geographic comparison of water policy lessons has practical as well as intellectual value.' We follow James Wescoat jr (2005) in recognising the practical and intellectual value of these cross-case comparisons. It is, however, important to note that several conceptual and methodological frameworks exist to structure comparison and accumulate learning; here, our goal is to describe the scope for comparison to motivate more systematic comparisons in future work.

In this chapter, we examine four dimensions of the public dialogue about the Basin Plan: a) balancing multiple objectives in water reform; b) defining environmental-water requirements; c) basin planning across jurisdictions; and d) adaptive learning in environmental-water recovery. We develop this contribution in four parts. In section two, we briefly review the evolving water-policy context in Australia to introduce key elements of the public dialogue about the Basin Plan. Section three presents the rationale for choosing the Colorado and Columbia basins as case studies of basin governance in the US west, and it describes their physical characteristics and water institutions. The fourth section compares the US basins along the focal dimensions summarised above. The fifth section summarises key lessons for the MDB in relation to the strengths and challenges associated with these four elements in the Basin Plan. We conclude by underscoring the scope for mutual learning and further comparative research.

Policy Evolution in Basin Planning

The Basin Plan exists within a wider context of federal policy reforms to address competition for scarce and variable water supplies in the MDB. The recent history of reforms to achieve integrated basin governance begins with the 1992 *Murray–Darling Basin Agreement* and the 1994 Council of Australian Governments (COAG) framework (Connell and Grafton 2008). These reforms established a framework and set of principles adopted through an intergovernmental agreement by the States, including recognition of environmental uses as legitimate and the establishment of limits on extractive uses. The 2004 *Intergovernmental Agreement on a National Water Initiative* built on the COAG framework and furthered its emphasis on market-enabling reforms, a return to environmentally sustainable levels of extraction, risk assignment, and future adjustment processes for water users and communities. The *Water Act 2007* aimed to optimise the economic, social and environmental outcomes of water use. The *Water Act* further stipulates the need to identify environmental-water requirements, establish SDLs to meet these requirements, and prepare a Basin Plan towards these ends.

The *Guide to the proposed Basin Plan* (MDBA 2010) was released on 8 October 2010. It includes six elements (MDBA 2010:36)

1. determining environmental-water requirements
2. assessing socioeconomic impacts
3. establishing SDLs
4. transition to the SDLs
5. implementing the Basin Plan
6. delivering outcomes.

Of these, the assessment of socioeconomic impacts, establishment of SDLs and design of transitional arrangements have attracted extensive community interest and political response. The Murray–Darling Basin Authority (MDBA) commissioned a series of socioeconomic studies in the lead-up to the development of the Basin Plan, and additional analysis is under way. Concerns about the social and economic costs of ecologically sustainable policy making are not unique to Australia or the MDB, however, and the western United States has confronted similar challenges in the context of drought and endangered species.

Governance in the Colorado and Columbia Basins

The Colorado and Columbia river basins bracket the diverse range of physical and institutional characteristics prevalent in the western United States. Both, however, are interstate and transnational rivers experiencing an increasing federal role in basin governance to coordinate multiple States and water-user associations to share the benefits, costs and risks of water-planning and allocation decisions.

Colorado

Physical Characteristics

The Colorado River Basin covers 629 100 sq km—roughly one-twelfth of the continental United States. The river descends 3658 m along its 2334 km length, and discharges into the Gulf of California. The average natural flow in the gauged record (1906–2008) at Lee Ferry is 18 500 gigalitres. This average obscures high stream-flow variability: the low and high flows (6930 GL and 31 200 GL) were recorded in 1977 and 1984, respectively. Flow variability is buffered by the large storage to run-off ratio in the basin, which approximates 4:1. Much of the 67 300 GL reservoir capacity is in Lakes Powell and Mead. The Colorado supports 30 million people, varied ecosystems, irrigates 800 000 ha of crops, generates hydroelectricity and provides recreation.

Water Institutions

The Colorado is managed by the Secretary of the Department of the Interior, who has designated the US Bureau of Reclamation as its lead agency. It is a multi-jurisdictional watershed: seven States and the United Mexican States (Mexico) share the river. In the United States, the basin was divided into two (largely in accordance with hydrology) in the *Colorado River Compact* of 1922. The Upper Division States (UDS) include Colorado, New Mexico, Utah and Wyoming and the Lower Division States (LDS) include Arizona, California and Nevada. Although the interests of each State within the UDS and LDS differ, they do tend to negotiate as a block. Mexico has a separate entitlement as per a 1944 treaty.

Columbia

Physical Characteristics

The Columbia River Basin (668 000 sq km) is comparable in size with the Colorado but not in volume (see Table 23.1). The river discharges into the Pacific

Ocean west of Portland, Oregon, after descending 820 m over 2000 km from its headwaters in British Columbia, Canada. It has an average volume at the Dalles Dam of 234 945 GL—an order of magnitude higher than the Colorado. Stream flow is characterised by spatial and seasonal variability. Chronic seasonal water deficits in the tributaries occur in late summer when peak agricultural use coincides with natural low flows. The Columbia supports a population of seven million, 2.8 million ha of irrigated agriculture, and a salmon fishery with high ecological, cultural and economic significance. A series of 31 dams generates hydroelectric power through the Federal Columbia River Power System (FCRPS).

Water Institutions

The Columbia is managed by a 'patchwork quilt' of laws, policies and jurisdictions, with allocation authority vested at the State level (Schoessler et al. 1997). The Columbia traverses parts of eight States in the United States, and Canada. The Columbia differs from the Colorado because it lacks a lead federal agency in water planning and an interstate compact to apportion water rights between the States. The FCRPS establishes a federal overlay in water management for power and conservation activities under the 1980 *Federal Power Act*. Further, a 1961 international treaty shares flood control and hydropower benefits between Canada and the United States. Tribes have substantial property rights, including 1855 *Stevens Treaty* rights to salmon harvests.

Table 23.1 Physical and institutional dimensions

	Dimensions				
Basin	**Physical**		**Institutions**		
	Annual inflow	*Size*	*Basin regime*	*Federal role*	*Objectives*
Colorado	18 500 GL	629 100 sq km	Interstate compact	US Bureau of Reclamation	Multidimensional 'Law of the River'
Columbia	234 945 GL	668 000 sq km	Northwest Power Act	Northwest Power and Conservation Council	Multidimensional 'patchwork quilt'
Murray–Darling	32 800 GL	1 061 469 sq km	Cwlth Water Act	Murray–Darling Basin Authority	'Optimise economic, social and ecological water use'

Four Elements of Sustainable Water Reforms

The MDB Plan reflects a commitment to reconcile multiple objectives, comprehensively specify environmental-water requirements, integrate planning at the basin scale, and implement transitional arrangements to comply with the

SDLs established through this process. Neither the Colorado nor the Columbia has established comparable commitments, although the Columbia has invested in substantial institutional innovations and infrastructure.

Balancing Multiple Objectives of Water Reform

Colorado

Water reform in the Colorado Basin has been patchy and piecemeal under a framework of laws, regulations, court cases and treaties collectively known as the 'Law of the River'. The first of these—the 1922 *Colorado River Compact*—sought to 'provide for the equitable division' of the Colorado River. The Compact was negotiated during wet conditions relative to the long-term average inflows over the 100-year gauged stream-flow record. A large research effort in the early 1990s—the 'Severe and Sustained Drought Study'[1]—sought to better understand the risks to water-supply reliability using both gauged and tree-ring-based stream-flow records and basin-wide hydrological models. Policy recommendations were not enacted largely as a result of political complacency following high stream flows in 1995 and 1997. In fact, as the current drought intensified, the efforts of the basin States and the Secretary of the Department of the Interior were focused on how to share surplus flows, as decided in the Interim Surplus Guidelines (USDOI 2001). The severe drought that began in 1999 provided the 'focusing event' (Pulwarty and Melis 2001) for the Interim Guidelines for Lower Basin Shortages and the Coordinated Operations for Lake Powell and Lake Mead (USDOI 2007)[2] and later for the first whole-of-basin approach, initiated in January 2010 with the Bureau of Reclamation's Colorado River Basin Water Supply & Demand Study (hereafter, Colorado Basin Study). The track record of reconciling multiple objectives has been limited, but capacity has grown. The deliberate adaptive-management approach in the Colorado has provided the basin States with sufficient flexibility to modify system management to changing conditions; the Bureau of Reclamation has engaged with States and tribes to develop technical capacity and cultivate stakeholder participation in the planning efforts.

Columbia

Unlike the Colorado, the Columbia lacks an interstate water compact to govern water sharing across the States. State water codes establish the contemporary

1 For studies relating to this effort, see 'Coping with a severe and sustained drought on the Colorado River', *Journal of the American Water Resources Association*, Special Issue, vol. 31, no. 5 (1995).

2 The 2007 Interim Guidelines incorporate: delivery cuts to LDS based on trigger elevations in Lake Mead; rules for the joint operation of Lakes Powell and Mead over the full range of storage conditions; and an innovative conservation mechanism: Intentional Created Surplus (ICS). ICS is currently available only to LDS but it is the probable mechanism for delta recovery in Mexico.

objectives of water reform to resolve imbalances among water uses for hydropower, consumptive purposes, and fish. For example, the *Washington Water Resources Act* of 1971 seeks to protect and use water for the greatest benefit of the people of the State, and it explicitly includes the water needs for fish habitat; it also prioritises local solutions to water-resource management.

The recovery of migratory salmon fisheries impacted by hydropower generation and surface-water diversions has provided a primary impetus for regional coordination. The *Northwest Power Act* of 1980 established the Northwest Power and Conservation Council (NPCC), which released its first fish and wildlife plan in 1982, with the most recent amendments in 2009. Recovery activities implement the fish and wildlife program strategy developed by the NPCC using more than $8 billion from 1999–2009, financed and administered by the Bonneville Power Administration, a federal power utility with mitigation obligations (BPA 2010).

The Yakima River of the Columbia Basin includes one of the first three basin studies funded under the 2009 *Secure Water Act*; the Deschutes River Basin is another tributary of the Columbia that has received funding in the 2010 round of grants under the Act. The study scope is illustrative in contrast with the Colorado. The Yakima River Basin Study and Associated Basin Restoration Implementation Plan include a characterisation of current and future water needs (both in-stream and off-stream), an assessment of climate impacts, and an evaluation of alternatives for meeting in-stream and off-stream needs and integrating them with other restoration strategies. This study effort to rebalance human and environmental water uses at the basin scale comes closest to integrated management across multiple, competing objectives and stakeholder values.

Defining Environmental Water Requirements

Colorado

From the viewpoint of the Colorado River Basin, it is striking to read the explicit environmental-flow requirements for ecosystem function and specified ecosystem assets in the MDB Guide. It is not that there is an absence of environmental concern in the Colorado,[3] but rather that system water is not explicitly set aside for environmental flows. The Colorado Basin Study will attempt to quantify

3 Prominent examples include the Glen Canyon Dam Adaptive Management Program, including experiments to restore sandbanks in the Grand Canyon National Park (see <http://pubs.usgs.gov/fs/2010/3009/fs2010-3009.pdf>) and ecosystem restoration efforts encompassed in the Lower Colorado River Multi-Species Conservation Plan.

environmental-flow requirements for targeted areas. The location of the Colorado River delta in Mexico, however, means that water requirements for this ecosystem are outside the scope of the Basin Study. Neither the United States nor Mexico ensures minimum base flows or flood pulses to maintain the health of the delta ecosystem; the delta receives flows and sediments only during extreme flood events, such as in 1983. There are ongoing bi-national conversations on how to restore the delta; base flow requirements of 61.7 GL with a flood pulse four or five times this volume every five years are often cited.[4] The responsibility for securing these environmental flows would be shared equally between three partners: the United States, Mexico and a consortium of non-governmental organisations (NGOs). Such an agreement would provide a mechanism for Mexico and the NGOs to access storage in the United States.

Columbia

The Columbia River Basin has among the most advanced scientific knowledge and policy framework for meeting environmental needs in the western United States. Like the MDB, in the Columbia Basin, the recognition of environmental needs occurred after rivers had become oversubscribed. Methods for prescribing environmental flows first emerged in the western United States in the late 1940s (Tharme 2003), and efforts in the Columbia (Tennant 1976) were motivated by legal and regulatory salmon recovery efforts. Each State has adopted different methods for assessing and meeting environmental requirements based on minimum flows required for salmon migration. Oregon established the 1955 Minimum Perennial Streamflows program to preserve base flows. Washington established a similar program in the late 1960s, and Idaho and Montana followed suit in the 1970s.

A comprehensive framework for assessing and defining environmental needs has been elusive (NRC 2004); however, the listing of 13 salmon and steelhead runs under the *Endangered Species Act* in the early 1990s triggered more coordinated efforts to identify and mitigate factors constraining salmon recovery, including flow limitations. The 2000 and 2008 Biological Opinions on the FCRPS established flow-augmentation requirements on the main stem of the Columbia. Local and State processes have guided the development of flow requirements on the tributaries where salmon reproduce on small systems vulnerable to late-summer depletion during peak irrigation use. The lack of a comprehensive basin-wide assessment of environmental-flow needs in the Columbia akin to the effort being attempted in the MDB Plan has made it difficult to establish priorities and assess trade-offs among competing targets for water recovery.

4 These flow requirements were provided by personal communications with federal officials, but they are not yet 'official'. The authors are unsure whether these flow requirements are based on an assessment of ecosystem needs or political acceptability.

Basin Planning across Jurisdictions

Colorado

The Colorado River Basin States are significant players with the Bureau of Reclamation in long-range planning, particularly with respect to proposed modifications to system management. Garrick et al. (2008) identified a key opportunity for stakeholder learning about the Reclamation Bureau's long-range planning model—specifically, the Colorado River Simulation System (CRSS)—in the almost sequential process of designing the 2001 Interim Surplus Guidelines (USDOI 2001) and the 2007 Interim Guidelines for Lower Basin Shortages and the Coordinated Operations for Lake Powell and Lake Mead (USDOI 2007). In the Environmental Impact Statement for the 2007 Interim Guidelines, the 'Basin States Alternative' was one of six reasonable operational alternatives analysed. The five alternatives (the sixth was the status quo) were developed collaboratively with NGOs, basin States, tribal and other stakeholders, in a process that depended on the transparency of CRSS and the ease and relative simplicity of the RiverWareTM software in which CRSS is implemented (Jerla et al. 2010). The 2007 Interim Guidelines will be up for renegotiation no later than 31 December 2020, at which time the Bureau of Reclamation, the basin States and other interested parties will evaluate the effectiveness of the new measures in increasing system management flexibility before extending them beyond 2026. Meanwhile, the Colorado Basin Study is the first authorised under the *Secure Water Act*; it builds on nearly a century of basin-scale conflict and coordination. A defining feature of this effort is federal coordination with the basin States, environmental NGOs and the tribes in assessing the long-range water supply and demand.

Columbia

In the absence of an interstate river compact, the NPCC and the federal agencies coordinating salmon recovery provide the primary opportunities for basin-scale integration in long-range basin planning. The move towards catchment-scale governance and whole-of-basin integration, however, emerged from the bottom up at the local and State levels as an alternative to reliance on courts for conflict resolution in response to salmon declines (Benson 1996). The 2000 amendments to the NPCC Fish and Wildlife Program delineated 62 sub-basins and called for the development of fish and wildlife plans for the main stems of the Snake and Columbia river basins and 58 sub-basin plans to cover the tributaries. The sub-basin plans emphasised 'locally developed, integrated' plans to establish biological objectives and implementation strategies through a technical assessment, an inventory of past recovery efforts, and a management plan (NPCC 2005). The NPCC ensured basin-wide consistency with the Fish and Wildlife program by ratifying the sub-basin plans as part of the 2009 program amendments.

The sub-basin planning process dovetails with State-level programs to manage water resources along tributaries affected by groundwater pumping. In contrast with the statutory provisions establishing state-coordinated, local catchment planning in Oregon and Washington, catchments in Idaho and Montana have developed voluntary, bottom-up approaches to integrated planning. The Blackfoot Challenge of Western Montana is a noteworthy example, where a multi-stakeholder forum has coalesced since the early 1970s. The forum developed scientific, planning and drought-response measures, which culminated in voluntary diversion reductions to meet fisheries requirements established by State-held in-stream water rights that were historically unmet in dry years.

Adaptive Learning in Environmental-Water Recovery

Colorado

There are bi-national conversations on how to restore the Colorado delta, but for now, the environmental-flow requirements that have gained traction in the bi-national dialogue are for the Ciénega de Santa Clara, Mexico. The Ciénega is the largest wetland in the Sonoran Desert and an important rest stop along the Pacific Flyway; its restoration is also the unintended consequence of a temporary solution to a bi-national water-quality issue. In 1973 in response to unacceptably high salt loads, Minute 242 to the 1944 treaty was signed. In 1974 the US Congress passed the *Colorado River Salinity Control Act*, which authorised the Bureau of Reclamation to construct pipelines and the Yuma Desalting Plant (YDP). A bypass drain was constructed both as an interim measure to divert saline agricultural drainage outflow to the Santa Clara Slough and as the future conveyance of the YDP brine stream. Until 2009, US compliance with Minute 242 had been met by bypassing saline agricultural backflow to the Ciénega. Bypass water is not counted in Mexico's entitlement but rather is water in excess of LDS entitlements that is released from basin reservoirs.

In 2009, a decision was taken to both run the YDP at one-third capacity to conserve system flows *and* replace reduced bypass flows. The replacement bypass water has to be conserved or transferred within the system. The United States delivered its portion of these replaced flows (USDOI 2009:23); they represent the first dedicated environmental flows in the Lower Colorado River within the Colorado River Accounting System. This is a significant milestone and was not without its detractors; some argued that a better experiment would have been to monitor how the Ciénega responded to reduced flows and higher salt loads. Replacement flows are most likely to be met by contracting with an irrigation district to forbear a proportion of its irrigation entitlement, which is then transferred to one of the three responsible parties.

There are a number of long-term, large-volume and also pilot irrigation forbearance agreements in the US south-west that provide price points for such environmental transfers. The Bureau of Reclamation has engaged in five pilot projects to learn how best to design, monitor and implement these agreements and numerous lessons have been learned (Colby and Bark 2010). The 2004 pilot program was cancelled after Arizona raised objections to it. In the 2006–07 pilot program, the Bureau of Reclamation partnered with the Metropolitan Water District (MWD) of Southern California and the Palo Verde Irrigation District (PVID) Land Management Crop Rotation and Water Supply Program. This supplementary agreement with the Reclamation Bureau conserved 3.7 GL in 2006 and 8.6 GL in 2007. The 2008 and 2009 pilot programs were in Arizona and conserved 4.3 GL, while the 2010 program conserved 4.6 GL. Responding to lower agricultural prices, program costs fell from US$97 285/GL in 2008–09 to US$72 964/GL in 2010. Longer-term agreements, such as the 35-year agreement signed in 2004 between MWD and PVID, transfer larger volumes—in this case, up to 160.4 GL/yr. Irrigators are paid a combination of a sign-up payment and an exercise cost by MWD. The preferred solution is temporary transfers rather than permanent entitlement buy-outs.

Columbia

A number of market-based approaches to reallocate water for salmon recovery have been designed and tested in the Columbia Basin (Hardner and Gullison 2007). Enabling conditions have developed at the State level or within pilot catchments within the States. Legal reforms establish caps or formal basin closures to establish overall limits to appropriation and to provide legal certainty about the extent, validity and relative priority among competing rights in the consumptive pool. The establishment of caps on water use is, however, incomplete and uneven across the Columbia, and diversion limits have traditionally been based on historical use patterns that already exceed sustainable diversion levels. Although environmental water needs were recognised as beneficial uses as early as the 1940s, and transfers were authorised to convert existing water rights to in-stream purposes, environmental-flow needs have yet to be comprehensively specified or prioritised to guide efficient acquisitions.

The *1987 Oregon Instream Water Rights Act* is a point of departure in the development of markets for environmental flows. The Oregon experience demonstrates an iterative and adaptive process. A diverse set of organisational actors has emerged to implement and manage water transactions for environmental recovery. A series of state-wide or catchment-level programs began to reallocate water rights through temporary and permanent acquisitions and irrigation efficiency savings. Non-profit water trusts, quasi-governmental basin organisations, and government-run water-acquisition programs have become coordinated by nested governance arrangements to achieve basin-wide integration and accountability (Garrick et al. in review). The Columbia Basin Water Transactions Program (CBWTP) formed in 2002 to integrate these nascent efforts at the basin scale by coordinating and administering $4 million of annual funding for local partners. Two of 13 sub-basins within CBWTP have achieved their flow targets—the Deschutes of Central Oregon (66.9 GL) and the Lemhi River (15.4 GL) of Idaho—and, by extension, have achieved threshold-level impacts by restoring sufficient habitat to reintroduce salmon species extirpated from the region in the case of the Deschutes (Garrick and Aylward 2010). The environmental-water buyer is the primary source of competition for water along over-allocated tributaries as demand remains low for agricultural transfers and urban growth has been limited until the past decade. As a result, pricing information has been established through two mechanisms in the absence of active water markets for consumptive rights: reverse auctions and case-by-case appraisals using an income cost replacement method that compensates farmers for income forgone.

The implementation experience over the past 15 years offers a group of key lessons about adaptive learning in market-based environmental recovery efforts. The policy reform underpinning market-based recovery has evolved over five decades in response to crises associated with drought, shifting water needs and ecological degradation. It has been necessary to harness these larger drivers to establish institutional changes conducive to the formation of water markets for environmental flows. The adaptive approach has guided the design of transactional tools to address issues of scale and scope. Water transactions initially targeted small tributaries where relatively small net increases in in-stream flows provide ecological benefits through river connectivity to aid salmon reproduction. Implementation relied on 'seed' leases to establish proof of concept and to assuage irrigator concerns about the impacts of environmental water transactions on agricultural viability. The scope of transactions has expanded to consider other factors limiting salmon recovery. Another lesson is the reliance on cooperation and engagement with local stakeholders to design water-recovery projects with careful attention to irrigator fears, incentives and capital infrastructure needs. Adaptive learning has been the crosscutting theme through proactive monitoring, evaluation and feedback processes.

Table 23.2 Comparative perspective on the Murray–Darling Basin from the western United States: scope for mutual learning

1. Balancing multiple objectives of water reform		
Strengths	*Challenges*	*Lessons from the western United States*
The Commonwealth Water Act of 2007 and the Murray–Darling Basin Authority provide explicit objectives for optimising the socioeconomic and ecological benefits of water use and prescribe whole-of-basin planning.	Engagement with local communities presents a challenge in the specification of sustainable diversion limits (SDLs) and the design of transitional arrangements.	The Colorado Basin has excelled recently in structuring long-range planning to engage States and tribes in response to prolonged drought. The coordination on shared challenges has established the technical capacity for local communities and diverse stakeholders to engage in complex planning processes and build trust for addressing a complex and comprehensive basin-wide study of long-range supply and demand. The Columbia Basin has specified tribal water rights and undertaken salmon recovery by coupling local, sub-basin planning with basin-wide coordination. The planning and implementation process has generated bottom-up capacity for policy mandates and accountability mechanisms administered by basin-wide and regional authorities.
2. Defining environmental-water requirements		
Strengths	*Challenges*	*Lessons from the western United States*
The comprehensive specification of environmental-water requirements in the Murray–Darling provides the basis for risk-based decision making about water recovery and delivery for ecosystem functions, environmental assets, the productive base and key environmental outcomes.	Current efforts seek to assess socioeconomic impacts to guide trade-offs under different scenarios of SDLs. The use of long-term averages to develop SDLs underscores the importance of State water plans to translate SDLs into water requirements.	The Colorado Basin lacks adequate specification of environmental needs due to international boundaries. Degradation of flow-dependent assets along the main stem and in the delta has already occurred beyond thresholds for long-range ecological health. The Columbia Basin has developed flow requirements through inconsistent methodologies but has identified flow components needed to target water-recovery efforts for dynamic environmental needs.

3. Basin planning across jurisdictions		
Strengths	*Challenges*	*Lessons from the western United States*
The MDBA has developed a consultation process to solicit feedback in development of the Basin Plan. This process has included the addition of the Guide to the Plan to engage regional and urban centres and inform development of the draft and final versions of the Plan.	The fears and concerns of irrigator communities remain high in areas confronting proposed reductions, which has led to new initiatives to assess socioeconomic impacts and tailor transitional arrangements.	The technical capacity developed by stakeholder groups in the Colorado in planning for drought has equipped the States, environmental groups, tribal governments, researchers and consultants to engage in basin planning for long-range system management. Dialogue with Mexico is ongoing on ecological concerns, shortage and storage sharing to provide flexibility for addressing more complex problems. The bottom-up approach of cross-jurisdictional integration in the Columbia has empowered local communities to pursue substantive planning efforts coupled with basin-wide coordination and integration.

4. Adaptive learning in environmental water recovery		
Strengths	*Challenges*	*Lessons from the western United States*
Transitional arrangements offer incentives Transitional arrangements offer incentives for water recovery and infrastructure through the Restoring the Balance program. This financing along with the establishment of the Commonwealth Environmental Water Holder (CEWH) provides substantial capacity to secure and manage environmental water.	Uncertainty exists about the willingness of irrigators to participate in the transitional arrangements; the portfolio of water that the CEWH may acquire; and the appropriate mechanisms for managing and delivering the environmental water holdings. Water-recovery efforts raise the corresponding governance challenge of managing the holdings—spatially, temporally, and volumetrically—to achieve ecological outcomes.	The Colorado Basin has not been active in the development of environmental water recovery; however, its experience with forbearance programs has demonstrated the use of pilot arrangements and temporary agreements to test and adapt mechanisms for drought-year flexibility. The Columbia is a laboratory for institutional innovation in the development of market-based programs to recover water for environmental flows. Experience in the Columbia demonstrates the value of nested governance arrangements to combine local institutional capacity with regional and federal financing and oversight.

Lessons for the MDB Plan

This chapter provides a preliminary comparison and seeks to motivate further opportunities for mutual learning between the western United States and Australia as both regions adapt to a water-scarce future. Table 23.2 elaborates key elements of the MDB in terms of the strengths (that is, lessons it imparts to the western United States), challenges, and potential lessons from the western United States.

Conclusions

What does this preliminary comparison with the western United States offer in the Murray–Darling context? And how could more structured comparison and exchange promote mutual learning about basin governance across multiple jurisdictons? The Colorado and Columbia provide insight on different dimensions of the MDB experience. As the Basin Plan progresses through future development phases and into implementation, the western United States and other international audiences will follow the Murray–Darling as it confronts challenges that international experience suggests are among the hardest to address: coordination of multiple objectives at the national level; comprehensive specification of environmental-water requirements at a whole-of-basin scale with consistent technical criteria to guide operational environmental watering decisions; cross-jurisdictional integration through an independent, basin-wide authority; and transitional arrangements to comply with SDLs. Although the western US reforms are halting and uneven, Australia might find relevant lessons from trial and error in the United States to: engage with and build technical capacity in governmental and non-governmental stakeholders at multiple levels; specify dynamic environmental needs to approximate time- and place-specific flow regimes and to respond to drought-year impacts; promote adaptive learning in market-based environmental recovery through pilot projects, monitoring and evaluation; utilise temporary agreements to build towards permanent protection; and rely on local and State governmental and non-governmental actors to liaise with irrigation communities without losing accountability to basin-scale objectives. The US experience in negotiating tribal water rights is also instructive as Australia seeks to recognise native-title rights to water.

The preliminary comparison developed here underscores the conclusion that the scope for exchanging lessons between the Murray–Darling and the western United States might be broader than previously recognised. Opportunities for mutual learning stem from shared challenges and varying types and levels of experience with adaptive river basin management strategies. The development of the Basin Plan provides an impetus for sustained and more systematic comparisons between Australia and the western United States.

Bibliography

Benson, R. 1996, 'A watershed issue: the role of streamflow protection in northwest river basin management', *Environmental Law*, vol. 26.

Bonneville Power Administration (BPA) 2010, *BPA invests in fish and wildlife*, Fact Sheet, Environment, Fish & Wildlife, Bonneville Power Administration, Portland, Ore., <http://efw.bpa.gov/IntegratedFWP/FACT_SHEET_Invests_fish_wildlifeFINAL.pdf>

Colby, B. and Bark, R. 2010, 'Inter-sectoral water trading as a climate change adaptation strategy', in Q. Grafton and K. Hussey (eds), *Water Resources Planning and Management*, Cambridge University Press, UK.

Connell, D. and Grafton, R. Q. 2008, 'Planning for water security in the Murray–Darling Basin', *Public Policy*, vol. 3, no. 1, pp. 67–86.

Garrick, D. and Aylward, B. 2010, Transaction costs and adaptive efficiency in emerging markets for environmental flows: lessons from the Columbia Basin, Economics and Environment Network Symposium, The Australia National University, Canberra.

Garrick, D., Jacobs, K. and Garfin, G. 2008, 'Models, assumptions, and stakeholders: planning for water supply variability in the Colorado River Basin', *Journal of American Water Resources Association*, vol. 44, pp. 381–98.

Garrick, D., Lane-Miller, C. and McCoy, A. (forthcoming), 'Institutional innovations for governing environmental water in the western US', *Economic Papers*.

Hardner, J., and Gullison, T. 2007, *Independent external evaluation of the Columbia Basin Water Transactions Program (2003–2006)*, Hardner and Gullison Consulting, Portland.

Jerla, C., Morino, K., Bark, R. and Fulp, T. 2010, 'The role of research and development in drought adaptation on the Colorado River Basin', in Q. Grafton and K. Hussey (eds), *Water Resources Planning and Management*, Cambridge University Press, UK.

Langford, J., Briscoe, J. and Porter, M. 2010, 'Creating wealth from water', *The Australian*, 1 November 2010, <http://www.theaustralian.com.au/creating-wealth-from-our-water/story-fn6nj4ny-1225945844874>

Murray–Darling Basin Authority (MDBA) 2010, *Guide to the proposed Basin Plan*, Murray–Darling Basin Authority, Canberra.

National Research Council (NRC) 2004, *Managing the Columbia River: Instream flows, water withdrawals, and salmon survival*, National Academies Press, Washington, DC.

Northwest Power and Conservation Council (NPCC) 2005, 'Deschutes subbasin plan', in *Columbia River Basin Fish and Wildlife Program*, Portland, Ore.

Pilz, D. (2010), 'Lessons in water policy innovation from the world's driest inhabited continent: using water allocation plans and water markets to manage water scarcity', *University of Denver Law Review*.

Pulwarty, R. S. and Melis, T. S. 2001, 'Climate extremes and adaptive management on the Colorado River: lessons from the 1997–1998 ENSO event', *Journal of Environmental Management*, vol. 63, pp. 307–24.

Schoessler, M. A., Blumm, M. C., Swift, B. M., Northwestern School of Law, Northwest Water Law and Policy Project and Western Water Policy Review Advisory Commission 1997, *A survey of Columbia River Basin water law institutions and policies*, Report to the Western Water Policy Review Advisory Commission.

Tennant, D. L. 1976, 'Instream flow regimens for fish, wildlife, recreation, and related environmental resources', *Fisheries*, vol. 1, pp. 4, 6–10.

Tharme, R. E. 2003, 'A global perspective on environmental flow assessment: emerging trends in the development and application of environmental flow methodologies for rivers', *River Research Applications*, vol. 19, pp. 397–441.

US Department of the Interior (USDOI) 2001, 'Record of Decision. Colorado River Interim Surplus Guidelines', *Final Environmental Impact Statement*, January, Department of the Interior, Washington, DC, <http://www.usbr.gov/lc/region/g4000/surplus/surplus_rod_final.pdf>

US Department of the Interior (USDOI) 2007, *Record of Decision. Colorado River Interim Guidelines for Lower Basin Shortages and the Coordinated Operations for Lake Powell and Lake Mead*, December, Department of the Interior, Washington, DC, <http://www.usbr.gov/lc/region/programs/strategies/RecordofDecision.pdf>

US Department of the Interior (USDOI) 2009, *Colorado River Accounting and Water Use Report Arizona, California, and Nevada Calendar Year 2009*, Department of the Interior, Washington, DC, <http://www.usbr.gov/lc/region/g4000/4200Rpts/DecreeRpt/2009/2009.pdf>

Wescoat, J. jr 2005, 'Water policy and cultural exchange: transferring lessons from around the world to the western United States', in *In Search of Sustainable Water Management: International lessons for the American west and beyond*, Edward Elgar, UK.

White, G. 1957, 'A perspective of river basin development', *Law and Contemporary Problems*, vol. 22, no. 2, pp. 157–87.

24. What Can the Murray–Darling Basin Plan Achieve? Will it be enough?

Neil Byron

The Murray–Darling Basin Authority (MDBA) has been criticised by many for what it has done so far, and how it has managed and communicated its extraordinary task. Debate is raging about whether the MDBA should reduce the amount of water irrigators are allowed to extract from rivers across the whole Murray–Darling Basin (MDB): not at all (as many irrigators assert): by 3000–4000 gigalitres (as the MDBA is currently considering); or by up to 7600 GL (as many environmental non-governmental organisations are demanding).

For comparison, according to the MDBA's recent *Guide to the proposed Basin Plan* (2010), under the current rules, the long-term average aggregate extraction plus interceptions is 13 677 GL or *43 per cent* of the estimated long-term natural flows. The long-term average environmental use is 13 996 GL or *44 per cent* of the estimated long-term natural flows. The balance—a long-term average of 5105 GL or 15 per cent of the estimated long-term natural flows—reaches the Murray River mouth. As an example, one of the MDBA's scenarios—reducing the total extraction by 4000 GL below current levels—would make these percentages 30 per cent for irrigators, 48 per cent for the environment and 24 per cent for the river mouth.

The purpose of this chapter, however, is not to debate the 'right' level of reduction in average levels of water extraction by irrigators. The central argument is that successive Federal Governments' entire MDB 'strategies' have not been well designed or well implemented. The Guide is just a product of the MDBA, which is a product of the *Water Act*, which in turn is a product of the (hastily produced) 2007 *National Plan for Water Security*. Errors and false assumptions from the 2007 press release are progressively compounding and accumulating. This series of errors, assumptions and simplifications makes it quite unlikely, in my view, that the Commonwealth can 'fix the MDB' *just* by following the current path.

In the 2010 Productivity Commission report on recovering water in the MDB, I concluded: Australians generally want to see an environmentally sustainable Murray Darling system (including the wetlands and riparian lands as well as the rivers per se) supporting viable, sustainable, world-class agriculture and viable dynamic communities that are good places to live and raise a family.

I believe that is still true.

But Australia is now three years into implementation of the 10-point Turnbull–Rudd 'strategy to fix the MDB' in 10 years for $10 billion, but with little confidence that it is going to deliver that outcome. Certainly, the $10 or $13 billion allocated—probably more—will all be spent over the decade of the 'strategy'. The Commonwealth will own some thousands of gigalitres of water entitlements. But it is still debatable whether the MBD ecosystems, including World Heritage Areas, Ramsar-listed wetlands, national parks and all the small unnamed wetlands, will be in much better condition than they were last year or 10, 20 or 50 years ago. Even if individual elements of the landscape are improved, improving the health of the *whole connected landscape* must include agriculture. And the social and economic impacts on industries and communities in the Basin will be non-trivial, even though most will probably manage to adapt and survive, and some will prosper.

Campbell (2010), Ison (2007) and Myers (2010) have also commented that the problem has been constructed in an unhelpful way: as a technical issue, in which a small number of scientists will work out the 'right' answer, announce what the rest of society has to do then all interested parties will do as instructed. This is very deterministic and authoritarian—akin to NASA planning a lunar landing. Campbell, Ison and I have argued elsewhere, at various times, for a bottom-up adaptive strategy as a more effective path for dealing with water resources as a wicked problem.[1]

Resolving and unscrambling the legacy of diverse and inappropriate MDB arrangements are not things a 'plan' can do. Although the National Water Initiative (NWI) and the *Water Act* are national in a broad sense, Australia still cannot agree on a plan because the policy has not been debated. In effect, we are having the policy debate now, *within* the plan. Working out the policy consensus is not an appropriate task for an 'independent authority' to do.

A *real* policy—preceding the technical planning—would have diagnosed precisely what the problems are in the Basin, ensured a feasibility study and thorough analysis of all major outlays of public money, and then directed the available resources at ensuring sustainable use of the Basin, a sound economy, and a viable rural society not dependent on subsidies.

It is not too late to use all of the high-quality science that has been done—*in a different way*. Governments could start engaging seriously with people at local

1 'Wicked problems require thinking that is capable of grasping the big picture, including the interrelationships among the full range of causal factors underlying them. They often require broader, more collaborative and innovative approaches. This may result in the occasional failure or need for policy change or adjustment. They go beyond the capacity of any one organisation to understand and respond to, and there is often disagreement about the causes of the problems and the best way to tackle them.'

and regional scales who know the details of how their ecosystems, agricultural systems and social systems interact. Without the commitment, energy and knowledge of these people, any externally imposed plan is unlikely to work. *Connectedness* across the landscape is crucial, not isolating everything into components—this wetland, that clump of trees, that stretch of riverbank, this species of frog—and agriculture is crucial to that connectedness. Environmental, social and economic dimensions must be integrated, instead of treating them as stand-alone silos. Landholders and catchment management authorities (CMAs) do this all the time (Robins and Dovers 2007), but governments consistently find it very difficult to operate any way other than in departmental silos (Productivity Commission 1999).

Critiques of the MDBA's *Guide to the proposed Basin Plan* abound. Paul Kelly (editor-at-large of *The Australian*), economist Henry Ergas and Andrew Campbell (the former head of Land & Water Australia and one of the founders of Landcare) are among those who have written from the author's own perspective, and they each criticise different aspects of the current proposals. Such specific criticisms can be pieced together to seriously question the whole undertaking. Most importantly, the aim of this chapter is to suggest how the strategy could be redeemed.

Four Judgments about the Existing MDB Strategy

The Commonwealth's Murray–Darling Basin strategy to reverse environmental deterioration in the Basin was not well designed. If the problems in the Basin are to be dealt with effectively, cost effectively and equitably, we need a well-thought-out policy and the current strategy will need to be overhauled and extended.

Having presided over a Productivity Commission inquiry into the design and implementation of the strategy, and observing for the subsequent year, I have reached four conclusions.

1. Legislatively, in the form of the *Water Acts 2007* and *2008*, the 'strategy' is poorly designed.
2. Administratively, the 'strategy' has been set up on an ill-conceived basis with an inappropriate institutional structure.
3. Financially and economically, the 'strategy' could prove wasteful.
4. Underlying all of these deficiencies, the plan is fundamentally confused about causes and effects, and about means and ends. The problem is seen as a technical one to be solved by experts, rather than a social learning, adaptive process—an inclusive, long-term national conversation.

I do not intend to discuss the politics, except to restate a few well-known facts: a $10 billion policy commitment was announced before the 2007 election. No-one has ever explained where the figure of $10 billion came from, or why 60 per cent of the $10 billion was allocated to subsidising irrigation infrastructure. To my knowledge, no-one has even claimed that a detailed feasibility or benefit–cost analysis exists, let alone has tabled one. Parliament passed the bill in 2007 unanimously in both houses without dissent.

During the 2007 election campaign, Kevin Rudd committed Labor to an *extra* $3 billion for the same policy. Parliament passed the renewed bill under the new Labor government in 2008—again unanimously, in both houses, without dissent.

If anyone had asked in the Commonwealth Parliament in 2007–08 '*Who wants to save the Murray–Darling Basin*', all hands would have risen. But if the question was '*Does anyone understand the nature of the problems facing the MDB, how we got into this mess, the options for getting out of the mess, how long that will take, how much it will cost, and whose cooperation do we need to succeed*' then I suspect no hand would have been raised. But no-one will ever know, because that question was not asked until now. Our purpose here is not to discuss politics, but rather to concentrate on the science, economics, and governance.

Why do I assert that the legislation itself is poorly designed?

Legislatively, the strategy has three flaws, in my view.

a) At its heart, the *Water Act* requires that key environmental assets and processes must not be compromised. Lawyers might argue forever about how to define 'key' and what changes, if any, amount to being 'compromised'. The estimated number of key environmental assets ranges from the 16 internationally listed wetlands, to an MDBA list of more than 33 000 'significant wetlands' (including some unnamed dry watercourses) based on the International Union for Conservation of Nature (IUCN) list of rare and endangered species. The MDBA Guide refers to 2442 sites, but then focuses on 18 Hydrological Indicator Sites (HIS). If the Gwydir Wetland, for example, is a key environmental asset (and most people agree it is), how much of it must be 'not compromised'—200 000 ha, 120 000 ha or just the small area (about 862 ha) that is Ramsar listed?

b) On most readings, the legislation deliberately and explicitly *prevents the MDBA from taking social or economic costs into account* in estimating the environmental water requirements and setting the sustainable diversion limits (SDLs) for each river. The legislation mandates reduction in extractive use of water to whatever level is required to protect all environmental values (apparently without taking social or economic costs into account). The Water

Minister, Tony Bourke, obtained a new 'legal opinion' to the effect that the *Water Act* permits social and economic issues to be taken into account in setting SDLs. Either way, it is unavoidable that there will be collateral costs associated with returning more water to the environment, and these costs are likely to impact on the agricultural economy of the MDB. They are likely to generate some political backlash against *any* government that implements an environmental plan regardless of social and economic disruption. A better piece of legislation might have carefully included such considerations.

c) Economists recognise that the quantity of water demanded by any irrigator depends on its price. But in contrast, the legislation sees the quantity of water demanded for the environment as totally independent of price—whether water has a market value of $1 or $1000 per megalitre, a 'key environmental asset' *must* have 'X' ML if scientists agree that it *needs* X ML/year (on average) to ensure that its ecological integrity is not compromised. Many believe that the environmental benefits will be very great and/or that the economic and social costs and disruptions will be small and manageable. That might be true. It is, however, inescapable that trade-offs must be made. *No process has been created to explicitly enable Australian society to have that national conversation and make those judgments.*

This is a political debate about the differing values and priorities within Australian society so it must ultimately be a *social consensus*; environmental science, hydrology and economics can and should inform the political decision, but reciting facts cannot make such trade-offs. Any organisation that is required to make recommendations based on the benefits of restoring more water to the environment (but without considering the social and economic costs *and* benefits) would probably generate just the sort of acrimony now evident.

So what is wrong with the way the strategy is being administered?

The States' repeated failure to deliver a viable water policy over many years, and to reduce over-allocations to irrigation (as they had promised in 1994 and again in the 2004 National Water Initiative), was seen as confirmation that the Commonwealth Government had to impose a top-down initiative.

But designing the strategy as a top-down Commonwealth initiative was unhelpful. Local and community groups were omitted in the way the strategy was conceived. The whole Landcare network was excluded from the planning and design processes, as Andrew Campbell has pointed out. The CMAs were also excluded from the design and content of the strategy. The only reference to CMAs in Volume I of the Guide is on pages 194–5:

> All basin states have existing mechanisms for integrating natural resource management at the regional level—for example through the

> implementation of regional NRM [natural-resource management] plans. To the extent that these regional plans relate to the management of water resources, it will be important that they are consistent with the directions and arrangements proposed under the Basin Plan.

But by ignoring such groups, the strategy fails to draw upon their data, experience and local environmental and agricultural knowledge. To substitute for such inputs has been extremely costly in time, effort and compensation for errors. Without these contributions, the strategy is virtually guaranteed to be, at the very least, highly contentious and inefficient.

But leaving these groups out also means that they have little commitment to the strategy. Without their commitment, the strategy will lack the necessary cooperation at the grassroots level. This is evident now—an enormous amount of goodwill and social capital have been destroyed by trying to impose a plan developed in isolation by experts, regardless of how technically excellent that plan is (or appears to be now).

Is the strategy likely to be economically and financially sound?

The strategy as currently designed has the potential to be inefficient. It will be many years before taxpayers know whether they have received value for money for the 10-year, $10 billion 10-point strategy. It would be reassuring to have some early confirmation that it is likely to generate significant net benefits. The Governor of the Reserve Bank said in November 2010 (replying to a question about the National Broadband Network, or NBNCo) that *all* major public investments should be subject to feasibility studies and benefit–cost analyses—before implementation.

The Federal Government has purchased about 1000 GL of water entitlements (on paper), but it still is not clear what the Commonwealth Environmental Water Holder (CEWH) will do with this asset. Achieving the desired environmental benefits will depend on *how* the CEWH uses the endowment of water, not just on the size of the endowment. For example, it now seems that, in specific places, it will be impossible to deliver the sheer volumes of water that would be needed to create over-bank flows (required to deliver the intended amounts of environmental watering of a wetland) because of hydrological constraints and the amount of flooding that would be imposed on third parties. So, if it will be necessary to use engineering works and measures that can achieve the desired results with much less water (as is being discussed for Hattah Lakes and Chowilla), the SDL might not need to be as restrictive. Using water efficiently is important, whether watering a crop or a wetland.

Second, the feasibility studies have yet to result in engineering initiatives that have made an appreciable impact on the problem. Almost $6 billion was allocated over 10 years for infrastructure projects. As far as I can ascertain, none of the major infrastructure projects has yet passed even the most basic cost/benefit analysis.

The Australian National Audit Office (ANAO 2010), the Productivity Commission and independent economist Henry Ergas concluded that the infrastructure projects could be far more expensive per megalitre of water saved than buybacks of similar water. The taxpayer could end up paying up to $10 000/ML to obtain water rights with a market value of about $2000/ML. Australians might get better value by spending that money on water buybacks. But making this level of expenditure on buybacks across the Basin in a short time might cause the unit prices (and the social and economic consequences) to rise.

In fact, it might not be necessary to spend that $6 billion on either subsidising infrastructure or buying back water entitlements in the MDB. Perhaps a pause is possible (given the flooding) until we are confident about (or agree on) how much water is needed, when, where and how to achieve MDB environmental sustainability. It is far from clear that the best and smartest way to assist irrigators or rural communities is to offer them $6 billion of taxpayers' money on the condition that they may spend it only on water-engineering works. Governments could actively engage with *all* the people concerned (not just irrigators and conservationists) and seek their vision and advice on what would give them better social, environmental and economic outcomes. It seems arrogant and presumptuous to specify in advance (without even asking) that the money is only for water engineering.

Finally, why do I believe the strategy is conceptually confused?

Conceptually, the strategy is based on this perception of 'the problem': rivers, wetlands and riparian systems in the MDB have been under severe stress; the cause is excessive extraction of water from the rivers for irrigation. If the extraction of water from rivers for irrigation is reduced to SDLs, and the water so saved is reassigned for environmental purposes then the Basin will be restored to ecosystem health—a sustainable, working river system.

This perception seems self-evident to many, but it is simplistic. It is true that there has been a decline in the environmental quality of the MDB over several decades and that this decline has *coincided* with a continuous increase in the extraction of water for irrigation in the MDB (until the recent decade of drought). The strategy is based on this simple perception of causality. Not only is this simple notion pervasive that 'all the environment needs is more water' but there is a corollary: 'the more water added, the better the environmental outcomes will be.'

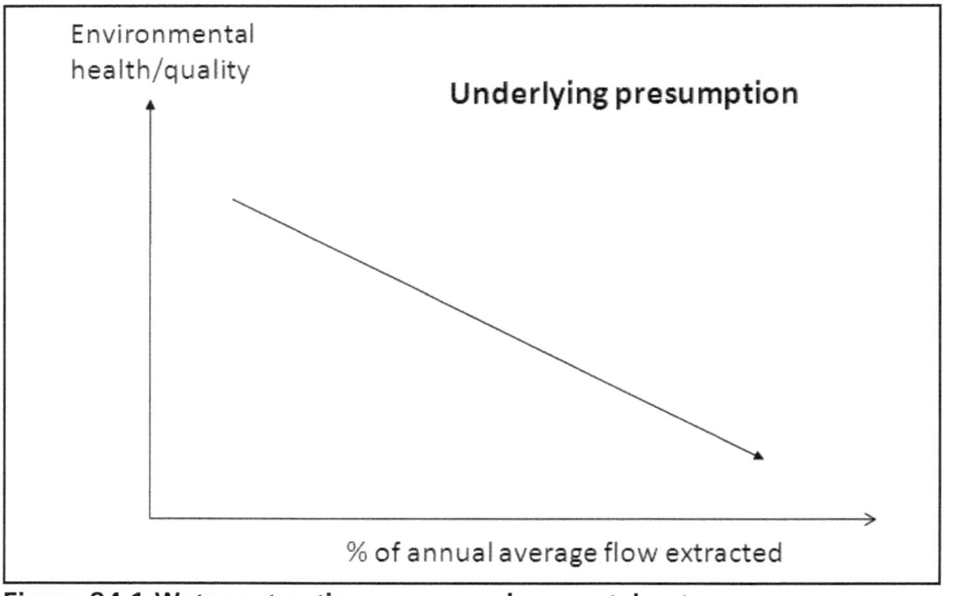

Figure 24.1 Water extraction versus environmental outcomes

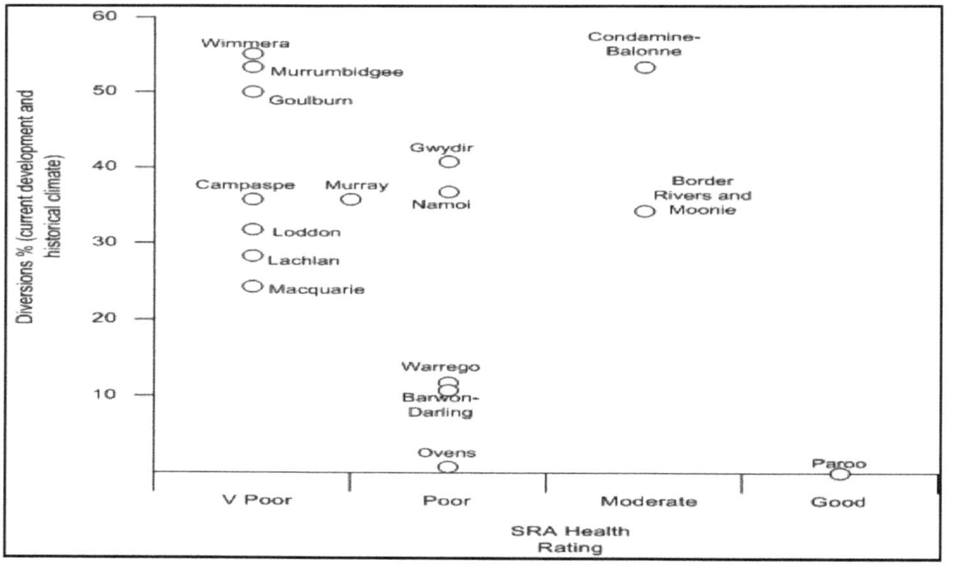

Figure 24.2 Where is the correlation?

While reducing extraction for irrigation might be *necessary* in many (but not all) of the rivers in the Basin, it is unlikely to be *sufficient*. The problems are more complex than that. Excessive extraction of water by irrigators is not the *sole* (and might not even be the greatest) threat to ecosystem health and sustainability in the MDB.

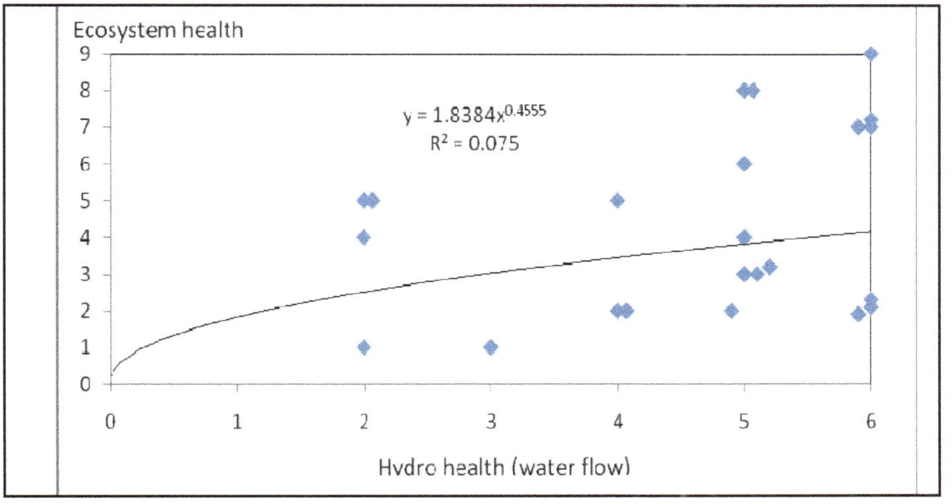

Figure 24.3 Ecosystem health

Note: The rating of 6 for hydrological health is the highest and 0 is the lowest. The rating of 9 for ecosystem health is the highest and 0 is the lowest. The ratings are altered slightly to ensure data for each river are visible. This will alter the relationship but only very slightly.

Source: MDBA (2010: Tables 2.2 and 2.3).

The highly variable outcomes for the ecosystem health of rivers or reaches that are rated the same in terms of flow conditions in the figure above suggest the need for caution. River flows seem a poor indicator of ecosystem health in highly regulated and so highly altered river systems. Unfortunately, the Guide seems to make the obvious (but possibly misleading) assumption that all that is needed for ecosystem health is to 'just add water, and the more water you add, the better it will be'.

At least *four factors* other than the quantity of water extracted from the rivers have contributed to environmental decline in the MDB.

1. The *seasonal timing* of water flows might be a more important contributor to environmental decline in the southern connected MDB than the actual *quantity* of flows.
2. *Land-use patterns* that intercept water before it ever reaches the rivers have contributed to the situation.
3. *Natural-resource management* issues have contributed to the problem in ways that have little to do with extraction of water from the rivers for irrigation.
4. It could be the *variation* in watering that matters—the timing, depth, duration, and frequency of flooding in wetlands, and the variation in river heights for in-stream and riparian biodiversity—not just the average flow per year, in the '*land of droughts and flooding rains*'.

So getting the average flow per year right is an important first step but the results depend on *how* the water is applied. That will not be known until the Environmental Watering Plan is finalised, and detailed on-the-ground implementation practices have been agreed on. This crucial part of the whole exercise has barely been discussed yet.

First is the matter of the *timing of water flows* compared with the quantity of water extracted from the rivers. Most non-specialists do not understand that the timing of water flows in the *southern Basin* is the major contributor to environmental decline, because water flows in the southern Basin have been re-engineered to reflect the *seasonal* requirements of irrigated agriculture downstream. In the headwaters of the Murray River, winter rains and snow are stored in dams and then released during the summer when needed for agriculture. This re-engineering has been a significant contributor to environmental decline in the southern MDB, because the flora and fauna of the MDB had evolved to flourish in rhythm with the natural seasonal flows, which have now been inverted. As the MBDA (2010:610) states:

> The ecological values of the Barmah Millewa forests have been threatened by several factors, but the *main* impact has been through river regulation, particularly through a decrease in medium-sized spring floods and an increase in small summer floods (MDBC 2006). Similarly GHD (2009) report that river regulation has been implicated in the steady decline of ecosystem health in the NSW Central Murray State Forests over the past 75 years—*inappropriate flood regimes have been the main adverse impact on the ecological character of the site*. [Emphasis added]

Second, land-use patterns that intercept water *before it reaches the rivers* are a factor in environmental decline in the MDB—in both the northern and the southern Basin. Farm dams are filled by intercepting rainwater that would have flowed into rivers. Overland flows are captured in southern Queensland and northern New South Wales. 'Stock and domestic' (S&D) extractions in most of the Basin are not metered, or charged for. So as water has become scarcer, the incentive for riparian users to take as much free water as possible under the S&D provisions becomes stronger. Plantation forestry in the upper catchments is also intercepting significant quantities of water before it ever reaches the rivers. The Guide recognises this—the MDBA has attempted to include 'interceptions' in its technical calculations—but the focus remains on reclaiming water for the environment from those who extract it from the rivers for irrigation. A problem for the MDBA (and the whole Commonwealth strategy) is that it has very limited powers under the *Water Act* to control interception and land use; it remains up to the States. Hopefully, the MDBA will require the States to address this between 2014 and 2019, to gain approval of their next State plans.

Third, natural-resource management problems are a factor in environmental decline in the MDB—quite independently of the extraction of water by whatever means. Failure to take this factor into account has often meant that environmental problems have been attributed to the extraction of water for irrigation, when they might be due to other problems altogether. Many of the rivers assessed as having poor or very poor environmental health are subject to little extraction for irrigation, but are infested with feral animals, such as European carp or trout—as in the Ovens River in the figures above. Feral animals have also caused widespread damage in wetlands—for example, pigs and goats in the Macquarie Marshes. The greatest threat to the three Hydrological Indicator Sites in the Lower Lachlan River (such as the Booligal Wetlands) could be salinity, not lack of river flows. The SDL 3000 scenario in the Guide proposes to reduce irrigation extraction in the Lachlan by 44 GL, from current diversion levels of 302 GL to 258 GL—that is, from 17 per cent of estimated long-term average flow to 14.7 per cent; and it would increase the environmental flow by the same amount—that is, from 65 per cent to 67 per cent of long-term average flows. The extra 2 per cent is unlikely to 'save the wetlands' if nothing is done about the salinity threat.

A strategy that overlooks these other factors is incomplete. In fact, such a strategy will simply not be able to reverse the environmental decline in the MDB if it is not actually targeting all of the real causes of the decline. Some hope that the strategy will reverse enough of the environmental decline to make a significant impact. Perhaps, but no systematic and systemic studies that quantify the different impacts of the various causal factors underpin the current strategy. Indeed, the current strategy could actually exacerbate the environmental decline in certain parts of the MDB; by increasing water flows in some rivers, it could cause further proliferation of pests, or erosion, or salinity, which are the actual problems. In fact, for at least half of the 18 HIS that the MDBA focuses on, lack of flow volume is not the only or major threat. There is no available assessment yet of whether lack of flow is the greatest threat to the 2442 wetlands the MDBA has identified as the 'key' sites that may not be 'compromised'. Hopefully, it has not simply been assumed.

All these quibbles about the MDBA's plans could be just reactionary hype by rent seekers and lobbyists, as some say, including Henry Ergas. Proponents of the current strategy could argue that it is not intended to be 'cost effective'. But proponents should be able to argue convincingly that the strategy will be effective, and that perverse outcomes can be avoided.

Conclusions

This chapter does not criticise the MDBA for its recent Guide; it was doing precisely what it was instructed to do by the *Water Act*. It is not an 'attack' on environmental scientists or conservation groups who want to see healthier ecosystems in the MDB; I share that goal. It does not 'attack' irrigators or rural interests concerned about their future (even if claims of impending catastrophe are exaggerated).

The central tenet is that this whole '*Save the MDB*' strategy was based on a glib assumption that reducing irrigators' extractions from rivers in the MDB was necessary *and sufficient* to achieve a healthy, sustainable MDB. It certainly is necessary (but not equally in every valley) but it is certainly not sufficient.

If Australian governments and communities are serious about achieving an environmentally, economically and socially sustainable MDB (and I believe we all should be) then we could be wasting our collective time, effort and money on this ill-conceived strategy. I believe that the goal is quite achievable, and could even cost less than $10 billion. But it would have to be implemented in a much smarter way, particularly by working very closely with the people on the ground who understand the complexities of the system and know what is feasible and necessary. This would probably include the design of ecosystem-service institutions capable of delivering environmental benefits as well as livelihood outcomes.

It is not too late to improve the Turnbull/Howard/Rudd/Wong strategy for the MDB to ensure it will indeed address all the differentiated, complex problems in different parts of the Basin. I argue that the key to achieve this is to start with the knowledge and energy of the people on the ground, engaging them in a serious, continuing exploration of what is feasible, workable and acceptable. This is not to suggest every valley or CMA may just do whatever it wishes. Rather governments should agree on a broad national framework within which the local and regional communities may propose workable solutions, supported by the best scientific advice they can get, from all sources.

Bibliography

Australian National Audit Office (ANAO) 2010, *Performance audit: administration of the Water Smart Australia program*, Auditor-General Audit Report No. 21 2009–10, Australian National Audit Office, Barton, ACT, <http://www.anao.gov.au/uploads/documents/2009-10_Audit_Report_21.pdf>

Australian Public Service Commission 2007, *Tackling Wicked Problems: A public policy perspective*, Australian Public Service Commission, Canberra, <http://www.apsc.gov.au/publications07/wickedproblems.pdf>

Campbell, A. 2010, 'Rethinking the basin plan', *Crikey*, 16 November 2010, <http://Blogs.crikey.com.au/rooted/2010/11/16/Murray murmurings>

Ergas, H. 2010, 'Water plan will cost a tsunami', *The Australian*, 12 November 2010, p. 16.

Ison, R. 2007, 'New plan, same old thinking', *New Matilda*, 7 February 2007, <www.newmatilda.com>

Kelly, P. 2010, 'Mighty river of trouble heading Gillard's way', Editorial, *The Australian*, 13 October 2010.

Murray–Darling Basin Authority (MDBA) 2010, *Guide to the proposed Basin Plan.Volume 1: Overview*, Murray–Darling Basin Authority, Canberra.

Myers, P. 2010, 'War over water will produce no winners', *Sydney Morning Herald*, 4 December 2010.

Productivity Commission 1999, *Implementation of ESD by Commonwealth Departments and Agencies*, Productivity Commission, Canberra.

Productivity Commission 2010, *Market Mechanisms for Recovery of Water in the Murray Darling Basin*, Productivity Commission, Canberra.

Robins, L. and Dovers, S. 2007, 'Community-based NRM boards of management: are they up to the task?', *Australasian Journal of Environmental Management*, vol. 14, no. 2, pp. 111–22.

25. Planning as Performance: The Murray–Darling Basin Plan

Ray Ison, Philip Wallis

The Emerging Performance

The performance that is emerging following the release of the *Guide to the proposed Basin Plan* at 4 pm on Friday, 9 October 2010 has been scripted since the creation and passage of the *Water Act 2007* (Cwlth) and the associated creation of the Murray–Darling Basin Authority (MDBA).[1] What is unfolding is a tableau that is the product of the structural determinism of its design.[2] In this chapter, we first characterise certain features of the emerging performance following the release of the planning Guide. These features are then set within some of the structural determinants of the current policy/planning design. We then take a step back and address the question of relevance of framing planning as a 'performance'. Arguments are then mounted for utilising this framing in a climate-change world—that world in which the enactment of the Basin Plan has to proceed and in which future cycles of planning will happen. We conclude with some suggested policy and practice initiatives that, from our choice of framing, seem warranted right now.[3]

Even by Australian standards, the media has been in a mild frenzy since the release of the Guide. It is not our purpose to draw out the significant threads of all that has been said already—nor do we wish to fully recapitulate the main features of the Guide. We leave that to others. Instead, we employ the metaphor of planning as performance as an analytical device to make sense of what is happening. Our organising image of a performance (Morgan 1986) is that of an orchestra or jazz band, though other forms of performance might offer equally relevant insights. We also employ a metaphor analysis—albeit selective—to drill down into the musical score to gauge some of the underlying notes that give rise to the melody. As all metaphors reveal and conceal, so the metaphors we employ have the same caveat (McClintock et al. 2003, 2004).

1 The current scene from the performance actually started days earlier with leaks to the media and other bodies.

2 A tableau is a striking scene or picture; structural determinism refers to the understanding that whatever happens to a system is a result of its present structure and is determined by it.

3 Frames are used to negotiate the complexity of the world by determining what requires attention and what can be ignored. A frame is the context through which a person interprets the world.

The most significant feature of the new performance is that the music has been reset to favour one set of players over others—those who have historically played the tune. The new arrangement privileges those who play for the environment over those who have historically played for agriculture, the economy or other human needs. Importantly, though, the environment is generally anthropomorphised and, in language at least, seen as a player in its own right (a form of ecological determinism?). Of course, the preferences and values of public servants, members of both parties in the Federal Parliament, interest groups, scientists and so on have composed the new arrangement and put it into play. In addition, just because a new arrangement has been created does not guarantee it will ever be performed, let alone performed well. To mix metaphors, there is still a lot of water to go under the bridge.

The performance that has emerged in the weeks following the release of the Guide is characterised by hyperbole, extravagant rhetoric, genuine fears, contestation, incipient conflict and poor listening. The public forums are characterised more by debate (literally, to put down) than dialogue (when meaning runs through) (Kersten and Ison 1998). Jamie Pittock (2010), on one side of the argument, claims that the Basin Plan does not go far enough in returning water to the environment. He suggests that 'if the basin plan does not faithfully implement Australia's obligations under the [Ramsar] convention, wetland conservation activists could seek redress in the High Court', thus hinting at conflict to come. Images of nooses around irrigators' necks were a feature of the first public meeting at Shepparton (Ker 2010), as was the burning of the Guide outside a packed public meeting in Deniliquin. The Australian Conservation Foundation (ACF 2010) argues that 'a healthy river needs more water, but the same rule doesn't always apply for agriculture. Between 2001 and 2007 dry conditions meant irrigation industries used about 70 per cent less water, and the economic value of irrigation production fell by only 0.12 per cent'.[4]

The other main feature of this phase of the performance is the extent to which 'consultation' is being pursued as the main—perhaps only—form of stakeholder engagement.[5] This has been largely structured into the act—though more creative engagement processes might legitimately have been pursued. The theoretical implications and limitations of consultation are discussed by, amongst others, Collins and Ison (2009).

Some performance metaphors can be gauged from an article by Ross Gittins (2010) in the Fairfax press. Recognising that a metaphor takes the form X *as* Y or X *is* Y then some of the main metaphors in this article can be deduced

4 The point has been made that these average data mask trends and local variations and ignore the impacts of debt loads.

5 Reports about the framing of community engagements vary—it is described in some reports as consultation but in others as 'information provision sessions'; this could reflect a lack of clarity about purpose.

(Table 25.1). All metaphors have theoretical entailments that can be understood as taken-for-granted assumptions that are held within a metaphor-in-use. Some of these are described in Table 25.1. Entailments can be enabling or disabling but it is only by making them explicit that we can discern the role they play and how this differs with context.

Table 25.1 Some of the main metaphors employed by Gittins in his article entitled: 'Don't think you can keep on neglecting me, Darling'

Key concept	Metaphors	Entailments
Sustainability as	Dangerous	Threatens assumptions
	Irresistibly attractive	?
	Having a wonderful ring to it	?
	Dripping with virtue	?
Environment as	Able to fail	Something static
	Sustainable	Something static?
	Natural	Humans are not natural
	Abusable/saveable	A product of human design
Ecosystems as	Healthy	Sickness and health knowable
	Having tipping points	Behave as complex adaptive systems
	Like flogging a horse	Can be killed
Politics as	Heads in the sand	Not open and adaptive
	Exaggerating claims	Rhetorical practice
Country towns as	Declinable	Viability knowable

By unpacking our performance metaphor further, it makes sense to ask: what are the elements that, together, give rise to a performance? Well, the instruments could be classed as: i) governance mechanisms (for example, regulations, legislation, market mechanisms, consultation, education, information provision); ii) institutions/social technologies (see Ison 2010b) (examples include the Basin Plan, the Guide, the *Water Act*, the minister); iii) organisations, such as the Murray–Darling Basin Authority (MDBA), the Council of Australian Governments (COAG), themselves made up of networks of institutions; and iv) theories used knowingly, or not—sometimes taking the form of ideologies. The performers include public servants, water professionals, scientists, economists, researchers, modellers, MDBA board members, politicians (though, as seen later, some deny this at present), interest groups and, depending on your perspective, the consulted.[6] Citizens participate vicariously, mediated by a plethora of theatre critics who write the reviews or offer their perspective on the media.

6 Richard Price points out that the media—that is, journalists—could also be considered a major player.

Who might be regarded as the audience? To be effective, most performances require audiences who continue to be satisfied. Those affected but not involved could be seen as part of the audience—the voting public perhaps—differentiated into rural citizens and city citizens perhaps, as well as international observers.

Structural Determinants of the Current Performance

Asking who the conductor is, or might be, reveals some of the main structural determinants of the current performance. One observation is that there is clearly a lot of deflection of responsibility between the Government and the MDBA over the Basin Plan. In the Guide, the MDBA (2010: iii) repeatedly asserts how it perceives its role:

> While the Murray–Darling Basin Authority (the Authority) is charged with developing a Basin Plan for the Minister's consideration, this occurs within the framework of the Water Act 2007 (C'wlth). The Commonwealth Parliament in 2007 and 2008 clearly laid out the general objectives of the Water Act, and prescribed how the Basin Plan was to be developed.

The MDBA then further spells out that the Government has twice decided to develop a plan for the Murray–Darling Basin (MDB), with support from both sides of Parliament and the Basin States. It also repeatedly points out the prescriptive nature of the *Water Act 2007* (Cwlth). This is especially important in setting the sustainable diversion limit (SDL) options, as they consider less than 3000 gigalitres or more than 4000 GL back to the environment 'will not meet the requirements of the Water Act' (MDBA 2010:xxi). Any less than 3000 GL would not serve environmental needs, while any more than 4000 GL would not optimise economic and social outcomes. In this case, the 'scientific' range is 3000–7600 GL. The proposal in the Guide is summarised in gross terms in Figure 25.1.

Meanwhile, the Water Minister, Tony Burke, chose to highlight the independence of the MDBA (*Lateline*, ABC TV, Friday, 8 October 2010):

> [T]he decision on the release is made by an independent authority. No minister tells them what to do. No minister should tell them what to do. It's their role to conduct an independent consultation and what they've brought out today is not 'the Basin Plan'. What they've brought out today is a guide to a draft of the Basin Plan.

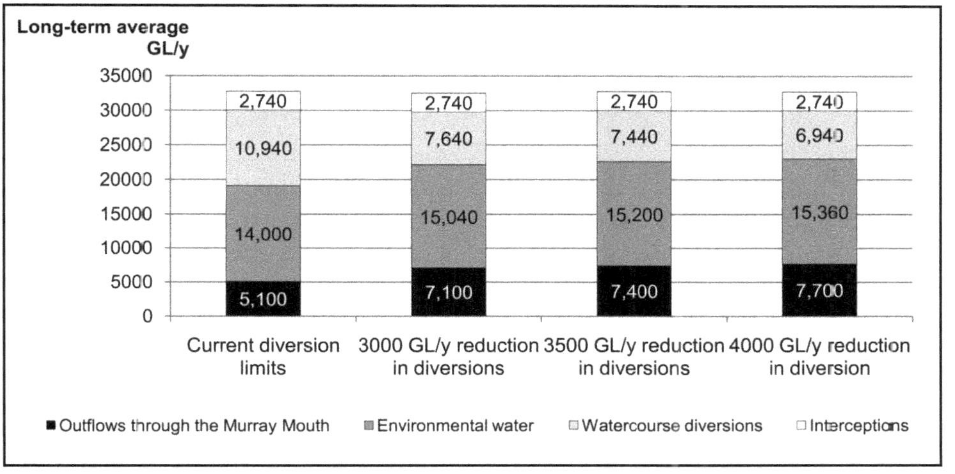

Figure 25.1 Murray–Darling Basin Guide proposals for whole-of-basin adjustments

Source: Adapted from The Weekly Times, 13 October 2010

This position, it can be argued, shows a lack of clear leadership (and responsibility) from the Government. And the MDBA is hardly being the frank and fearless public service that former National Water Commission head Ken Matthews argued for upon his retirement (Keane 2010; Mathews 2010). At the time of the Guide's release, the media gave a lot of oxygen to 'angry farmers', as controversy sells papers. We argue that these are more than superficial issues. They are structural issues connected with the original *Water Act* and thus the MDBA design.

And in the wings await the various State interests. Victorian Water Minister, Tim Holding, has called 'for an end to simply stripping basin communities of their most valuable asset…We have concerns with what has been proposed by the MDBA and the impact this would have' (*The Weekly Times*, Wednesday, 12 October 2010:1). SA Premier, Mike Rann, has been reported welcoming the Plan, saying that it will 'overturn a century of greed' (Jones and AAP 2010).

The portrayal of what is at stake in Figure 25.1 masks the performance element of the ongoing governance of the MDB even under existing arrangements—the implementation of the Plan if you like. In performance terms, water is clearly not divisible into different forms of water—for example, a particular release could contain both irrigation and environmental water. And presumably, environmental water will become subject to the same regimes as other forms of water in terms of efficiency, monitoring, and so on. The institutional complexity that could arise might be enough to undermine the whole performance.

There are also valid questions that can be asked about the nature and boundaries of the theatre: is it a biophysical boundary, an economic boundary, a sustainable-population boundary, a rural-livelihoods boundary, or some combination of these? In the Australian context, the 'performance space' is made complex by the historical as well as contemporary aspects of federalism and the need to coordinate multiple performances across horizontal and vertical spatial dimensions, not to mention temporal dimensions for which current organisations and institutions are poorly designed (Figure 25.2).

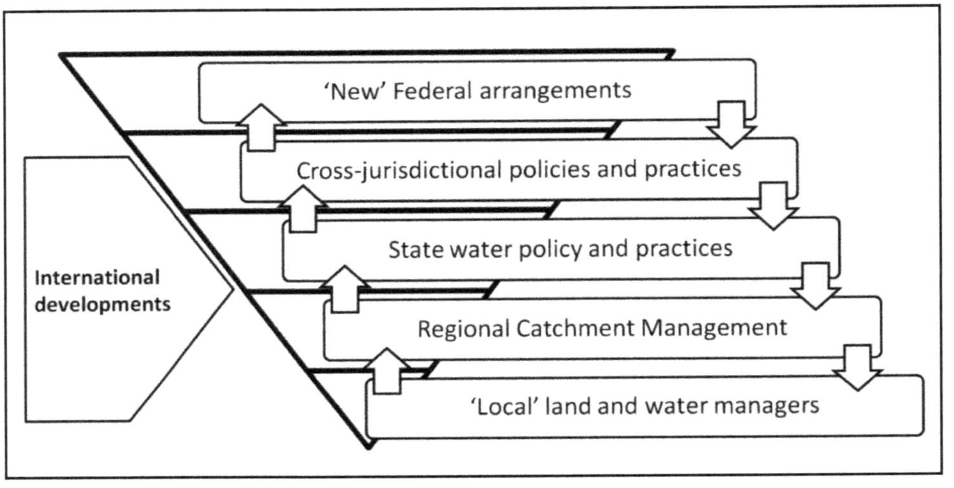

Figure 25.2 A model of the 'performance space' in which the Murray–Darling Basin Plan has to be enacted

Drawing on Ison et al.'s (2007) analysis of typical environmental-governance arrangements, it is apparent that the *Water Act 2007* (Cwlth) is framed on an assumption that there is a known or knowable problem that will remain relatively static over time. Further, it is assumed that such problems are best addressed by regulation (usually understood as command and control), fiscal or market mechanisms and the provision of information in attempts to educate stakeholders through largely one-way consultation processes. Historically, 'stationarity'—the idea that natural systems fluctuate within an unchanging envelope of variability—is a foundational concept that permeates training and practice in water-resource engineering, but as Milly et al. (2008) argue, 'stationarity is dead and should no longer serve as a central, default assumption in water-resource risk assessment and planning. Finding a suitable successor is crucial for human adaptation to changing climate.' They further argue that 'climate change undermines a basic assumption that historically has facilitated management of water supplies, demands, and risks' (Milly et al. 2008:573–574). Climate-change adaptation—if framed as a 'wicked problem' (APSC 2007)—also undermines a position that assumes that water governance and management are problems of the known or knowable type and thus raises questions about the ongoing utility of traditional governance mechanisms (Godden and Ison 2010; Ison et al. 2007).

Performances built on stationarity and fixed knowledge forms give rise to systematic (that is, linear, step-by-step) practice rather than systemic practice that is relational, recursive and circular, and characterised by learning and adaptation (Ison 2010b).

Why Focus on Performance?

Performances, if they are effective, are intrinsically systemic because they give rise to a relational, in contrast with a linear, dynamic—for example, performer with audience, with other performers, with conductor, and so on. Performances also necessitate the building of relational capital, which results from the interactions between the other forms of capital, including natural, social, artificial and human—all of which are systemically connected.

Relational capital is precious; it is hard to build but easy to destroy. The practices of the Australian Public Service (APS) traditionally run counter to the cultivation and conservation of relational capital. Examples are legion within the public sector of the undermining of joined-up practice by the intentional and unintentional undermining of relational capital of the sort that creates ongoing effective performances and that is central to social learning (Ison and Wallis 2009; Ison et al. 2007).

There is also the question of performance in the APS. In July 2009, Lynelle Briggs, the then APS Commissioner, argued the need for[7]

- removing unnecessary obstacles to innovation, to improve the quality of outcomes in complex and uncertain policy areas
- developing more variegated accountability and performance management arrangements, better suited to new modes of policy implementation

She also made the case for developing more horizontal accountability mechanisms (a form of horizontal governance) and the need for skills and capabilities for APS staff in

1. problem framing and boundary setting
2. generating fresh thinking on intractable problems
3. working across organisational and disciplinary boundaries
4. making effective decisions in situations with high levels of uncertainty

7 See <http://www.apsc.gov.au/media/briggs150709.htm>

5. being able to tolerate rapid change in the way problems are defined
6. engaging stakeholders as joint decision makers (not just providers or recipients of services).

These capabilities seem in short supply at the moment and the institutional arrangements far from conducive to their enactment (Ison and Wallis 2009). Westminster-style governance performance leaves a lot to be desired. For example, Ringen (2009) reported on a major study looking at what the UK New Labour Party achieved in terms of its own social-policy objectives over the period 1997–2007.[8] He studied the flagship policies of child poverty, education, social justice and health and found that they had achieved 'absolutely nothing'. His study provides strong evidence for the systemic failure of UK governance by highlighting the problems that emerge when governments adopt a command-and-control approach and fail to mobilise citizens or stakeholders in policy development and implementation. His sobering conclusion is that no UK Government, of any political persuasion, can currently get done what it is elected to do.

Ringen's findings illustrate a situation that can be understood as a 'structure-determined system'. Not only governments are constrained by the system in which they operate. Take, for example, utility companies that deliver social goods, such as water or energy. Most now have as a main measure of performance the profit derived from sales of water or energy. The system is not structured to recognise that in today's world the main social benefit from water and energy comes from how little water or energy is used and the efficiency of its use. We create measures of performance that conserve particular structural relations that give rise to certain forms of organisation. Only by inventing new organisations, comprising different structural relations, can we break out of the constraints of particular structural determinisms.

Performance begins at the interpersonal level; it entails choreography of the emotions (Russell and Ison 2005). For example, research into group functioning and effectiveness has shown that informal contracting prior to starting group processes enhances performance and sets a more positive emotional dynamic. The following type of informal contract has been effectively used in participatory research (Ison et al. 2009) and has the potential to be applied more broadly to public participation in the MDB.

- Provide others with the experience of being listened to.
- Adopt behaviour that checks out your own understandings and assumptions first.
- Appreciate the diversity of experiences and perspectives.

---

8 See <http://www.youtube.com/watch?v=AHcfNy1_zqA&feature=related>

- Feel comfortable asking questions or saying you do not know.
- Agree that who says what stays here.
- All participants take responsibility for monitoring this contract.

This process needs to be adapted to context and this particular set of agreements will not be valid in all settings. An analysis of Barack Obama's approach to political practice suggests some other design considerations for more effective interpersonal performances (see Ison 2010b, based on Freedland 2008)

- encountering of the other as a legitimate other
- predisposition to learning (which in itself is a way of abandoning certainty)
- capacity for listening—such that he creates for those in the conversation the experience of being actively listened to
- capacity and technique of 'mirroring back' his understanding of the position of others
- understanding and valuing multiple perspectives in respect to a situation or issue of concern
- ability to move between different levels of abstraction and to synthesise different strands of an argument
- awareness that change comes through relationships
- ability—knowingly or not—to be both systemic and systematic
- use of diagrams as a 'mediating object' in his practice.

Better performances need to be designed and sustained for living and governing in a climate-change world.

Recasting the Current Performance: Some options

What should we turn to if there is a systemic failure of this public-policy process? We suggest the following.

- Reframe the problematique, as a central part of inventing systemic and adaptive governance for managing a co-evolutionary dynamic, as a purposefully designed 'learning system' organised as an ongoing systemic inquiry (Ison 2010ab). To do this involves framing the Murray–Darling Basin and its future as a coupled socio-ecological system.
- Reorganise future water governance as a contribution to innovation in 'horizontal governance' (see Ison 2010a).

- Invest in social learning (Collins and Ison 2009) as a means to generate ongoing (real-time) effective performances.

Following Giddens (2009:8), who argues that 'to develop a politics of climate change, new concepts are needed', Ison (2010b) puts forward *systemic inquiry* as both a practice and a potential institution better able to be employed in situations of complexity and uncertainty. Systemic inquiry is designed for the governance and management of uncertainty. One might equally add that, to develop a praxis of water governance as part of climate-change adaptation, other new, conducive institutional arrangements are also needed.

In the face of the uncertainties, complexities, interdependencies and multiple stakeholdings in the MDB, an approach to its governance and management is needed that is adaptive and contingent. In such situations, a national, systemic inquiry could have been chosen as an alternative governance mechanism instead of a traditional regulatory, legislative and planning approach. An effectively constituted systemic inquiry—after all, the issues are unlikely to go away in the short to medium term, if ever—could become a vehicle for the deployment of social-learning approaches and the adoption of systems practices.

As noted earlier, current forms of governance in most Western democracies—despite their many strengths—are not well suited for managing long-term, complex issues (Ison 2010a). Helen Ingram (2008), who has long experience of water governance, argues:

> Attempts to design improved water resources management and institutions must attend to context. Standardised reforms have failed time after time...In general, clumsy solutions that embrace multiple perspectives and appeal to different kinds of logic are preferable... mixed strategies that appeal to different ways of knowing are likely to be more effective (p. 17).

Given the governance we have at the moment what would we recommend starting from current circumstances? What next?

- Create a cabinet-level, interdepartmental 'water committee'—make water everyone's business—that is better able to deal with the complexities of continuing water-governance reform (and remain cognisant that water is and will remain a key strategic issue on the Australian continent).
- Institutionalise a national, *systemic inquiry* into water and energy governance in a climate-changing world. Such an innovation would address the limitations of the three-year electoral cycle, which is too short for long-term natural-resource and climate-change issues. Such an innovation is warranted in what is now a period new to human history (Ison 2010a). Features of the

UK Climate Committee and the UK Royal Commission on Pollution could help in designing such an institution.

- Use the current political refocus on regional Australia to address the policy vacuum that has developed in Australia around rural and regional futures. This needs to be understood as a process of exploring systemic opportunities framed as 'livelihoods' and not industries, sectors, farms, and so on.
- Invent a range of new ecosystem services that adds to the livelihood mix of current and future rural inhabitants.
- Reform—on a coherent, national basis—catchment management authorities (CMAs) as institutions able to manage a coupled 'socio-ecological system'. Managing catchments as coupled socio-ecological systems requires recognition of the systemic interconnection of humans to their environment if an ongoing effective performance is to be created. In this context, planning is a form of social technology that mediates these connections.

Bibliography

Australian Conservation Foundation (ACF) 2010, Giving a lifeline to our Murray–Darling, Email to supporters, 13 October 2010, Australian Conservation Foundation, Melbourne.

Australian Public Service Commission (APSC) 2007, *Tackling Wicked Problems. A public policy perspective*, Australian Government/Australian Public Service Commission, Canberra.

Collins, K. B. and Ison, R. L. 2009, 'Jumping off Arnstein's ladder: social learning as a new policy paradigm for climate change adaptation', *Environmental Policy & Governance*, vol. 19, no. 6, pp. 358–73.

Foerster, A., Ison, R. L. and Godden, L. 2008, Systemic and adaptive water governance: reconfiguring institutions for social learning and more effective water managing?, Unpublished paper for a program of research prepared for a seminar to explore the role of social learning in water policy and law, Friday, 5 December 2008.

Freedland, J. 2008, 'The Obama story. The improbable journey', *The Guardian*, Thursday, 6 November 2008, pp. 1–8.

Giddens, A. 2009, *The Politics of Climate Change*, Polity, Cambridge, UK.

Gittins, R. 2010, 'Don't think you can keep on neglecting me, Darling', *National Times*, 13 October 2010, <http://www.smh.com.au/opinion/politics/dont-think-you-can-keep-on-neglecting-me-darling-20101012-16ho4.html>

Godden, L. and Ison, R. L. 2010, 'From water supply to water governance', in M. Davis and M. Lyons (eds), *More Than Luck: Ideas Australia needs now*, Centre for Policy Development Limited, Haymarket, NSW, <http://morethanluck.cpd.org.au/about/>

Ingram, H. 2008, Beyond universal remedies for good water governance: a political and contextual approach, Paper presented at the Sixth Biennial Rosenberg Water Policy Forum on Water for Food: Quantity and Quality in a Changing World, Zaragoza, Spain, <http://rosenberg.ucanr.org/documents/V%20Ingram.pdf> pp.1-21.

Ison, R. L. 2010a, 'Governance that works. Why public service reform needs systems thinking', in M. Davis and M. Lyons (eds), *More Than Luck: Ideas Australia needs now*, Centre for Policy Development Limited, Haymarket, NSW, http://morethanluck.cpd.org.au/more-than-luck/governance-that-works/

Ison, R. L. 2010b, *Systems Practice: How to act in a climate-change world*, Springer, London.

Ison, R. L. and Wallis, P. 2009, Submission to the Advisory Group on the Reform of Australian Government Administration, Department of the Prime Minister and Cabinet, Canberra, <http://www.dpmc.gov.au/consultation/aga_reform/pdfs/0074a%20SAGP%20Monash%20University.pdf>

Ison, R. L., Röling, N. and Watson, D. 2007, 'Challenges to science and society in the sustainable management and use of water: investigating the role of social learning', *Environmental Science & Policy*, vol. 10, no. 6, pp. 499–511.

Ison, R., Russell, D. and Wallis, P. 2009, *Adaptive water governance and systemic thinking for future NRM: action research to build MDBA capability*, Monash Sustainability Institute Report 09/4, Melbourne.

Jones, H. and Australian Association Press (AAP) 2010, 'Rann: time to end the greed', *Border Mail*, 9 October 2010, p. 11.

Keane, B. 2010, 'Public service "docile and unassertive"', *Crikey*, Thursday, 7 October 2010, <http://www.crikey.com.au/2010/10/07/time-to-shake-up-our-docile-and-unassertive-public-service/>

Ker, P. 2010, '"Rope around our neck", cry frustrated farmers', *The Age*, 13 October 2010, <http://www.theage.com.au/victoria/rope-around-our-neck-cry-frustrated-farmers-20101012-16huc.html>

Kersten, S. and Ison, R. L. 1998, 'Listening, interpretative cycles and dialogue: process design for collaborative research and development', *The Journal of Agricultural Education & Extension*, vol. 5, pp. 163–78.

McClintock, D., Ison, R. L. and Armson, R. 2003, 'Metaphors of research and researching with people', *Journal of Environmental Planning and Management*, vol. 46, no. 5, pp. 715–31.

McClintock, D., Ison, R. L. and Armson, R. 2004, 'Conceptual metaphors: a review with implications for human understandings and systems practice', *Cybernetics and Human Knowing*, vol. 11, no. 1, pp. 25–47.

Mathews, K. 2010, There's a telegram for you..., Valedictory address by Ken Matthews AO on his retirement from the Australian Public Service, 6 October 2010, <http://www.nwc.gov.au/resources/documents/Oct_6_-_VALEDICTORY_2010.pdf>

Milly, P. C. D., Betancourt, J., Falkenmark, M., Hirsch, R. M., Kundzewicz, Z. W., Lettenmaier, D. P. and Stouffer, R. J. 2008, 'Stationarity is dead: whither water management?', *Science*, vol. 319, no. 5863, pp. 573–4.

Morgan, G. 1986, *Images of Organization*, Sage, London.

Murray–Darling Basin Authority (MDBA) 2010, *Guide to the proposed Basin Plan*, Murray–Darling Basin Authority, Canberra.

Pittock, J. 2010, 'A watery plan of compromises', *The Drum Unleashed*, Australian Broadcasting Corporation, Sydney, <http://www.abc.net.au/unleashed/40026.html>

Ringen, S. 2009, *The Economic Consequences of Mr Brown: How a strong government was defeated by a weak system of governance*, The Bardwell Press, Oxford.

Russell, D. B. and Ison, R. L. 2005, 'The researcher of human systems is both choreographer and chorographer', *Systems Research and Behavioural Science*, vol. 22, pp. 131–8.

26. Enhancing Collaborative Management in the Basin

Katherine A. Daniell

> Appropriate policy in a democracy is determined through a process of political debate. The right course of action is always a matter of choice, never of fact.
>
> — Davidoff (1965:331)

Introduction

This chapter examines the potential benefits and costs of enhancing collaborative planning and management practices in the Murray–Darling Basin (MDB), relative to the current approach being taken at a federal level since the creation of the *Water Act 2007* and the Murray–Darling Basin Authority (MDBA). In particular, the opportunities and challenges of increasing stakeholder engagement in water reform and planning processes will be investigated. A brief analysis of knowledge and expertise available for organising and implementing engagement activities to support effective collaborative planning and management practices will be provided, along with suggestions on how such knowledge and expertise might best be used to enhance the current reform process.

Challenges of Water Management: The need for collaborative approaches

Each person values water in different ways. Many sources of water are likely to be valued concurrently for providing basic life-support functions such as for drinking, sanitation and food production, as well as for maintaining ecosystem health, a range of economic livelihoods and personal, cultural and spiritual wellbeing. Given these multiple and competing values for water—in particular, in a world of rapidly increasing population, environmental degradation and globalised telecommunications—the management of water is an increasingly political process. Recognising the political and value-based nature of water management is the key to successfully developing institutions and types of governance that allow the complex, uncertain and conflict-ridden situations seen in today's water basins to be successfully navigated (Syme and Hatfield-Dodds 2007). The political nature of contemporary water management is often characterised by a process of deciding how available water should or could be used and shared between a variety of stakeholders and the environment, and conflict resolution amongst these stakeholders[1] (Delli Priscoli and Wolf 2009).

Under such conditions, traditional forms of centralised technocratic or 'command-and-control' management and reliance on engineering solutions and technologies are now typically considered insufficient for most water systems (Gleick 2000). One of the key reasons for this is that it is increasingly rare that individuals or governments have the capacity to make and implement their own water-management decisions without the help of other stakeholders, as power and resources for managing water systems are increasingly distributed. For this practical reason and other key reasons such as promoting thriving democracy (Dryzek 1990; Fischer 1990, 2000), it is imperative, rather than optional, that the development of water policies and their management and implementation plans be developed in an inclusive and collaborative way with a full range of management agencies, community stakeholders and members of the scientific community (Dietz et al. 2003; Loucks 1998; Thomas 2004). In this way, plans can be based on the best available scientific and stakeholder knowledge, with values-based decisions and management agreements collectively negotiated to ensure sufficient stakeholder 'buy-in' and capacity to successfully implement them.

This need to acknowledge the key roles that values play in water management and the need for a participatory or collaborative approach are reiterated in almost every recent well-known international water document, such as the *Dublin Statement*, which outlined that '[w]ater development and management should be based on a participatory approach, involving users, planners and policy makers at all levels' (ICWE 1992). Improving the uptake of this approach has been hampered in some quarters, however, by technocracies unwilling to give up some of their power for the benefit of water systems, and by a range of other issues such accountability, legitimacy and lack of understanding of how 'participatory' or collaborative approaches can be implemented (Tan et al. 2008). As Ingram and Schneider (1999: 27) stated more than 10 years ago, 'The most fundamental flaw in contemporary water policy is that many value questions in which ordinary citizens have a great interest, are being framed as technical questions.' Failing to understand the importance of developing inclusive stakeholder water-management processes and plans that are not only scientifically validated but broadly stakeholder legitimated (Landry et al. 1996)—and enabling institutional infrastructure and facilitating environments to support these processes—has been seen as one of the main routes to inadequate plan implementation, continued water-system damage and conflict (Delli Priscoli 2003).

---

1 Stakeholders are considered as people, institutions or organisations that have a stake in the outcome of decisions related to water management, as they are directly affected by the decisions made or have the power to block or influence the decision-making process (Nandalal and Simonovic 2003).

Recent Reforms in the Murray–Darling Basin: A move to centralised control

The bipartisan-supported federal water-reform process linked to the *Water Act 2007*, the creation of the MDBA and the $12.9 billion Water for the Future plan presents a historic opportunity for investing in improved water-management practices, infrastructure and monitoring across Australia. Following a long line of management reforms in the MDB (see Connell 2007)—the more recent of which have made efforts to address issues of over-allocation of basin water resources and to restore environmental health to key environmental assets—these new reforms seek, in part, to overcome perceived impasses and protracted negotiations that occurred in the Murray–Darling Basin Ministerial Council (the political forum invested with powers to make decisions for the Basin as a whole under the 1992 *Murray–Darling Basin Agreement*). To this end, decision-making power for the acceptance of the Murray–Darling Basin Plan, which will set sustainable diversion limits for water withdrawals and interceptions across the Basin, has been invested in a single federal minister. The minister is informed by the work and recommendations of the independent MDBA, a Basin Officials Committee, a Basin Community Committee and the new slimmed-down version of the Murray–Darling Basin Ministerial Council, which the federal minister also chairs.

The first incarnation of these reforms under the Howard Government—in particular, the National Plan for Water Security—marked a significant shift from the direction of previous reforms that had moved to treat the environmental, economic and social issues of water, land and environmental management in the Basin in an integrated manner through both collaborative and market mechanisms. The Howard reforms put the focus instead on centralised authority, infrastructure efficiency improvements and using economic instruments—including water buybacks—in an attempt to ensure water-supply security and, in the process, Australian economic security. With the instalment of the Rudd–Gillard Government in 2008 and its Water for the Future plan, many of the initiatives of the Howard plan were maintained, but the priorities were re-badged as 'taking action on climate change, using water wisely, securing water supplies and supporting healthy rivers' (Wong 2008). Attempts were also made to alter the tone of the reforms from a technocratic and directive federal management intervention to one that reflected a more integrated and cooperative management style. For example, as stated by the then Minister for Climate Change and Water, Penny Wong (2008), at the presentation of the Plan: 'It is imperative for Commonwealth, state and local government[s] to share a common understanding of the problems in water and respond in a comprehensive and coordinated way.'

Working under the *Water Act 2007*—drafted during the Howard administration—the MDBA has, however, taken the decision to pursue a more minimalist approach to cooperation and collaboration with other levels of government, regional management groups and local stakeholders. It appears that under the strict time pressures for the development of the Murray–Darling Basin Plan and the need to base the plan on the 'best available science', the MDBA worked to commission and collate large numbers of studies from consultants and academics, gather data, reports and models from State governments and to go about the synthesis work, analyses and choice of options for the plan in-house, with a minimal amount of external consultation.[2] An objective of such a process design could have been to limit the power of certain stakeholder or lobby groups opposed to any significant change. Whether such an approach is actually capable of achieving desired changes is, however, highly questionable if such groups have enough political power to block or upset the process in other ways (Daniell 2008; Fischer 2000).

Understanding Preliminary Reactions to the MDBA Planning Approach

This centralised technocratic approach that has appeared in practice—although supported by some stakeholders, including environmentalists, economists and government officials, who consider that the only way of restoring health to the Basin's ecosystems is by imposing cuts to water allocations and not allowing 'lobby' groups to intervene in the process—is far from supported by all. In particular, the lack of transparency of the planning process in clarifying underlying assumptions on which synthesis and planning decisions are made, and the common lack of openness to engage in discussions about these assumptions—so that community members and other land and water managers at different administrative levels can understand them—have received widespread criticism from, and distressed many of, the MDB's stakeholders.[3]

Until these reforms, many of the natural-resource management (NRM) processes in the MDB—especially over the past couple of decades—have aimed to be of a predominantly cooperative or collaborative nature (Bellamy et al. 2002; Boully 2004; Margerum 2008; SCEH 2000; Tan et al. 2008). In particular, at a local level (for example, through Landcare and integrated catchment management groups) and at a regional level (for example, through the NRM regional bodies), stakeholders have been encouraged through higher-level management groups (for example, the Community Advisory Committee linked to the former MDB Ministerial Council)

2 See the Basin Plan Knowledge and Information Directory for an overview of commissioned and collated information that has informed the development of the Basin Plan: <http://thebasinplan.mdba.gov.au/bpkid/>

3 See, for example, the comments stemming from the community information session tour: <http://www.mdba.gov.au/communities/latest-news>

and funding programs (for example, the Federal Government's Natural Heritage Trust) to become actively involved in developing knowledge of, support for, and implementation and monitoring of reform programs such as the National Action Plan for Salinity and Water Quality and The Living Murray First Step. None of these programs has been immune to criticism, and a number of authors have suggested how the engagement mechanisms and governance arrangements could be further improved (Bellamy et al. 2002; Crase et al. 2005; Marshall and Stafford Smith 2010). Criticism came as many of these processes did not reach all of the 'ideal' outcomes that these collaborative approaches aim for—such as: increased social and political capital; agreement on information and shared understandings; ending stalemates; developing high-quality agreements; cost-effective decision making; inciting learning and change beyond the original participating stakeholders; driving innovation; creating a cascade of changes in attitudes, behaviours and actions; and fostering institutions and practices that involve flexibility and networks (Connick and Innes 2001). Nevertheless, many positive advances were still made with relatively meagre resources relative to the magnitude of the issues facing the Basin, and the governance arrangements were often used in international discussion as a best-practice example of cooperative or collaborative trans-boundary water management (Delli Priscoli and Wolf 2009).

The large injection of funds coupled with the recent reforms (the *Water Act 2007* and the Water for the Future plan) to transfer certain powers to the Federal Government was therefore welcomed by many stakeholders, who saw it as a potential means for strengthening their existing capacity to effect positive change in their regions and communities. Perceptions were quick to change, however, when stakeholders realised that the MDBA was planning to develop the plan largely 'in-house' with seemingly little regard for understanding the existing local, regional and State knowledge, networks and experience. This exclusion process and perceived disrespect for stakeholder knowledge and management experience were key drivers of a range of negative feelings and criticism directed at the MDBA and the Federal Government. For example, Leith Boully, Chair of the previous Murray–Darling Basin Ministerial Council's Community Advisory Committee, in a public lecture at the Australian Academy of Sciences for the Water Management Options for Urban and Rural Australia series in July 2010, expressed her feelings:

> I'm angry that the Canberra machine does not seem to care about the impacts of water reform on real people in real communities. As an irrigator, and member of a community in the Lower Balonne, I no longer feel that the contribution I make to society is valued. Rather, I'm intensely aware of the disdain that the city's chardonnay set treats us with. I'm powerless to do anything about that.

She is certainly not alone, as other State government officials and catchment managers, who are less able to publicly express their opinions, also state in private their concerns and feelings of disempowerment experienced from being largely excluded from the Basin planning process that they care about. In particular, they express their disappointment in not being able to contribute their knowledge or energy to enhance the reform process, and not being able to work through the uncertainties and challenges associated with the new Basin Plan with their constituents to build community understanding.

Curious observers and researchers in a range of water-management, governance and stakeholder-engagement areas have also been watching this reform process unfold with intense interest, as the move back to a technocratic 'command-and-control' type approach and the separation of water from land management—together with the use of market mechanisms—appear very different from currently promoted 'best-practice' approaches to water governance and management outlined at the beginning of this chapter.

A Preliminary Attempt by the MDBA to Enhance Engagement

One of the intriguing aspects of the MDBA's work to date has been the perceived gap between the stated principles of its approach to stakeholder engagement (see Box 26.1)—which appears to coincide with some aspects of what a cooperative or collaborative approach might involve—and what has occurred to date in practice.

Box 26.1 Principles underpinning the MDBA's approach to stakeholder engagement

Our approach to stakeholder engagement is guided by the following principles upon which we will operate and to which we will be accountable.
We are committed to ensure that our stakeholder engagement is:

- **transparent** — we will engage with transparent purpose, goals, accountabilities, expectations and constraints
- **inclusive and targeted** — we will seek to engage with individuals and organisations that represent the full diversity of those who will be affected by the Basin Plan. We will seek to engage Indigenous people and people who have English as a second language in culturally appropriate ways. We will provide opportunities for people with disabilities, including vision and hearing impaired.
- **appropriate and adaptive** — we will use levels and methods of engagement that suit the group being consulted and our strategy will be adaptive to feedback
- **accessible and innovative** — we will provide clear, accessible and comprehensive information to people in order to help them understand their engagement with us
- **respectful** — we will conduct engagement activities in a manner that fosters mutual respect and trust by listening to feedback and responding where possible. We will treat comments and submissions as well as collect and store information in accordance with the Privacy Act 1988(Commonwealth)
- **supportive** — we will be sensitive to how the changes resulting from the Basin Plan impact individuals.

Source: MDBA (2009: 5)

Perhaps sensing the growing frustration of stakeholders who were able to access little information about the advances being made on the Draft Murray–Darling Basin Plan, the MDBA made the decision to release a 'Guide' to the Draft Plan. This initiative—a step not originally planned under the *Water Act*—was seen by the MDBA as an additional voluntary mechanism for engaging stakeholders and giving them a chance to comment on the synthesis and planning option definition it had carried out, as well as to improve its work, before the formal consultation process on the Draft Murray–Darling Basin Plan was held. As stated by the MDBA in its *Guide to the proposed Basin Plan*:

> The objectives of the engagement process for the Guide and proposed Basin Plan are to:
>
> - provide information about the Guide and the proposed Basin Plan
> - give opportunities for people to provide feedback on the Guide, to ensure the proposed Basin Plan is based on the best available information
> - give opportunities for people to provide feedback on, and input to, the proposed Basin Plan, including through a formal submission process. (MDBA 2010)

The objectives of this process were therefore seemingly aimed at providing a basis for simple information exchange, rather than more interactive or collaborative forms of engagement. The manner in which the main part of this information and consultation program was carried out—the town-hall meeting tour and associated call for submissions—has, however, yet to quell many stakeholders' concerns about the planning process. It rather seems to have added to the frustration and anger of some stakeholders, with scenes of people burning the Guide outside a couple of meetings (Franklin 2010). During the meetings many people also spoke emotionally about their fears and anxieties about the way the planning is being carried out, including that the meetings have done little to enhance their understanding of the science and analysis that underlie the planning propositions presented in the Guide. Many found it particularly disappointing that the technical volume of the Guide was not released prior to the meetings, with this 'hiding' of information considered a reason to distrust the science. Some potential reasons for why this particular initiative of information meetings seemed to receive so many hostile reactions will be investigated later in the chapter after a brief exploration of how collaborative approaches can be developed and the choice of stakeholder-engagement methods can be aided.

Existing Knowledge of Developing Effective Collaborative Management and Stakeholder-Engagement Approaches

Considering the need for, and importance of, collaborative approaches and engaging a range of stakeholders from the policy, public and scientific spheres in water planning and management, much research in this domain has occurred over the past three to four decades. Lessons and knowledge about what kinds of processes work in what contexts have been developed from extensive experience and assessment of collaborative management approaches and stakeholder-engagement processes around the world. Throughout a range of academic disciplines and management practices, there has been growing concern about working across boundaries (organisational, cultural, political, administrative, and so on) and developing inclusive and stakeholder-informed decision-making processes. For example, there has been a long history of developing 'public participation' in many spheres of public policy, or more deliberative versions of political involvement of citizens in decision-making and knowledge-creation processes—for example, through studies of 'deliberative democracy' or 'social learning'. Breaking down barriers between professionals and other stakeholders—including service users in both the public and the private spheres—as well as between organisations has also led to much knowledge and expertise that can be found under a range of appellations, including 'stakeholder participation', 'participatory processes', 'institutional coordination', 'adaptive management', 'collaborative advantage', 'inter-organisational management', 'multi-stakeholder platforms', 'soft systems approaches', 'conflict management' and 'modelling with stakeholders', although many of the key findings are surprisingly similar (Daniell 2008; Huxham 1996; Lynham et al. 2007; Renger et al. 2008; Tan et al. 2008; Voinov and Bousquet 2010). Specifically in the water domain, about 10 years ago, practitioners and scholars were still investigating what was desired from participatory processes in different contexts and what methods could be used to achieve these goals (Dovers 2000). More recently, however, through extensive field testing of methods, a much better understanding of these issues has emerged, and individual designers have been able to engineer processes to reach their desired outcomes (Hare et al. 2006). Numerous guidelines, books and papers on the design and use of participatory methods now also exist (for example, Aslin and Brown 2004; Beierle and Cayford 2002; Creighton 2005; Stern and Fineberg 1996) to help process organisers to achieve a range of desired outcomes. Such guides tend to provide in-depth information on

- *stakeholder analysis*, including how to identify and select or invite stakeholders to participate
- *decision analysis*, including how to identify and select the issues to be examined and gather existing knowledge on them

- *participation planning*, including the selection of methods for different stages in the decision-making or planning cycle and logistics planning for stakeholder-engagement events (von Korff et al. 2010).

Very recent research and practical processes have focused more strongly on how to increase the efficiency, effectiveness and innovation potential of large-scale or multi-level participatory water-management processes, as well as how to more successfully manage and transform water conflicts. Useful results from such work include a need to focus efforts on constructing effective organisation teams or inter-institutional networks for collectively managing or 'co-engineering' participatory processes to achieve multiple goals, and on the potential for individuals in both formal and informal participatory processes to encourage transitions in governance to more collaborative and adaptive arrangements (Daniell et al. 2010b; Meijerink and Huitema 2010; Moellenkamp et al. 2010). Linked to the importance of a diversity of values present in water-management debates, important work on integrating or adhering to appropriate ethics is also appearing as an aid for preventing and resolving water-related conflicts; and the need to find balanced perspectives will enable more effective cooperation and collaboration for water management to occur (Daniell et al. 2009; Delli Priscoli and Wolf 2009). In particular, for effective and efficient water planning to occur, in-depth reflection and debate over at least two different levels of process and content should ideally occur, as outlined in Table 26.1.

Table 26.1 Two sets of questions to investigate for collaborative water management

Stakeholder process for managing water systems	Project organisation process for managing the participatory process
Why ought a water plan be created?	Who ought to be responsible for organising and managing the participatory process?
What ought to be the goals of the water plan?	How ought the scope and purposes of the water-management plan be decided?
What ought to be the actions to achieve these goals?	How ought the decision be made on who ought to participate and when?
Who ought to be responsible for funding, resourcing and implementing these actions and when?	Which participatory methods ought to be used and why?
How ought progress towards these goals be measured?	Who ought to design, implement or facilitate the use of these methods with the participants?
How ought the plan be adjusted based on these evaluations?	Who ought to analyse and synthesise the results stemming from the participatory process?
	How ought the evaluation of the process take place and who ought to be allowed access to the raw data and final results?

Source: Daniell et al. (2011).

The questions in the first column of Table 26.1 are ideally used to guide discussion and eventually decisions on these issues throughout a collaborative stakeholder process for managing water systems. The questions in the second column of Table 26.1 then refer to how this collaborative stakeholder process might itself be organised, with these questions typically guiding the negotiations and decisions of an organising team, which could include some stakeholders. Further explanation of this differentiation of participatory processes—one with water-planning decisions as a key outcome, and the second having organisational decisions on the development of a collaborative stakeholder process—is provided in Daniell et al. (2010b). Examples of ethical dilemmas and methods of working through them based on some of the questions in Table 26.1 are presented in Daniell et al. (2009). Further tools for working through conflicts and consensus building are provided in Delli Priscoli (2003).

Failing to resolve misunderstandings or conflicts surrounding each of these questions could lead to serious challenges in developing successful collaboration processes and achieving better outcomes through water-management plans. In particular, it is important to note that the answers to the questions are often made unconsciously, with little explicit discussion by those wishing or needing to engage. This 'setting the terms' of engagement—if it is not carried out in a collaborative or at least an open manner—can be one of the main reasons why certain stakeholder groups such as 'the public' or 'citizens' choose to resist invitations to engage. In this chapter, space is not available to discuss typical aids that can be used to elicit and discuss potential answers to all these questions, so just a couple that could be relevant for understanding the current angst over the MDB planning process will be briefly examined in the next section.

Choosing Stakeholder-Engagement Methods and Participants

The first issue to understand for aiding the choice of what methods can be most effectively used with certain participants is that there is a sliding scale of what can be considered 'engagement' or 'participation'. Many classifications are based on the level of power sharing between decision makers and other stakeholders (for example, Arnstein 1969; Mostert 2003), considering that only processes that have the potential for significant alteration of decision-makers' viewpoints or mutual learning (such as the interactive processes of co-thinking or co-deciding) should be called 'participation' or 'engagement'. Other processes, which limit interactive communication and the impact of stakeholders' views on decision making—such as information provision or consultation—are considered to be lesser forms of participation or 'tokenism' (Arnstein 1969). Often due to time and resource constraints, hard decisions on how to limit the extent of participation, but maintain the potential for success, must be made.

Just one classification aimed at facilitating this choice is presented in Figure 26.1. The classification considers that if, for example, relative to a particular object of interest (for example, a study on a particular topic; implementation planning for specific management actions), stakeholders have potentially high levels of resources available (for example, knowledge, implementation capacity, authority, finance) and a high stake in the decisions related to the object of interest (for example, they will be strongly affected by or could make or block the decision) then they should be actively involved in the 'decision-aiding' process associated with that object. Further explanation of this classification and its use in the design of participatory processes can be found in Mazri (2007) and Daniell et al. (2010a).

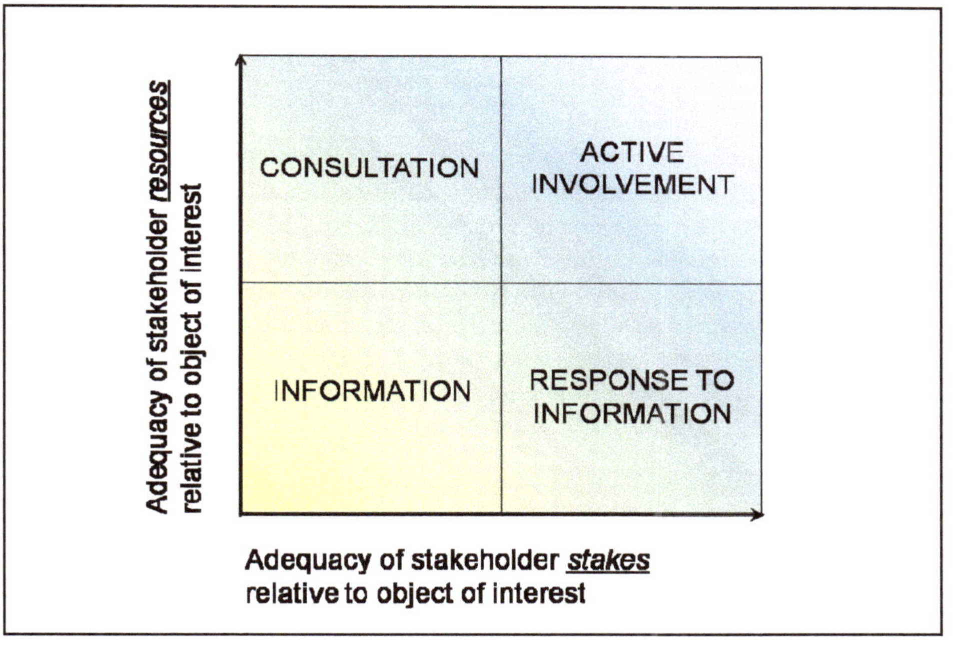

Figure 26.1 Understanding when to involve stakeholders at different levels of engagement

Source: Mazri (2007).

A second issue to consider prior to choosing methods and participants is the importance of determining the objectives of the engagement, the resources available to support the stakeholder-engagement process, and the phase of the decision-making process in which the methods are to be applied. A number of these are outlined in Table 26.2. The choice of objectives should ideally be strongly linked to the results of reflection on the questions outlined in Table 26.1.

Table 26.2 Potential objectives, resources and decision-making phases that can impact on appropriate choice of methods and participants

Objectives	Resources	Decision-making phases
• Information provision • Education • Improving two-way communication • Social learning • Enhancing legitimacy of decisions • Enabling democratic governance • Conflict resolution • Legal/organisational requirements • Building personal relations and social capacity • To achieve a better water-management outcome	• Time and finance • Skills in designing and using methods • Organisational will and leadership • Existing trust levels and relationships • Power to make and implement decisions • Knowledge of the policy area • Stakeholder interest and capacity in engaging (both agencies and communities)	• Identifying and structuring issues and values • Situation analysis • Eliciting preferences • Developing and assessing management options • Negotiating choices • Implementation planning • Monitoring and evaluation • Policy and plan adjustment

Depending on the objectives, the resources available, and the decision-making phase targeted for the engagement process, different methods can be selected. A number of methods—ranging from less interactive to more interactive—are outlined in Table 26.3, along with what they might ideally be used for, the key challenges associated with the method, its potential cost and how many people can participate. Further discussion of these methods and many more are available in a range of publications (for example, Aslin and Brown 2004; Chambers 2002; Creighton 2005; Forester 1999).

Some methods in Table 26.3 allow for participants to be chosen at will (for example, workshops, Delphi, mail-outs), some are typically open to all (for example, broadcasts, town-hall meetings, some online gaming or forums) and others have specific methodologies for the selection of participants (for example, citizens' juries, consensus conferences or some surveys). Most methods require careful design, implementation and monitoring to ensure that they have the best possible chance of meeting their planned objectives. It is very common for a suite of different methods to be employed with different participants for separate stages of the decision-making process. A range of documents (for example, Aslin and Brown 2004; Daniell 2008; Tan et al. 2008; von Korff et al. 2010) explains how this might be done. A range of expertise is available across Australia and internationally in water management, community development and business that can be mustered for these processes, including facilitators, mediators, decision analysts, communications experts and participatory-process management specialists. Nevertheless, despite the best intentions, there are some barriers that can prevent effective stakeholder engagement and collaborative approaches from occurring, and these need to be understood and managed. Examples of barriers for decision makers and scientific experts, and stakeholders and the public, are presented in Table 26.4.

Table 26.3 Common 'less interactive' and 'more interactive' stakeholder engagement methods with some of their properties

	Methods	Ideal use	Key challenges	Cost	Participant numbers
Less interactive	Mail-outs/press articles/ broadcasts	Broadscale information distribution/awareness raising	Tailoring information to audience; finding attractive format	$-$$$$	Few to very large numbers
	Information stands/ roadshows	Providing overview information and/or people to explain information	Making information easily understandable to people; finding knowledgeable people able to answer questions	$-$$$$ (depending on length/ size of roadshow)	Potentially large numbers
	Town-hall/public meetings	Providing overview information and/or people to explain information	Can heighten conflict if information is contentious or disputed; difficult to hear many voices	$-$$	Dependent on size of meeting hall. Typically < 2000
	Public presentations/ Q&A sessions	Providing information of interest and encouraging some debate	Requires a good facilitator to maintain a positive Q&A session	$-$$$	Dependent on room size unless televised. Typically < 200
	Mail, phone and in-person surveys	Eliciting information from a targeted population	Obtaining expertise to develop and administer a useful and well-constructed survey	$$-$$$$	Dependent on survey design and resources to carry it out. Potentially large numbers
	Delphi analysis (typically experts)	Developing a structured expert view on an issue	Facilitation of method use and choice of experts	$-$$$	Three to many (especially web-based Delphis)
	Consultation by written submission	Eliciting feedback with a view to considering new information and differing opinions in decision making	Not being a superficial process; synthesis and treatment of submissions	$-$$$$ (depending on synthesis work)	Potentially large numbers

	Citizens' juries/ consensus conferences	Developing judgments on controversial or little publicly examined topics	Organisation of the events; having political buy-in to considering decisions/ recommendations	$$-$$$$	Approximately 10–150
	Participatory modelling	Developing shared representations as a basis for joint investigations and informing decisions	Managing organising team and participant dynamics/ effectively structuring complex information	$-$$$	Approximately 5–50
	Facilitated workshops/ focus groups	Encouraging dialogue and collaborative work, including making trade-offs through use of techniques such as multi-criteria decision analysis	Establishing agreed workshop aims and finding effective facilitators who can work with participants to achieve them	$-$$	Approximately 10–30 per workshop. More can be handled in parallel by multiple facilitators
More interactive	Games (role-playing, simulation, online)	Developing understanding of a specific situation and impacts of actions	Finding resources for game development; having access to appropriate props/technology	$-$$$$ (depending on game development costs)	Variable depending on game and platform
	Problem structuring methods	Aiding decision making in complex, uncertain and conflict-ridden situations	Finding facilitators with a working knowledge of the required methods	$-$$$	Approximately 5–50
	Visioning/ scenario building/ search conference	Developing and assessing potential futures	Finding facilitators with a working knowledge of the required methods	$-$$$$	Approximately 5–150
	Participatory planning/GIS	Jointly developing action plans and spatialising information provided by participants	Finding effective facilitators and having access to GIS technology/ maps or spatial models	$-$$$$	Variable depending on scope of planning process. Large numbers can participate through online GIS systems

More interactive	Participatory evaluation	Encouraging participant reflection and learning with a view to applying lessons in the future	Finding facilitators and evaluation specialists with a working knowledge of the required methods	$-$$$$	Variable depending on scope of evaluation
	Discussion forums (online, TV, in person)	Encouraging dialogue, debate and mutual learning	Finding discussion facilitators and mediators; developing appropriate platforms for interaction if online or on TV	$-$$$$	Variable depending on media. Effective interaction likely with relatively small numbers
	Multilateral negotiations	Developing joint agreements or treaties	Finding effective mediators, chairpeople and information synthesisers/ drafters	$-$$$$$	Variable. Typically 3–200
	World cafés	Developing conversations and collective understanding of multiple interconnected issues	Organisation of the event; continuing the conversations after the end of the event and acting on knowledge	$$-$$$$	Variable depending on aims. Can range from 12 to hundreds or thousands
	Advisory panels	Developing and synthesising knowledge and opinions to inform decision-making processes	Selecting a broad range of advisors who will be able to work effectively together and add value to decision making	$-$$$$	Variable depending on aims. Can range from 5 to hundreds or thousands

Table 26.4 Barriers to stakeholder engagement and collaborative approaches

For decision makers and scientific experts	*For stakeholders/the public*
Lack of will to involve others in decision and science processes	Lack of interest or time to become involved in such processes (especially in an 'out-of-crisis' period)
Lack of organisational support and leadership	Previous bad stakeholder-engagement experiences
Inadequate resources, including time, finance, knowledge and a lack of training as facilitators and with participatory methods	Lack of other resources to participate (for example, knowledge, technology, financial support)
Lack of personnel continuity (difficulties building and maintaining relationships and trust)	Mistrust in coordinators
Inability to manage stakeholder expectations and conflict	Scepticism that participation will make a difference
Legal, security or other institutional constraints	Nothing obvious in it for them

At this point it is worth acknowledging that if barriers cannot be overcome, it is often better not to try to engage stakeholders than to convene substandard participatory processes that are likely to disappoint stakeholders or lead to 'over-consultation', as more damage than good is likely to result (for further discussion on this point, see Barreteau et al. 2010). In some cases, however, where certain stakeholder inputs and cooperation are required to achieve key desired outcomes of water reform, perseverance to engage and negotiate will be necessary.

Enhancing Collaborative Management and Stakeholder Engagement in the Basin

In light of this brief analysis of the need for collaborative approaches and some of the methods of stakeholder engagement that can be used to support them, it is now possible to postulate a number of reasons why there appears to be so much anger, frustration and anxiety felt by many stakeholders across the Murray–Darling Basin, and how altering the current management approach could lead to improved outcomes.

Managing Stakeholder Anxiety, Frustration and Anger

In any important reform process, fear of the unknown typically makes people feel anxious and often defensive towards change. This anxiety is a natural emotional response to a situation in which negative consequences are anticipated (Stephan and Stephan 1985). Stakeholders across the MDB recognise that these new water reforms—and in particular the development of the Murray–Darling Basin Plan—represent potentially massive change and could likely threaten their current way of living. Without any convincing and encouraging information to the contrary provided to them before the release of the Guide and all of its supporting documentation, many stakeholders were therefore harbouring a range of negative emotions directed predominantly at the MDBA and the Government. In such a situation—in which stakeholder perceptions of the reforms appear to be deteriorating and associated potentially harmful mental health impacts could be envisaged in the short and longer term—what can be done to improve the reform process?

First, in times of change, leadership is the key for providing a vision of the future that can be positively anticipated, even if short-term pain is part of that vision as being a necessary part of the transition. Showing the importance of the gap between the vision and what maintaining the status quo could represent can help to galvanise support for action and planning to work towards a preferred future. Linked to this process, anxiety over uncertainty can be mitigated by providing more certainty and attempting to build trust. One way of working on this uncertainty reduction and trust building is to provide and explain the consequences of information as soon as it comes to hand, so that people can understand the reasons for the situation. Also, knowing that support will be available to help manage any negative consequences can reduce anxiety, as it is also a means of understanding options and reducing uncertainty. The key here is 'understanding'. A good example of how to do this was provided by the Queensland Premier, who held regular press conferences with other leaders from the Police and the Army in the lead-up to, and during, the recent flood and cyclone crises, to explain the situation, relay information from scientists, clarify what actions were being taken to manage the situation and what affected people could do, and empathise with people and reassure them that everything possible would be done to help them cope with any consequences of the extreme events. As with anxiety, the effective management of anger, frustration and disappointment typically requires people to be able to 'let out' their emotions in a safe environment and have the opportunity to be understood. Building mutual understanding of these emotions and being shown the respect and willingness to work through the root causes of these emotions will eventually build more positive emotions and capacity to cope with the situations presented. So, to what extent was the choice made by the MDBA to release the Guide and couple it with a series of town-hall meetings likely to succeed in improving the reform?

From Table 26.3, it can be seen that although town-hall meetings can be used for informing relatively large numbers of people about a limited range of issues at low cost (compared with other methods), they present the challenge of potentially heightening conflict if the information is controversial or disputed. They are also not typically appropriate for fostering two-way conversations, understanding and trust that are required to overcome negative emotions. Informed participatory-process analysts would therefore most likely have predicted that, given the mood of stakeholders in the Basin before the release of the Guide, the proposed engagement process was likely to inflame rather than calm tensions. If, however, the sessions were led exceptionally well by charismatic, convincing communicators with an attractive vision of the Basin and in-depth knowledge of the science underlying the planning decisions, some improvements in stakeholder sentiment could have occurred.

In terms of other relatively low-cost stakeholder-engagement approaches that could lead to more favourable outcomes, there are some options available. For example, given that the anxieties are in part linked to the Basin Plan and lack of understanding of the knowledge and assumptions underlying it, one approach potentially more adapted to this specific issue could include a 'roadshow' with MDBA scientific staff manning a stand/caravan for a period in each key regional centre where they have all of the data, information and studies to help stakeholders to gain understanding of the issues that interest them most and to build trust with the MDBA on a more personal basis. This would, however, be only a small part of a larger process needed to rebuild and further enhance all stakeholders' confidence, understanding and collective capacity to work together across a range of geographical and administrative scales. Developing a potentially valuable and acceptable collaborative-management approach, with a range of appropriate stakeholders and stakeholder-engagement methods employed, would require much thought, concerted effort, finance and, more importantly, engagement with a range of stakeholders to determine their needs, constraints and aspirations for the current and future phases of the MDB planning process.

Looking to the Future: Strengthening leadership and collaborative practice

Hope remains across the Basin that the recent reforms will lead to a brighter future for the Basin and the communities it supports, even if this hope is currently tempered by a range of negative emotions, and many communities are trying to cope with the impacts of floods after many years of drought. It is well known that positive changes can take place when people feel the need to act, and are respected and encouraged in their efforts. The 2010–11 flood crisis across large parts of Australia has shown us that with good leadership, timely and continuous information provision, empathy, trust in others' capacities and putting efforts into coordination and support, many Australians will rise to a challenge and are more than willing to volunteer and to work together through extremely difficult and potentially devastating situations. We should learn from this experience and determine how it can be reapplied to build a positive future for the Basin. Generosity and goodwill in aiding adjustment processes can soften the blow of loss and provide a foundation for resilient and positive communities and individuals to get back on their feet after tumultuous change. Lessons and knowledge available from other MDB, Australian and international experiences can also be gleaned on how communities can use collaborative values-focused processes to plan for the future, including large cuts to water allocations (see, for example, Richardson et al. 2011, (chapter 22)), estuarine risk management (Daniell et al. 2008) and flood and drought risk management (Daniell et al. 2011).

With a new Chairman of the MDBA having just been appointed and the Parliamentary Inquiry drawing to a close, a window of opportunity is available for the Federal Government to lead the MDB planning process in a new direction. This direction would ideally set the enabling governance conditions that would allow: 1) a clear vision for the Basin to be articulated; and 2) communities, regions, and State and Federal governments to come together in a broad collaborative effort to propose ways of restructuring the Basin for the future. Informally starting development of the State water-resource plans before the Basin Plan is finalised could provide a concrete basis for assessing the potential basin-wide impacts of planning decisions. The keys to this endeavour would be to

- acknowledge that decision making associated with the MDB Plan is based on values, objectives and specific visions for the future, as well as science
- lead a well-resourced collaborative approach to the next phase of plan development to actively engage all stakeholder groups, including all existing management agencies, in an inclusive and respectful manner, where real dialogue can occur on a vision for the Basin, achievable plan objectives, scientific and stakeholder knowledge, development of management options and their impacts under a range of scenarios, potential adjustment packages and monitoring plans
- work to engage stakeholders in ways they want to be engaged (as far as possible), with a commitment made to value their input and support them through the decision-making and implementation phases of the reforms.

This collaborative approach would provide a basis for a Basin Plan that could: a) adequately value the river-basin system along with the people managing and relying on it for their livelihoods; b) better ensure that the Plan is based on the best available scientific and stakeholder knowledge; and c) allow stakeholders to develop ownership of the Plan and prepare for changes through its implementation.

Conclusions

This chapter has outlined that for the desired outcomes of the Murray–Darling Basin reform process to be reached, the centralised technocratic approach of the MDBA with minimal levels of interactive engagement with stakeholders is inappropriate. Rather, investment is required in a more collaborative approach to the reforms where the best available science and experience from practice on participatory processes should be drawn upon. Town-hall meetings should not be the main engagement method used in a collaborative approach; rather, a range of interactive and less interactive methods needs to be employed. Such an approach could include a range of regional and local panels, facilitated workshops that include visioning activities and participatory planning based on scenarios of potential futures. These would complement the information gathering and consultation already carried out through the responses to the *Guide to the proposed Basin Plan* and the current Parliamentary Inquiry. Learning from the recent flood and cyclone crises in Queensland, holding regular official press conferences with the relevant ministers and the Chair of the Murray–Darling Basin Authority, and potentially using larger panels of community, government and scientific leaders for Q&A sessions might also aid in showing leadership, rebuilding trust and demonstrating the priority accorded to this important national endeavour. Acknowledging that the Government and communities are working together to reduce the negative impacts of the reforms and that support will be made available to those affected through potentially difficult transitions could help people to rebuild a sense of pride in being a part of these significant new reforms. It would provide the impetus to many stakeholders to engage or drive the process at their level, allowing them to offer their knowledge and energy to make the new measures a success in securing a viable and sustainable future for the Basin.

Acknowledgments

Thank you to the Fellows, Friends and supporters of the Peter Cullen Trust for many thought-provoking discussions, which have inspired a number of reflections presented in the chapter, as well as to an anonymous reviewer for comments and suggestions that have improved its structure and clarity.

References

Arnstein, S. R. 1969, 'A ladder of citizen participation', *Journal of American Institute of Planners*, vol. 35, pp. 216–24.

Aslin, H. J. and Brown, V. A. 2004, *Towards Whole of Community Engagement: A practical toolkit*, Murray–Darling Basin Commission, Canberra, <http://publications.mdbc.gov.au/download/towards_whole_of_community_engagement_toolkit.pdf>

Barreteau, O., Bots, P. W. G. and Daniell, K. A. 2010, 'A framework for clarifying "participation" in participatory research to prevent its rejection for the wrong reasons', *Ecology and Society*, vol. 15, no. 2, art. 1, <http://www.ecologyandsociety.org/vol15/iss2/art1/>

Beierle, T. C. and Cayford, J. 2002, *Democracy in Practice: Public participation in environmental decisions*, Resources for the Future, Washington, DC.

Bellamy, J., Ross, H., Ewing, S. and Meppem, T. 2002, *Integrated catchment management: learning from the Australian experience for the Murray–Darling Basin*, Report prepared for the Murray–Darling Basin Commission, CSIRO Sustainable Ecosystems, Canberra, <http://www2.mdbc.gov.au/__data/page/911/ICM_Learning_from_Australian_Experience.pdf>

Boully, L. 2004, Participatory governance: intra and inter governmental consultation and community engagement in the Murray–Darling Basin Initiative, Presented to Seventh Annual Corporate Governance in the Public Sector Conference, Canberra, 20–22 April 2004.

Chambers, R. 2002, *Participatory Workshops: A sourcebook of 21 sets of ideas and activities*, Earthscan, London.

Connell, D. 2007, *Water Politics in the Murray Darling Basin*, The Federation Press, Leichhardt, NSW.

Connick, S. and Innes, J. 2001, *Outcomes of Collaborative Water Policy Making: Applying complexity thinking to evaluation*, Institute of Urban and Regional Development, University of California, Berkeley.

Crase, L., Dollery, B. and Wallis, J. 2005, 'Conceptualising community consultation in public policy formulation: the case of the Living Murray debate in the Murray Darling Basin of Australia', *Australian Journal of Political Science*, vol. 40, pp. 221–37.

Creighton, J. L. 2005, *The Public Participation Handbook: Making better decisions through citizen involvement*, Jossey-Bass, San Francisco.

Daniell, K. A. 2008, Co-engineering participatory modelling processes for water planning and management [*Co-ingénierie des processus de modélisation participative pour la planification et la gestion de l'eau*], (2 vols), PhD Thesis, The Australian National University, Canberra, and AgroParisTech, Montpellier, France.

Daniell, K. A., Coad, P., Ferrand, N., White, I., Jones, N., Guise, K., Marvell, C., Burn, S. and Perez, P. 2008, Participatory values-based risk management for the water sector, Paper presented at the Proceedings of the Water Down Under 2008 Conference, Adelaide, 14–17 April 2008.

Daniell, K. A., Mazri, C. and Tsoukiàs, A. 2010a, 'Real world decision-aiding: a case of participatory water management', in D. Rios Insua and S. French (eds), *e-Democracy: A group decision and negotiation perspective*, Springer, Dordrecht, pp. 125–50.

Daniell, K. A., Ribarova, I. S. and Ferrand, N. 2011, 'Collaborative flood and drought risk management in the Upper Iskar Basin, Bulgaria', in R. Q. Grafton and K. Hussey (eds), *Water Resources Planning and Management*, Cambridge University Press, UK, pp. 395–420.

Daniell, K. A., White, I., Ferrand, N., Riborova, I. S., Coad, P., Rougier, J.-E., Hare, M., Jones, N., Popova, A., Perez, P. and Burn, S. 2010b, 'Co-engineering participatory water management processes: theory and insights from Australian and Bulgarian interventions', *Ecology and Society*, vol. 15, no. 4, art. 11, <http://www.ecologyandsociety.org/vol15/iss4/art11/>

Daniell, K. A., White, I. and Rollin, D. 2009, 'Ethics and participatory water planning', *Proceedings of the 32nd Hydrology and Water Resources Symposium: 'H2009', 30 November – 3 December 2009, Newcastle, Australia*, pp. 1476–87.

Davidoff, P. 1965, 'Advocacy and pluralism in planning', *Journal of American Institute of Planners*, vol. 31, no. 4, pp. 331–8.

Delli Priscoli, J. 2003, *Participation, Consensus Building, and Conflict Management Training Course (Tools for Achieving PCCP)*, United Nations Educational, Scientific and Cultural Organisation—International Hydrological Programme prepared for the WWAP, Paris.

Delli Priscoli, J. and Wolf, A. T. 2009, *Managing and Transforming Water Conflicts*, Cambridge University Press, UK.

Dietz, T., Ostrom, E. and Stern, P. C. 2003, 'The struggle to govern the commons', *Science*, vol. 302, pp. 1907–12.

Dovers, S. 2000, 'Beyond everythingcare and everythingwatch: public participation, public policy, and participating publics', *Proceedings, International Landcare 2000 Conference*, Department of Natural Resources and Environment, Melbourne.

Dryzek, J. S. 1990, *Discursive Democracy: Politics, policy and political science*, Cambridge University Press, New York.

Fischer, F. 1990, *Technocracy and the Politics of Expertise: Critical perspectives on the managerial and policy sciences*, Sage Publications, Newbury Park, Calif.

Fischer, F. 2000, *Citizens, Experts, and the Environment*, Duke University Press, Durham, NC.

Forester, J. 1999, *The Deliberative Practitioner: Encouraging participatory planning processes*, MIT Press, Cambridge, Mass.

Franklin, M. 2010, 'Heated backlash forces Murray inquiry', *The Australian*, 15 October 2010.

Gleick, P. H. 2000, *The World's Water 2000–2001: The biennial report on freshwater resources*, Island Press, Washington, DC.

Hare, M. P., Barreteau, O., Beck, M. B., Letcher, R. A., Mostert, E., Tàbara, J. D., Ridder, D., Cogan, V. and Pahl-Wostl, C. 2006, 'Methods for stakeholder participation in water management', in C. Giupponi, A. J. Jakeman, D. Karssenberg and M. P. Hare (eds), *Sustainable Management of Water Resources: An integrated approach*, Edward Elgar, Chichester, UK, pp. 177–225.

Huxham, C. (ed.) 1996, *Creating Collaborative Advantage*, Sage Publications, London.

Ingram, H. and Schneider, A. 1999, 'Science, democracy, and water policy', *Water Resources Update*, vol. 133, pp. 21–8.

International Conference on Water and the Environment (ICWE) 1992, *The Dublin Statement on Water and Sustainable Development*, International Conference on Water and the Environment, Dublin.

Landry, M., Banville, C. and Oral, M. 1996, 'Model legitimisation in operational research', *European Journal of Operational Research*, vol. 92, no. 3, pp. 443–53.

Loucks, D. P. 1998, 'Watershed planning: changing issues, processes and expectations', *Water Resources Update, Universities' Council on Water Resources*, no. 111 (Spring), Carbondale, Ill., pp. 38–45.

Lynham, T., de Jong, W., Sheil, W., Kusumanto, T. and Evans, K. 2007, 'A review of tools for incorporating community knowledge, preferences and values into decision making in natural resource management', *Ecology and Society*, vol. 12, no. 1, art. 5, <http://www.ecologyandsociety.org/vol12/iss1/art5/>

Margerum, R. D. 2008, 'A typology of collaboration efforts in environmental management', *Environmental Management*, vol. 41, pp. 487–500.

Marshall, G. R. and Stafford Smith, D. M. 2010, 'Natural resources governance for the drylands of the Murray–Darling Basin', *The Rangeland Journal*, vol. 32, pp. 267–82.

Mazri, C. 2007, *Apport méthodologique pour la structuration de processus de décision publique en contexte participatif. Le cas des risques industriels majeurs en France*, Université Paris Dauphine, France.

Meijerink, S. and Huitema, D. 2010, 'Policy entrepreneurs and change strategies: lessons from sixteen case studies of water transitions around the globe', *Ecology and Society*, vol. 15, no. 2, art. 21, <http://www.ecologyandsociety.org/vol15/iss2/art21/>

Moellenkamp, S., Lamers, M., Huesmann, C., Rotter, S., Pahl-Wostl, C., Speil, K. and Pohl, W. 2010, 'Informal participatory platforms for adaptive management. Insights into niche-finding, collaborative design and outcomes from a participatory process in the Rhine basin', *Ecology and Society*, vol. 15, no. 4, art. 41, <http://www.ecologyandsociety.org/vol15/iss4/art41/>

Mostert, E. 2003, 'The European Water Framework Directive and water management research', *Physics and Chemistry of the Earth*, vol. 28, nos 12–13, pp. 523–7.

Murray–Darling Basin Authority (MDBA) 2009, *Stakeholder Engagement Strategy*, Murray–Darling Basin Authority, Canberra, <http://www.mdba.gov.au/files/publications/Stakeholder-Engagement-Strategy-brochure.pdf>

Murray–Darling Basin Authority (MDBA) 2010, *Guide to the proposed Basin Plan. Volume 1: Overview*, Murray–Darling Basin Authority, Canberra.

Nandalal, K. D. W. and Simonovic, S. P. 2003, *State-of-the-Art Report on Systems Analysis Methods for Resolution of Conflicts in Water Resources Management*, UNESCO-IHP.

Renger, M., Kolshoten, G. and De Vreede, G. 2008, 'Challenges in collaborative modelling: a literature review and research agenda', *International Journal of Simulation and Process Modelling*, vol. 4, pp. 248–63.

Standing Committee on Environment and Heritage (SCEH) 2000, *Co-ordinating Catchment Management: Report of the Inquiry into Catchment Management*, Standing Committee on Environment and Heritage, House of Representatives, The Parliament of the Commonwealth of Australia, Canberra.

Stephan, C. and Stephan, W. 1985, 'Intergroup relations', in G. Lindzey and E. Aronson (eds), *Handbook of Social Psychology*, [Third edition], Random House, New York.

Stern, P. J. and Fineberg, H. V. (eds) 1996, *Understanding Risk: Informing decisions in a democratic society*, National Academy Press, Washington, DC.

Syme, G. J. and Hatfield-Dodds, S. 2007, 'The role of communication and attitudes research in the evolution of effective resource management arrangements', in K. Hussey and S. Dovers (eds), *Managing Water for Australia: The social and institutional challenges*, CSIRO Publishing, Collingwood, Vic., pp. 11–22.

Tan, P. L., Jackson, S., Oliver, P., Mackenzie, J., Proctor, W. and Ayre, M. 2008, *Collaborative Water Planning: Context and practice literature review*, Land & Water Australia, Canberra, <http://lwa.gov.au/files/products/track/pn21213/pn21213.pdf>

Thomas, R. L. 2004, 'Management of freshwater systems: the interactive roles of science, politics and management, and the public', *Lakes and Reservoirs: Research and Management*, vol. 9, no. 1, pp. 65–73.

Voinov, A. and Bousquet, F. 2010, 'Modelling with stakeholders', *Environmental Modelling & Software*, vol. 25, pp. 1268–81.

von Korff, Y., d'Aquino, P., Daniell, K. A. and Bijlsma, R. 2010, 'Designing participation processes for water management and beyond', *Ecology and Society*, vol. 15, no. 3, art. 1, <http://www.ecologyandsociety.org/vol15/iss3/art1/>

Wong, P. 2008, Water for the future, Speech to the Fourth Annual Australian Water Summit, Sydney Convention and Exhibition Centre, Sydney, 29–30 April 2008.

27. Improving the Basin Plan: Options for consideration[1]

Mike Young

Re-Engaging with Basin Communities

Given the untidy release of the *Guide to the proposed Basin Plan*, the future depends on both the way the next phase of consultation processes are run and the policy decisions taken as the process runs forward. From an administrative perspective, many of the structural-adjustment and environmental-water management decisions lie beyond the remit of the Murray–Darling Basin Authority (MDBA). In recognition of this reality, this chapter begins by recommending that the new Chair of the Authority be given responsibility for coordinating development of a whole-of-government approach to the resolution of the Basin's problems and development of the Basin Plan.

There are many ways to implement the next consultation phase and restore community confidence; one way is to prepare a green paper. Rather than going over old ground, this green paper would identify the suite of policy choices that needs to be resolved in the course of preparing a Plan. Short, rather than long, this green paper could then be used to open dialogue and truly engage with the Basin. Those responsible for preparing this green paper would need to draw upon the advice submitted to the MDBA and to parliamentary inquiries that have considered the range of issues before basin communities. Once community confidence has been restored and they have been given time to respond, a white paper could be used to present the Government and the Authority's collective position on the policy changes that need to be made to restore health to the Basin.

Background

The *Guide to the proposed Basin Plan* released by the Murray–Darling Basin Authority (MDBA) represents one of the first attempts by a government to develop a management plan for a large river system that has its roots in scientific

1 The opportunity to discuss and develop the ideas presented in the chapter with Jim McColl is acknowledged with appreciation.

analysis. The approach taken was to trust the science and then present a strong set of recommendations to the community. As indicated elsewhere in this book, the approach taken by the Authority and the way it engaged with those who live in the Murray–Darling Basin failed to gain the trust and support of this community.

Rather than criticising the approach taken by the Authority and the communication mistakes made, this chapter focuses on institutional opportunities to develop and deliver a Plan that communities can embrace.

The recommendations made enable greater use of community skills and knowledge, encourage innovation in the management of environmental water and provide funding for adjustment. If these recommendations are adopted then the cost of fixing the Basin's problems in an equitable and efficient manner will be less.

In addition, it is recommended that the approach taken to the definition of sustainable diversion limits (SDLs) be changed. Consistent with the National Water Initiative, it is recommended that the environment's share of inflows into the system should be defined using an entitlement-based rather than a rules-based planning approach. The result is a regime that allows further adjustments to be made without the compulsory acquisition of water from entitlement-holders.

Ways Forward

Giving the Environment an Entitlement

The first opportunity to improve the Basin Plan identified in this chapter is to change the way it is proposed to define SDLs.

In the *Guide to the proposed Basin Plan*, the SDL is defined quantitatively as a long-term average—a volume.[2] The volume chosen to define the SDL is a long-term average amount that can be diverted, reduced by 3 per cent to adjust for the MDBA's assessment of that part of the predicted effects of climate change not included in regional plans.

---

2 At page 103, the Guide (MDBC 2010) says: 'Long term sustainable limits (SDLs) represent the volume of water that is available for consumptive use (irrigation, town water supplies, industry, etc) after the environment has received what it requires.' The amount the environment requires determines the limit and, hence, may not be changed by purchasing entitlements from a consumptive-water user.

If this approach is taken and once the implementation of the Basin Plan and regional plans is completed (in 2019), any further reduction of SDLs is possible only via the compulsory reduction of entitlements and the payment of compensation. [3]

This situation arises because, under the proposed Basin Plan, all water secured or held for the environment is defined as being outside the SDL. Further, Section 4 of the *Water Act 2007* defines

> the *environmentally sustainable level of take* for a water resource…[as] the level at which water can be taken from that water resource which, if exceeded, would compromise:
>
> (a) key environmental assets of the water resource; or
>
> (b) key ecosystem functions of the water resource; or
>
> (c) the productive base of the water resource; or
>
> (d) key environmental outcomes for the water resource.

Significantly—and if this quantitative approach is taken to the definition of a SDL—the words 'if, exceeded, would compromise' mean that water held for the environment is not part of the SDL. Once the Basin Plan is in place and regional plans approved, it will no longer be possible to reduce the amount of water being used by purchasing more water entitlements for the environment. The only way to increase the environment's entitlement or share of inflows is to revise the SDL for a region by either

- paying compensation for the reduction in the value of all entitlements in the system
- compulsorily acquiring a proportion of each entitlement.

The way out of this surprising—and presumably unintended—feature of the *Water Act* is to use an entitlement-based rather than a quantitative approach to the definition of SDLs. Fortunately, this is possible because Section 23(2) of the *Water Act* states that:

> A long-term average sustainable diversion limit for the Basin water resources, for the water resources of a particular water resource plan area or for a particular part of those water resources may be specified:
>
> (a) as a particular quantity of water per year; or

3 See the bottom of page 154 in the Guide (MDBA 2010) and Division 4 of the *Water Act*.

(b) as a formula or other method that may be used to calculate a quantity of water per year; or

(c) in any other way that the Authority determines to be appropriate.

Section 23(2)(c) makes it clear that an SDL may be defined in the most appropriate way the MDBA can identify. In this chapter, it is argued that the most appropriate way is to define the nature of the portfolio of entitlements that needs to be held by the environment in each region. There are a number of advantages of taking this entitlement-based approach.

1. Consistent with Section 23(2) of the *Water Act*, the Basin Plan would require that regional plans establish rules for the allocation of water to each type of entitlement in a region. The minimum proportion of each type of entitlement to be held by the environment would then be specified. In areas where water in one region is easily traded to another, this proportion should be specified in a manner that enables the goal to be reached by securing access to entitlements in other areas. The proportion specified should be a minimum. If subsequent research shows that the portfolio of entitlements held in the environment's interest is inadequate, the policy should state that more water entitlements should be acquired in a timely manner. No complex administrative revision of the SDL would be necessary.
2. The entitlement-based approach enables iteration towards a solution with a focus on ways to improve the efficiency of environmental-water use. If this approach is taken, the Basin Plan will need to indicate the minimum portfolio of shares to be acquired in each region, a time frame for its acquisition and a target portfolio to be acquired if no ways to improve the efficiency in the delivery of required environmental outcomes can be found.
3. The approach also makes it possible for environmental managers to manage supply risk more effectively. A South Australian environmental manager might, for example, conclude that the most cost-effective strategy would be to secure a portfolio of entitlements that included some general-security entitlements from New South Wales and some low-security entitlements from Victoria.[4] Similarly, an environmental-water manager might decide to sell some water allocation and use the money received to pay for the installation of a control gate that would allow the much more efficient use of environmental water.
4. An additional advantage of the property-right approach is the fact that accounting risks are distributed in proportion to the number of the entitlements held. If, for example, estimates of the amount of water being

4 This would have the additional advantage that it would enable an SA environmental-water manager to carry forward water from one season to the next.

intercepted by a forest are too conservative then all water users, including the environment, will have less water made available to them. As a result, environmental managers and irrigators have a joint interest in ensuring that this problem is solved. When a quantitative approach is taken, however, the SDL remains the same and the amount of water left to the environment is reduced by the size of the underestimate in the amount of water intercepted. In short, when a quantitative-planning approach is taken, the environment takes all the downside risk.[5] When an entitlement-based approach is taken, both sides have an interest in ensuring that water-accounting problems—such as failure to account adequately for the adverse effects of forest plantations or groundwater use—are fixed. The environment is given an entitlement equivalent to that given to other users. The approach forces governments to allocate water to the environment whenever they allocate water to any user holding an equivalent entitlement.

5. Amongst other things, this entitlement-based approach focuses discussion on the best way to manage environmental water and makes it difficult for people to add up the SDLs. Conceptually, one should add quantitative SDLs together only when the regions are closely connected and have similar flow distributions. Amongst other things, the Basin Plan needs to recognise that flows in the Darling River are many thousand times more variable than those in the Murray River system.

6. Another consideration is the fact that property-right-based approaches reverse the role of the entitlement system and regional plans. When a property-right approach is taken, regional plans do not need to be used to work out how to partition water between the environment and other entitlement-holders on a day-by-day basis. The environment's share is determined by the nature of the entitlements held not the SDL. As a result, regional plans can be made much more strategic and flexible in their orientation. Less prescription is necessary and more attention can be given to the nature of the outcomes required.

7. The proposed entitlement-based approach also deals with climate change and variability in a much more transparent manner. This is especially the case in the southern connected River Murray system, where the MDBA is required to establish a conveyance reserve to ensure that enough water is available to supply critical human needs throughout the system and generally maintain the river at a minimum level.[6] As a result of this arrangement, if an entitlement-based approach is taken, a built-in mechanism for managing

---

5 Irrigators will argue that they would prefer the existing quantitative regime, as this is what they understand. They also understand that when a quantitative regime is put in place, the impact of the majority of accounting risks falls on the environment.

6 Strictly, Section 86A(4) of the *Water Act* requires conveyance only as far as Wellington, South Australia.

the effects of adverse climate change is always in place; as the amount of water in the system reduces, a larger proportion of water is allocated to the environment.

In summary, by taking an entitlement-based sharing approach and defining the environmental water requirement (EWR) as the portfolio of entitlements to be held for the environment in each region, accounting risks can be managed more effectively, the need to compulsorily acquire water in the future can be avoided and a more flexible approach can be taken to the resolution of the Basin's problems.

Significantly, whenever more water is needed for the environment, the need to switch on the risk-allocation provisions of the *Water Act* that require the Commonwealth to pay compensation is avoided. The option to purchase entitlements for the environment from willing sellers is preserved. The compensation provisions set out in Section 77 of the *Water Act*, titled 'Payments to water access entitlement holders', would never need to be used.

Maximising Innovation: A regional approach

The next opportunity identified in this chapter is to enable greater use of local knowledge and skills in the management and use of environmental water.

Under the current administrative regime, water entitlements acquired for the environment are being transferred to the Commonwealth Environmental Water Holder (CEWH). The result is a regime in which, ultimately, the CEWH will have access to about 35 per cent of all water entitlements in the Basin. Once this process is completed, States and local communities will have to negotiate access to this water. They are not trusted to be able to get this right.

The alternative (recommended) approach is to transfer a significant proportion of the CEWH's entitlements to regional environmental-water trusts (see Young 2010). This approach builds upon the European concept of 'subsidiarity'—the notion that a central authority should be responsible only for those functions that cannot be handled more effectively at a regional level. While some environmental water needs to be held centrally to ensure that basin-wide trade-offs can be managed and opportunities to synchronise inter-regional initiatives pursued, in practice, much of this water will always be needed for use within each region.

In the case of environmental-water management, it is virtually impossible for any one person or entity to be aware of environmental needs and opportunities throughout the Basin. Local knowledge is needed and—as experience in Oregon's Water Trust in the United States has demonstrated—is best achieved by giving local trusts absolute control of water entitlements for use in their region. Using existing laws, the Commonwealth Government could decide to establish a set of regional environmental trusts and lease some of the CEWH's entitlement to these trusts on a long-term basis.

When a significant proportion of a region's environmental-water entitlement is held in a regional environmental-water trust, local managers can plan with confidence and explore innovative options. A local manager, for example, may decide to water an area for three weeks and set aside enough water to guarantee that this area can be rewatered the following year.

When a regional approach to environmental-water management is taken, there is greater opportunity to take advantage of local knowledge. Local pride in the extent of outcomes achieved per unit of water allocated to the environment emerges. Tension dissipates. More effort goes into the delivery of outcomes and less into negotiation with a centralised bureaucracy (Young 2010).

If knowledge about how to distribute environmental-water entitlements throughout the Basin is insufficient to do this on a permanent basis then, as a transitional arrangement, the Commonwealth could establish an environmental-water management trust in each region and instruct the CEWH to lease about 50 per cent of water being held by this entity to these regional trusts for, say, 10 years.

In summary, there is an opportunity to establish regional environmental trusts throughout the Basin, allocate entitlements to them and make them both responsible and accountable for wise use of this water.

Funding Adjustment

The next opportunity to consider is the question of how best to assist communities to adjust to a regime under which accounting and supply risks are shared more evenly between the environment and all other water users. It is current government policy to resolve the over-allocation problem by

- purchasing water entitlements for the environment only from people prepared to sell some of their entitlement at the current market prices[7]
- investing in projects that, by making irrigation more efficient, enable 50 per cent of the savings made to be transferred as an entitlement to the CEWH.

Typically, the price paid per megalitre of entitlement secured through an infrastructure project is two to three times higher than that paid for a megalitre of water purchased by the Commonwealth. Many criticisms have and will continue to be made of this approach.

7 It is current Commonwealth policy to try to purchase water entitlements without increasing market prices.

One of the criticisms made is that these two mechanisms do not address the interests of the communities likely to be adversely affected by the transfer of water entitlements from irrigators to the environment. Another criticism is the observation that government investment in projects that improve the efficiency of water use is best described as a 'subsidy' that disadvantages irrigators who have upgraded infrastructure at their own expense. Moreover, several people have observed that the $8.9 billion allocated for the purchase of water and investment in infrastructure is insufficient to secure the water needed to achieve the SDL limits proposed by the MDBA. There is, however, sufficient money to achieve the proposed SDLs and make some money available to assist communities to adjust if most of this money is used to acquire water entitlements for the environment.

Instead of directly funding infrastructure projects, the Commonwealth Government could decide to take a broader regional approach and plan to assist all communities to adjust to the new regime. If this decision is taken then it will be necessary to establish a set of regional-development funds and make contributions to these funds as water entitlements are acquired for the environment. Acting on the assumption that the impact on communities of water-entitlement acquisition for the environment is likely to increase as more and more water is purchased, the size of the contribution made to a fund per megalitre of water purchased could be increased as the number of entitlements purchased from a region increases.

Within broad guidelines, each region would then be free to determine how best to allocate this money and determine how much should be invested in projects that improve irrigation efficiency, how much should be invested in the restructuring of supply systems and how much in building the infrastructure needed to enable those adversely affected by the purchase of water for the environment to pursue new opportunities.[8] If this approach is taken, the resolution of over-allocation problems could be speeded up through the use of a reverse tender or other similar buyback arrangement that offers to pay more than the current very low prices being offered.

In summary, rather than investing in projects that improve irrigation efficiency, consideration could be given to the establishment of regional-development funds that give all members of a community an opportunity to apply for and receive financial assistance.

8 A warning needs to be issued: a review of Australian experience with 10 eras of adjustment experience found that, despite the best intentions, most adjustment-assistance programs have had severe unintended consequences that undermine most if not all of the benefits sought. Past programs have tended to impede rather than expedite change (McColl and Young 2006, 2007).

Carry Forward: Reducing the amount of water needed for the environment

Unfortunately, the last opportunity to improve the Basin Plan identified in this chapter would require amendment of the *Water Act* and agreement among all the governments involved in the Murray–Darling Basin. Given that the current Commonwealth Government does not hold a majority in either house of Parliament, any attempt to amend the *Water Act* is likely to be fraught with problems; nevertheless, this last opportunity is worth exploring. The opportunity is to amend the *Water Act* so that the only restriction on the amount of water that entitlement-holders and States can carry forward from one year to the next would be the presence of adequate storage space in dams. Amendment of the *Water Act* to allow all entitlement-holders and states to carry forward water from year to year would significantly reduce the amount of water that needs to be secured for the environment.

From the perspective of an environmental manager, water stocks need to be managed so as to enable the creation of many small and medium-scale flood-like conditions. In a regime under which allocations are made in proportion to the number of entitlements held, this is most efficiently achieved if the only restrictions on the amount of water that may be carried forward are those necessary to offset evaporative losses and manage dam spills. Amendment of the *Water Act* and of interstate water-sharing arrangements in the River Murray system is needed to do this.

In Queensland, carry-forward policies appropriate to environmental needs are already in place. In Queensland's regulated river systems, under what is known as a continuous accounting, entitlement-holders may carry forward as much water from year to year as they like—provided evaporative losses are accounted for and there is enough dam space. In Victoria and New South Wales, however, carry forward is allowed only up to 100 per cent of entitlement; stockpiling of more than one year's maximum allocation is not allowed. Under normal (Tier 1) conditions, South Australia is not allowed to carry forward water from one year to the next for anything other than urban water supply purposes. In addition, the amount of water that Victoria and New South Wales are required to release into South Australia is limited.

In summary, the Water Act *and interstate water-sharing arrangements could be amended to allow the unrestricted carry forward and transfer of water from one year to the next provided evaporative losses are accounted for and there is adequate dam space. In addition, the* Water Act *could be amended to allow South Australia complete freedom in deciding when to use any water that it has carried forward from one year to the next.*

Concluding Comments

In this chapter, four opportunities to improve the Basin Plan are identified. The first of these is to use a different mechanism to specify the SDL set for each region. The approach enables iteration towards a sustainable water-sharing regime for the Basin and removes all consideration of the need to compulsorily acquire water entitlements for the environment. The second opportunity recommends establishment of regional environmental-water trusts so that greater use can be made of local knowledge and opportunities to improve the management of environmental water are greater. The third identifies an opportunity to establish regional-development funds so that adverse effects of existing arrangements on the irrigation industry and on communities are avoided. The last opportunity would allow the more efficient inter-temporal management of storage for environmental purposes and, hence, reduce the size of the portfolio of water entitlements that needs to be secured for the environment.

In closing, it is important to draw attention to the fact that each of the recommendations made in this chapter would allow more effective use of the Basin's water resources and make it possible to achieve the same environmental outcomes with less water. As noted in the chapter by Connell (this volume), under existing arrangements, it is possible to acquire sufficient water to solve all the Basin's environmental problems.

The main purpose of the Basin Plan is to reduce the size of the portfolio of water entitlements that needs to be secured for the environment in each region and thereby reduce the cost of solving the Basin's problems.

Bibliography

McColl, J. and Young, M. 2006, 'Drought and structural adjustment', *Farm Policy Journal*, vol. 3, no. 2, pp. 13–21.

McColl, J. C. and Young, M. D. 2007, *Managing change: Australian structural adjustment lessons for water*, Report No. 16/05, CSIRO Land and Water, Canberra.

Murray–Darling Basin Authority (MDBA) 2010, *Guide to the proposed Basin Plan*, Murray–Darling Basin Authority, Canberra.

Young, M. D. and McColl, J. C. 2008, *A Future-Proofed Basin: A new water management regime for the Murray–Darling Basin*, University of Adelaide, SA.

Young, M. D. 2010, 'Managing environmental water', in J. Bennett, R. T. Kingsford, R. H. Norris and M. D. Young, *Making decisions about environmental water allocations*, Research Report, Australian Farm Institute, Surry Hills, NSW.

Contributors

Dr Onil Banerjee

Onil is a Postdoctoral Fellow with CSIRO's Natural Resource Economics and Decision Science Group. His research interests include economy-wide natural-resource policy modelling and development.

Dr Rosalind Bark

Rosalind is a resource/ecological economist with CSIRO Ecosystem Sciences working to synthesise socioeconomic research in policy analysis.

Henning Bjornlund

Henning Bjornlund holds two academic positions: at Canada Research Water Policy and Management at the University of Lethbridge, Alberta, and as an Associate Research Professor at the University of South Australia. He has researched water-management and policy issues in Australia since 1993 and in Canada since 2005.

Leith Boully

Leith Boully and her family operate a grazing, dryland farming and irrigation property near Dirranbandi in Queensland. She is an Adjunct Professor with the School of Agriculture and Food Science at the University of Queensland and has had a long involvement in water-resource management at the local, State and national levels, including as Chairwoman of the Community Advisory Committee to the Murray–Darling Basin Ministerial Council from 1999 to 2005 and as a founding member of the Wentworth Group of Concerned Scientists.

Dr Donna Brennan

Donna was an independent consultant and held an adjunct Senior Lecturer position at the School of Agriculture and Resource Economics, University of Western Australia, as well as an honorary position at the Crawford School of Economics and Government at The Australian National University. Previously she has worked at the Australian Bureau of Agricultural and Resource Economics, the University of Sydney, the Australian Centre for International Agricultural Research and CSIRO Land and Water. Her research focused on water-resource management, particularly in the Murray–Darling Basin. She passed away in 2010.

Dr Neil Byron

Dr Neil Byron is Adjunct Professor in Environmental Economics at The Australian National University's Crawford School of Economics and Government. Previously, he was Commissioner responsible for environmental, agricultural and natural-resource management issues in the Productivity Commission from 1998 to 2010; director of ANU's graduate program in Environmental Management and Development; and Assistant Director-General of the Centre for International Forestry Research in Indonesia.

Dr Jeremy Cheesman

Jeremy Cheesman is a Principal Economist at Marsden Jacob Associates, where he specialises in resource and environmental economics and policy. He obtained his PhD from the Crawford School, ANU. He has published widely in the areas of applied water, environmental and agricultural economics, including professional peer-reviewed journal papers, book chapters, technical reports, and conference papers.

Dr Daniel Connell

Daniel is in the Crawford School of Economics and Government at ANU. He published *Water Politics in the Murray–Darling Basin* in 2007 and now researches water-management institutions and their response to climate variability in the federal political systems of Australia, South Africa, the United States, Europe, India and China.

Dr Jeffery Connor

Jeff is a natural-resource economist in the CSIRO's Ecosystem Sciences, Social and Economic Science Group, who specialises in water-resource economics. He has worked and published extensively in Murray–Darling Basin water allocation and salinity economics and policy.

Dr Katherine Daniell

Katherine Daniell is a Research Fellow in the Centre for Policy Innovation at the ANU. Her research interests include water governance, participatory risk management, climate-change adaptation and multi-level decision-aiding processes. Katherine has received a number of awards and honours for her work including a General Sir John Monash Award and being elected as a Fellow of the Peter Cullen Water and Environment Trust.

Ray Evans

Ray Evans is a Principal Hydrogeologist at Sinclair Knight Merz. He has more than 35 years' experience in Australian hydrogeology and environmental geoscience. He has worked across all hydrogeological environments in Australia, and has been heavily involved in groundwater and salinity issues in the Murray–Darling Basin for the past 27 years. His experience ranges across the hydrogeology of regional aquifer systems and sustainable yield determinations.

Professor Max Finlayson

Max is the Director of the Institute of Land, Water and Society, Charles Sturt University, and a wetland ecologist with a strong interest in wetland management and communication. He is a past chair of the international Ramsar Wetland Convention's Scientific and Technical Review panel and continues to be involved with its work on wetlands and agriculture.

Professor Douglas Fisher

Douglas teaches and researches environmental and natural-resources law in the School of Law, Queensland University of Technology. His publications include *Australian Environmental Law* (2003), *Water Law* (2000), *Australian Environmental Law: Norms, Principles and Rules* (2010) and an international comparative analysis, *The Law and Governance of Water Resources: The Challenge of Sustainability* (2009).

Dr Dustin Garrick

Dr Garrick is a geographer based in Adelaide, South Australia, as a Fulbright Scholar in the CSIRO's Water for a Healthy Country Flagship and Centre for Regulation and Market Analysis, School of Commerce at the University of South Australia. As of June 2011, he is Postdoctoral Fellow of Water Security at the University of Oxford, with a focus on the design and performance of governance arrangements for trans-boundary rivers in Australia and the western United States.

Professor R. Quentin Grafton

Quentin is Professor of Economics and Director of the Centre for Water Economics, Environment and Policy (CWEEP) at the Crawford School of Economics and Government, ANU. He was Chair of the Socio-Economic Reference Panel of the Murray–Darling Basin Commission, 2008–09. In April 2010, he was appointed to hold the UNESCO Chair in Water Economics and Trans-Boundary Water Governance.

Dr Catherine Gross

Catherine is a Visiting Fellow at the Fenner School of Environment and Society, ANU, with research interests in climate-change adaptation and fairness and justice in environmental decision making.

Mark Hamstead

Mark Hamstead is a Water Policy Consultant with Hamstead Consulting Proprietary Limited. He has 25 years' experience in water-resource management, largely in government. He prepared State-wide guidelines for water planning and water trading in New South Wales and contributed to the development of several statutory water-sharing plans. Since leaving the NSW Government in 2005, Mark has been providing advice on water policy and planning to State and national government agencies.

Dr Glenn Harrington

Dr Glenn Harrington is a Principal Research Scientist with CSIRO Land and Water based in Adelaide. He has more than 15 years' experience in the groundwater industry, having worked for both government and the private sector in Australia, and in academia in Canada.

Dr Darla Hatton MacDonald

Darla is an environmental economist at CSIRO Ecosystem Sciences working on market-based instruments, institutional arrangements and non-market valuation problems in the area of water as part of the Water for a Healthy Country Research Flagship.

Professor Ray Ison

Ray works in Geography and Environmental Science and the Monash Sustainability Institute (Uniwater), Monash University. His research spans the biophysical and social disciplines (including organisations and institutions) and has been primarily interdisciplinary and collaborative in nature. His recent work includes research on 'social learning' and the implementation of the European Union's Water Framework Directive.

Dr Sue Jackson

Sue Jackson is a geographer with CSIRO's Division of Ecosystem Sciences. She has more than 15 years' experience researching the social dimensions of natural-resource management. She has conducted many consultancies for Aboriginal representative bodies in environmental and native-title policy arenas in the Northern Territory and the Kimberley, Western Australia. Her current research focus is Indigenous values associated with water and Indigenous participation in water-resource management. She is a research advisor to the Indigenous Water Policy Group formed by the North Australian Indigenous Land and Sea Management Alliance and a Research Associate at the Centre for Water Economics, Environment and Policy, ANU.

Qiang Jiang

Qiang Jiang has been a PhD student at The Australian National University since 2007. His thesis is about water management in the Murray–Darling Basin. Qiang has an interdisciplinary background with a Bachelor of Economics from Guangxi University, Master of Business Information Systems from the University of Wollongong, and three years' work experience with the Land and Water division of CSIRO.

Professor Jennifer McKay

Professor McKay is the inaugural director of the supported research centre — the Centre for Comparative Water Policies and Laws at the University of South Australia in Adelaide. She conducts socio legal research in Australia, USA, India, the Middle East and Spain. She was awarded a senior Fulbright fellow at UC Berkeley and has awards from the Australian Water Association.

Karlene Maywald

The Hon. Karlene Maywald has extensive experience in water politics. Karlene is a former Minister for the River Murray in the South Australian Government. She experienced first hand the tensions involved in the debate about how to implement environmentally sustainable river management when she was a National Party MP and a minister in the SA Labor Government (until March 2010).

Professor Chris Miller

Chris Miller is a professor in social work and social planning in the School of Social and Policy Studies, Flinders University. He has more than 35 years' experience in community development and social policy. He recently led a team of researchers that worked with six communities in the Murray–Darling Basin to assess the impacts of climate change.

Professor Mark Morrison

Mark is Associate Director of the Institute for Land, Water and Society at Charles Sturt University. Mark's current areas of research focus include non-market valuation of riverine resources, encouraging landholder participation in agro-environmental programs, climate-change communications and developing Indigenous businesses.

Dr Martin Mulligan

Dr Martin Mulligan is the Director of the Globalism Research Centre at RMIT University in Melbourne. He has a background in both environmental sociology and community development, and taught in the Social Ecology program at the University of Western Sydney before joining the Globalism Research Centre at RMIT. He has written extensively on the changing nature of community in the contemporary world and has a particular interest in ways of engaging the community in thinking about the future challenges of climate change. His books include *Ecological Pioneers: A social history of Australian ecological thought and action* (with Stuart Hill, 2001) and *Decolonizing Nature: Strategies for conservation in a post-colonial era* (with William Adams, 2006).

Professor Richard Norris

Richard is the Professor of Freshwater Ecology at the University of Canberra, and works on the biological assessment of rivers, including metal and coalmine effluents, heated water, agricultural effects, sewage effluents, siltation, environmental flows and predictive modelling. He played a central role developing Australia's National River Health Program and led the teams that undertook the Assessment of River Condition component of the first National Land and Water Resources Audit (2000) and the Snapshot of the Murray–Darling Basin River Condition (2001).

Dr Jamie Pittock

Jamie Pittock is at The Australian National University. He has a background in zoology and geography from Monash University, Melbourne, and, from 1989, he worked for various non-governmental environmental organisations. Jamie was Director of the Worldwide Fund for Nature (WWF) Global Freshwater Programme from 2001 to 2007, promoting sustainable river-basin management and representing WWF in international institutions. His research considers how our societies under climate change can better manage increasingly scarce and variable water resources to benefit people and nature.

Professor John Quiggin

John is an Australian Research Council Federation Fellow in Economics and Political Science at the University of Queensland. He is prominent both as a research economist and as a commentator on Australian economic policy. He has published more than 1000 research articles, books and reports on topics such as unemployment policy, microeconomic reform, privatisation, competitive tendering and sustainable management of the Murray–Darling system.

Stuart Richardson

Stuart Richardson is a groundwater specialist currently employed by Sinclair Knight Merz. He has 20 years of experience in salinity and groundwater management. Much of his work over the past several years has related to bringing sound science to the community and policy makers.

Dr Nick Schofield

Dr Nick Schofield is a Senior Executive Consultant with Sinclair Knight Merz specialising in water, climate change and natural-resource management, including strategic planning and futures. Nick was formerly Science Manager for Land & Water Australia.

Åsa Wahlquist

Åsa Wahlquist has been a rural journalist since 1984. She has a degree in Agricultural Science from the University of Adelaide. She has worked on ABC Radio's *Country Hour*, ABC TV's *Countrywide*, and as the rural writer for the *Sydney Morning Herald* and *The Australian*. She has won a dozen journalism awards, including a Walkley Award in 1996; the Australian Government Peter Hunt Eureka Prize for Environmental Journalism in 2005; and the European Community Journalist Award in 1994. She is the author *Thirsty Country* (2008).

Dr Philip Wallis

Philip is a Research Fellow at the Monash Sustainability Institute (Uniwater), Monash University, and a Fellow of the Peter Cullen Trust. He engages in collaborative, interdisciplinary research on social and ecological aspects of water and water governance.

Dr Jessica Weir

Jessica is the author of *Murray River Country: An Ecological Dialogue with Traditional Owners* (2009). Jessica's research focus is on the cultural dimensions of environmental issues, as well as the governance of native-title lands and waters in south-eastern Australia and the Kimberley. She works as a research fellow at the Australian Institute of Aboriginal and Torres Strait Islander Studies.

Dr Sarah Wheeler

Sarah Wheeler is a Senior Research Fellow of the Centre for Regulation and Market Analysis at the University of South Australia. Prior to graduating with her PhD in 2007, she worked in a variety of international and national organisations as a natural-resource economist and has published widely in the fields of organic farming, water markets, crime and gambling. She is currently the chief investigator of a large Australian Research Council linkage project on water markets: 'Improving water market outcomes through a better understanding of market behaviour.'

Dr John Williams

John is a founding member of the Wentworth Group of Concerned Scientists. He is one of Australia's most respected scientists, and has led the national debate about sustainable land management. Prior to joining the NSW Natural Resources Commission in 2006, he was Chief Scientist and Chair of the Department of Natural Resources Science and Information Board. He retired from CSIRO as Chief of Land and Water in 2004. In 2005, he was awarded the prestigious Farrer Memorial Medal for achievement and excellence in agricultural science. John is currently Adjunct Professor in Agriculture and Natural Resource Management at Charles Sturt University. He is also Director of John Williams Scientific Services Pty Ltd., which provides strategic advice and analysis in Agriculture and the Natural Resource Sciences.

Professor Mike Young

Mike is the Executive Director of the Environment Institute at the University of Adelaide. He is one of the country's leading water-policy experts, with a particular focus on the Murray–Darling Basin. He has been awarded the Land & Water Australia Eureka Award for Water Research and a Centenary Medal for his contribution to environmental economics.

Index

D

M

www.ingramcontent.com/pod-product-compliance
Lightning Source LLC
LaVergne TN
LVHW070906210926
848482LV00017B/1004
* 9 7 8 1 9 2 1 8 6 2 2 4 3 *